The Books Sānk and Pātanğal

Perspectives on Islamicate South Asia

VOLUME 2

The titles published in this series are listed at *brill.com/pisa*

The Books Sānk and Pātanğal

A Socio-cultural History of al-Bīrūnī's Interpretations of Sāṅkhya and Yoga

By

Noémie Verdon

BRILL

LEIDEN | BOSTON

The open access publication of this book has been published with the support of the Swiss National Science Foundation.

Cover illustration: drawing by Françoise Verdon.

Library of Congress Cataloging-in-Publication Data

Names: Verdon, Noémie, author.
Title: The books Sānk and Pātanǧal : a socio-cultural history of al-Bīrūnī's interpretation of Sāṅkhya and Yoga / by Noémie Verdon.
Description: Leiden ; Boston : Brill, 2024. | Series: Perspectives on islamicate South Asia, 2666-0202 ; volume 2 | Includes bibliographical references and index.
Identifiers: LCCN 2023048725 (print) | LCCN 2023048726 (ebook) | ISBN 9789004548534 (hardback) | ISBN 9789004680302 (ebook)
Subjects: LCSH: Bīrūnī, Muḥammad ibn Aḥmad, 973?-1048–Criticism and interpretation. | Bīrūnī, Muḥammad ibn Aḥmad, 973?-1048. Kitāb Bātanjal al-Hindī fī al-khalāṣ min al-irtibāk. | Bīrūnī, Muḥammad ibn Aḥmad, 973?-1048. Tārīkh al-Hind. | Sanskrit literature–Translations into Arabic. | Islamic Empire–Relations–India. | India–Relations–Islamic Empire.
Classification: LCC Q143.B5 V47 2024 (print) | LCC Q143.B5 (ebook) | DDC 500–dc23/eng/20231026
LC record available at https://lccn.loc.gov/2023048725
LC ebook record available at https://lccn.loc.gov/2023048726

Typeface for the Latin, Greek, and Cyrillic scripts: "Brill". See and download: brill.com/brill-typeface.

ISSN 2666-0202
ISBN 978-90-04-54853-4 (hardback)
ISBN 978-90-04-68030-2 (e-book)
DOI 10.1163/9789004680302

This book is printed on acid-free paper and produced in a sustainable manner.

Contents

Acknowledgements IX
List of Figures and Tables XI
Abbreviations, Editions and Other Usages XII

Introduction 1

1 **Cultural Contexts of al-Bīrūnī's Work and Writings** 13
1.1 Persian and Islamic Spheres of Influence 13
1.1.1 *Kāṯ, Ǧūrǧān, Ray and Ǧūrǧānīya* 13
1.1.2 *Ghazna and Kābul, the Gateways to Early Medieval India* 15
1.2 Al-Bīrūnī's Visits to al-Hind 21
1.2.1 *The Geographical Delimitation of al-Hind* 21
1.2.2 *Evidence from His Writings* 24
1.2.3 *Maḥmūd's Interests and His Court* 32
1.2.4 *The Various Sources of Information* 43
1.3 Elements of Culture of Gandhāra and Panjab 48
1.3.1 *Laghmān, Peshawar, Fort Rājagirī, Fort Lahūr and Fort Nandana* 48
1.3.2 *The Late Shahis and Their Religion* 53
1.4 Concluding Remarks 57

2 **The Social and Intellectual Contexts** 59
2.1 Building up Theoretical Knowledge on al-Hind 59
2.2 Collaborations and Multiculturalism at Royal Courts 68
2.3 Al-Bīrūnī and Indian Scholars 73
2.4 The Transmission of Living Traditions 80
2.5 Concluding Remarks 95

3 **Al-Bīrūnī's Translations within the Sāṅkhya-Yoga Traditions** 97
3.1 Sāṅkhya-Yoga Literature Predating al-Bīrūnī's Time 97
3.2 Tenets of Sāṅkhya and Yoga 103
3.3 Authorships and Titles 108
3.3.1 *The Sāṅkhyakārikā* 108
3.3.2 *The Kitāb Sānk* 110
3.3.3 *The Pātañjalayogaśāstra* 113
3.3.4 *The Kitāb Pātanǧal* 116

3.4 Intersections and Disagreements of the Two Philosophical Systems 121
3.4.1 *The Indian Sāṅkhya and Yoga* 121
3.4.2 *The Kitāb Sānk and the Kitāb Pātanǧal* 125
3.5 Concluding Remarks 126

4 Al-Bīrūnī's Interpretative Strategies 128
4.1 Al-Bīrūnī's Methods through the Lens of Translation Studies 128
4.2 Three Explicit Transformations 132
4.3 Al-Bīrūnī's Reshaping of the Original Texts 134
4.3.1 *Pedagogical Motives* 134
4.3.2 *Changes due to Logic* 140
4.4 Al-Bīrūnī's Transformations of Content 142
4.4.1 *Omission of Technical Notions and of Redundancies* 142
4.4.2 *Substitution Due to al-Bīrūnī's Cultural Background* 148
4.4.3 *Addition and Definition from Pre-existing Knowledge* 155
4.5 Concluding Remarks 157

5 On the Kitāb Pātanǧal and Its Sources 159
5.1 Scholarship Review 159
5.1.1 *Before the Edition of the Manuscript* 159
5.1.2 *After the Edition of the Manuscript* 166
5.2 The Commentary as an Integrated Part of the Kitāb Pātanǧal 168
5.2.1 *Explicit Integration* 168
5.2.2 *Tacit Integration* 173
5.3 A Problematic Laudatory Passage 190
5.4 Concluding Remarks 194

6 On the Kitāb Sānk and Its Sources 196
6.1 Scholarship Review 196
6.2 Methodological Considerations 199
6.3 Passages of the Kitāb Sānk and the Tradition of the Sāṅkhyakārikā 204
6.3.1 *The Innocent Man among Thieves, Birth Depending on Virtues and Vices, and Categories of Beings* 204
6.3.2 *The Soul as a Passer-by* 207
6.3.3 *Six Views on Action and Agent* 210
6.3.4 *States Different from Liberation and Four Levels of Knowledge* 218
6.3.5 *The Rainwater Which Varies in Taste* 225
6.4 Concluding Remarks 227

Conclusion 231

Appendix: Passages Related to the Kitāb Sānk Found in the Taḥqīq mā li-l-Hind 239
Glossary of Sanskrit Terms as al-Bīrūnī Interpreted Them into Arabic 251
Bibliography 254
Index 275

Acknowledgements

At the suggestion of Professor Johannes Bronkhorst, I let al-Bīrūnī into my life when I began my Master thesis with the aim of bringing together Islamic and Indological studies. Since that fateful moment, I have held an unrelenting appreciation for al-Bīrūnī's significance across disciplinary boundaries. Investigating his life and work has been more than satisfying, both for my inclination toward interdisciplinarity and my interest in exploring the obscure and mysterious. This research has strengthened my conviction that studying bridges between cultures is both enriching and essential for a global understanding of the past and of the present.

Much of this research is owed to the discussions generated in several talks I have had the opportunity to give during this investigation, and at the same time was inspired by many rich interactions with scholars from around the world. I would like to express my gratitude to all my colleagues and friends who have contributed to this research in one way or another. I also thank the Swiss National Science Foundation for supporting the open access publication of this volume.

First and foremost, I would like to thank my two thesis supervisors, Prof. Emeritus Johannes Bronkhorst and Prof. Emeritus Charles Genequand, who continually enriched the processes of my reflection, research and writing with invaluable insights and enthusiasm. I am also grateful to the two external experts who accompanied my doctoral research, Prof. Emeritus Michio Yano and Prof. Blain Auer, for their intellectual support, unyielding encouragement and willingness to give suggestions and constructive criticism.

Special thanks also go to Vladimir Lončar, who shared the instructive and extraordinary experience of creating a documentary movie on al-Bīrūnī with me while also providing me with the opportunity to relate my research on the distant past to contemporary field work.

I am especially grateful to Dr. Philipp André Maas who not only took part in my first experience co-writing an academic article, but also generously shared his scientific expertise as a philologist on the *Pātañjalayogaśāstra*. His constructive feedback on the early drafts of my work encouraged me to deepen my reflections and to posit my opinions with more confidence.

Stimulating discussions with Sara Cappelletti, Prof. Najaf Haider, Prof. Minoru Inaba, Prof. Abdur Rehman, Prof. Walter Slaje, Prof. Ingo Strauch and Dr. Marc Tiefenauer were also highly beneficial for my work, as they removed doubts, opened up pertinent questions and offered new perspectives.

I am greatly indebted to my mother Françoise Verdon, my late father Raymond Verdon, my sister Rachel Moret, and to my friends Noémi Knobel, Loreto Salazar, Elise Gasser, Gaël Zürcher and Sohan Prasad Shah who provided a steady stream of support and encouragement coming my way during my doctoral studies. I am thankful to my brother-in-law, Dr. Lionel Moret, with whom I shared my doubts during this process as he had also passed through the experience of writing a PhD dissertation. I also convey my gratitude to Petros Mapulanga, aka Sanka.

I also am thankful to Prof. Karin Preisendanz who took the time to share her valuable insights in a thorough work of revision.

Lastly, I would also like to thank Dr. Daniel Majchrowicz, Tabitha Spence, Dr. Juliette Vuille and Jennifer Teeter for proofreading my dissertation. Their valuable comments contributed to the removal of typos and grammatical errors, reduced French oddities and substantially helped me improve my academic writing in English.

The present book is a revised version of my PhD dissertation defended on December 16, 2015 (Department of South Asian Studies, University of Lausanne, Switzerland).

Figures and Tables

Figures

1 South Asian Subcontinent with visible natural borders (map prepared by the author) 24
2 Al-Bīrūnī's places of residence (map prepared by the author) 31
3 Example of Maḥmūd's bilingual coins (Cappelletti 2015: 150, ILLS 4B) 33
4 Temple A, Nandana (Meister 2010: fig. 52) 53

Tables

1 Synthesis of the text of the bilingual dirhams of Lahore, based on Cappelletti 37
2 Names of the months in Sanskrit, as transliterated into Arabic by al-Bīrūnī in *Al-āṯār al-bāqiya* 60
3 Names of the seven planets in Sanskrit, as transliterated into Arabic in *Al-āṯār al-bāqiya* 60
4 Names of the zodiacal signs in Sanskrit, as transliterated into Arabic by al-Bīrūnī in *Al-āṯār al-bāqiya* 61
5 Transliterations from Sanskrit to Arabic by al-Bīrūnī in the *Taḥqīq mā li-l-Hind* 67
6 List of references to the *Kitāb Sānk* in the *Taḥqīq mā li-l-Hind* 112
7 List of references to Pātanğal in the *Taḥqīq mā li-l-Hind* 118
8 Correspondences between question-and-answer sets in the *Kitāb Pātanğal* and the *Pātañjalayogaśāstra*, based on Pines' and Gelblum's annotations 137
9 Comparison of the wordings of the chapter-colophons of the *Kitāb Pātanğal* and *Pātañjalayogaśāstra* 174
10 Concordance of the passages on the different objects of concentration in the *Kitāb Pātanğal* and the *Pātañjalayogaśāstra* 178
11 Correlation of the questions from the *Kitāb Pātanğal* and the *Pātañjalayogaśāstra* 185
12 Passages related to the *Kitāb Sānk* and/or to Sāṅkhya concepts found in the *Taḥqīq mā li-l-Hind* 202
13 Correspondences in the six views listed on kā. 61 213
14 Summary of the analytical comparisons of the passages drawn from the *Kitāb Sānk* with the Sanskrit commentaries on the *Sāṅkhyakārikā* 228

Abbreviations, Editions and Other Usages

Abbreviations and Editions

Al-āṯār (1878)	*Al-āṯār al-bāqiya ʿan il-qurūn al-ḵāliya*. See Sachau 1878.
Al-āṯār (2001)	*Al-āṯār al-bāqiya ʿan il-qurūn al-ḵāliya*. See Azkaei 2001.
Ǧamāhir	*Kitāb al-ǧamāhir fī maʿrifa al-ǧawāhir.* Hakim Mohammed Said (ed.), Damas.
GPBh	*Gauḍapādabhāṣya*. See Sharma 1933.
JM	*Jayamaṅgalā*. See Vaṅgīya 1970.
kā.	*kārikā*(s) of the *Sāṅkhyakārikā*.
KP	*Kitāb Pātanǧal.* See Ritter 1956.
MP	*Maṇiprabhā*. See Śāstrī 2009.
MV	*Māṭharavṛtti*. See Vaṅgīya 1970.
PYŚ (1904)	*Pātañjalayogaśāstra* (including the *bhāṣya* and *sūtra* parts). See Āgāśe 1904 [part 1; *pāda*s II–IV].
PYŚ	*Pātañjalayogaśāstra* (including the *bhāṣya* and *sūtra* parts). See Maas 2006 [*pāda* I].
Al-qānūn (1955)	*Al-qānūn al-Masʿūdī. Abū l-Rayḥān Muḥammad bin Aḥmad al-Bīrūnī (d. 440 A.H.–1048 A.D.). Al-Qānūnuʾ l-Masʿūdī (Canon Masudicus). Vol. II (An Encyclopaedia of Astronomical Science). Edited by the Bureau from the oldest extant Mss*, Hyderabad 1955.
RM	*Rājamārtaṇḍa*. See Āgaśe 1904 [part 2].
RT	*Rājataraṅgiṇī*. See Stein 1892.
SK	*kārikā*(s) as printed in YD. See Appendix II in Wezler & Motegi 1998: 278–285
Tafhīm	*Kitāb al-tafhīm li-awāʾil ṣināʿa al-tanǧīm*. See Wright 1934.
Taḥdīd (1992)	*Kitāb taḥdīd nihāyāt al-amākin li-taṣḥīḥ masāfāt al-masākin*, P.G. Bulgakov, Fuat Sezgin, Imām Ibrāhīm Aḥmad (eds.), Frankfort 1992.
Taḥqīq (1887)	*Kitāb taḥqīq mā li-l-Hind min maqūla maqbūla fī l-ʿaql aw marḏūla li-Abū l-Rayḥān Muḥammad bin Aḥmad al-Bīrūnī*. See Sachau 1887.
Taḥqīq (1958)	*Al-Bīrūnī kitāb fī taḥqīq mā li-l-Hind min maqūla maqbūla fī l-ʿaql aw marḏūla*, Hyderabad 1958.
Taḥqīq (1983)	*Abū l-Rayḥān Muḥammad bin Aḥmad al-Bīrūnī. Kitāb fī taḥqīq mā li-l-Hind min maqūla maqbūla fī l-ʿaql aw marḏūla*, Beirut 1983.
TK	*Tattvakaumudī*. See Srinivasan 1967.
TVŚ	*Tattvavaiśāradī*. See Āgāśe 1904, part 1.
sū.	*sūtra*(s) of the *Pātañjalayogaśāstra*.

**Suvarṇasaptati*	*Suvarṇasaptati*. See Takakusu 1904b.
V1	*Sāṅkhyasaptativṛtti*. See Solomon 1973a [part text].
V2	*Sāṅkhyavṛtti*. See Solomon 1973b [part text].
Vivaraṇa (1952)	*Pātañjalayogaśāstravivaraṇa*. See Sastri & Sastri 1952 [*pādas* II–IV].
Vivaraṇa (1999)	*Pātañjalayogaśāstravivaraṇa*. See Harimoto 1999 [*pāda* I].
YD	*Yuktidīpikā*. See Wezler & Motegi 1998.

Other Usages

[ ] the author's additions and omissions in the English translations of Sanskrit and Arabic quotes.[1]

( ) the author's supplementary explanations in the English translations of Sanskrit and Arabic quotes.

〈 〉 the author's additions and emendations in the original Sanskrit and Arabic edited texts.

~ approximate correspondences between Sanskrit and Arabic passages.

Date conversions between Christian Era and Hegira are based on the converter available on the website of the Institute of Oriental Studies at the University of Zurich (https://www.aoi.uzh.ch/de/islamwissenschaft/hilfsmittel/tools/kalenderumrechnung/hegira.html) [accessed October 2023].

Arabic transliteration follows conventions of the Arab World Institute.

1 My interpretations of the Arabic passages drawn from al-Bīrūnī's *Taḥqīq mā li-l-Hind* and *Kitāb Pātanğal* presented in this book are much indebted to Sachau's, Pines' and Gelblum's valuable work.

Introduction

> I have translated two books into Arabic: the first of them is on fundamental elements and a description of what exists, named *Sānk*; the second on the liberation of the soul from the fetters of the body, known as *Pātanǧal*. These two [books] contain most principles around which their (i.e., the Indians) faith revolves, without the subdivisions of their religious laws.[1]

∵

A diverse body of evidence in the historical interactions between the Indian and Islamic worlds reflects the desire of ancient thinkers to share ideas and science across cultural and linguistic boundaries. Two periods of texts' transmission well illustrate these intercultural and intellectual exchanges. The second quarter of the eighth century CE saw the transfer of several Sanskrit works, primarily related to medicine and astronomy, to Muslim intellectuals. The Abbasid rulers in Baghdad, the capital of the Islamic territory at the time, encouraged these translations, notably through the impulse of administrators such as Yaḥyā al-Barmakī (d. 805), Ibrāhīm al-Fazārī (d. 777) or Yaʿqūb Ibn Ṭāriq (d. 796).[2] Thanks to the initiation of the latter two thinkers, portions of the Sanskrit *Brāhmasphuṭasiddhānta*, a text on astronomy written by Brahmagupta in 628 in Bhillamāla,[3] were, for instance, available to Arab Muslims as early as the eighth century CE. Other examples include the medical treatise *Carakasaṃhitā* and the *Pañcatantra*, a collection of Sanskrit fables translated into Arabic around the eighth century under the title *Kalīla wa Dimna*.[4]

---

1 *Taḥqīq* (1958), p. 6.1–4. Sachau 1910: I/8. On this quotation, see below p. 125.

2 On the Barmakids, see Elverskog 2010: 59–61 and Van Bladel 2011: 74–86 and 2012. Baloch (1973: 24–33) focuses on the roles of Ibrāhīm al-Fazārī and Yaʿqūb Ibn Ṭāriq in this process and highlights the connections between this intellectual development and the Islamic spread in Sind. See also Pingree 2012a. On al-Bīrūnī's knowledge of some of these translations, see below pp. 64.

3 On Brahmagupta and his works, see Pingree 1981: 254–257 and 1983.

4 In the ninth century CE, ʿAlī Ibn Sahl Rabbān al-Ṭabarī incorporated elements drawn from the *Carakasaṃhitā*, the *Suśrutasaṃhitā* or the *Aṣṭaṅgahṛdayasaṃhitā* in his *Firdaws al-ḥikma fī l-ṭibb* (*Paradise of Wisdom*) (Elverskog 2010: 61). Knowledge of the transmission of the *Carakasaṃhitā* into Arabic, however, remains limited, as there exists today no extant Arabic

The second broad movement of translation of Sanskrit texts started in the late thirteenth century. The context in which it occurred differed considerably from the preceding project. Works covering a large range of topics, from epics to treatises on medicine and science to Indian religious literature, were translated into Persian. The phenomenon was taking place at the courts of Muslim rulers established in north-western India. These translations include the *Ṭūṭī-nāma*, composed between the years 1313 and 1315 in Persian by ʿImād Ibn Muḥammad Ṯaġarī, a book, based on the Sanskrit *Śukasaptati*, which was dedicated to a sultan of Delhi, ʿAlā al-Dīn Ḵalǧi.[5] A few centuries later the Mughal emperor Akbar (1542–1605) also played a significant role in the transmission of Sanskrit literature into the Perso-Muslim cultural sphere. Notably, he had the *Mahābhārata*, known in Persian as the *Razmnama*, translated.[6]

Amid these two periods, al-Bīrūnī (973–ca. 1050) embodies the cross-cultural and intellectual interactions between the Indian and Islamic cultural spheres thanks to his work on al-Hind[7] and his Arabic interpretations of Sanskrit literature.[8] His contributions, however, contrast with the two aforementioned translation projects which were large-scale undertakings occurring in relatively stable political contexts. Al-Bīrūnī, on his part, lived in a context of regular political change and therefore worked for several patrons. Despite an early interest in Indian mathematics and astronomy, his comprehensive studies of the Sanskrit language and Indian sciences commenced when he came in contact with Maḥmūd of Ghazna (971–1030).

The present book focuses on al-Bīrūnī's transmission of Sanskrit texts into Arabic. As an introductory remark to his work, it must be noted that the Arabic verb *naqala* (نقل) signifies "to transfer", "to transmit" or "to translate." It does not necessarily convey the meaning of a literal translation in the modern sense of the term. If anything, early medieval Muslim thinkers generally carried out their work of translations by emphasizing the transmission of ideas found in their source-texts rather than the words themselves.[9] As I show in this study, al-Bīrūnī's contributions are to be counted within this intellectual tradition. Thus,

manuscript of the work. The *Pañcatantra* had been first translated into Pahlavi in the sixth century CE (Brockelmann 2012). ʿAbd Allah Ibn al-Muqaffaʿ (ca. 720–756), for instance, is among those who played a part in the transmission of the *Pañcatantra* into Arabic (Gabrieli 2012). Al-Bīrūnī mentions this author (*Taḥqīq* [1958], p. 123.10–15; Sachau 1910: I/159).

5 Beelaert 2008.

6 Athar 1992; Rice 2010.

7 The present book makes use of the term al-Hind to refer to the territory of India as al-Bīrūnī defined it in terms of cultural borders in the early eleventh century (see below Section 1.2.1).

8 See also Ernst 2003: 174–177.

9 On the early medieval tradition of translations see for instance Wisnovsky et al. 2011.

I use the English words "translations" or "interpretations" in the sense of "free translations" to designate al-Bīrūnī's productions.

Al-Bīrūnī conducted thorough and extensive research about India, which he communicates in his monograph on India, i.e., *Fī taḥqīq mā li-l-Hind min maqūla maqbūla fī l-ʿaql aw marḏūla* (في تحقيق ما للهند من مقولة مقبولة في العقل او مرذولة; *True Account on What the Indians Say, Both What is Accepted by Reason and What is Not*). This work composed approximately in 1030[10] is referred to hereafter as the *Taḥqīq mā li-l-Hind*. Al-Bīrūnī interpreted numerous portions of Sanskrit works on astronomy into Arabic, such as the aforementioned *Brāhmasphuṭasiddhānta*, which he referred to as *Brāhmasiddhānta*,[11] the *Pauliśasiddhānta* by Puliśa, the *Bṛhatsaṃhitā* by Varāhamihira, the *Karaṇatilaka* by Vijayanandin,[12] as well as purāṇic and epic literature, such as the *Viṣṇupurāṇa*, the *Ādityapurāṇa* and the *Bhagavadgītā*, entitled by him *Kitāb Gītā*.[13] He also quoted a few portions of the Arabic translation of the *Carakasaṃhitā*, referring to it as the *Kitāb Čaraka*.[14] The *Taḥqīq mā li-l-Hind* contains quotations of al-Bīrūnī's interpretations of these works.

In 1036, al-Bīrūnī composed a bibliography informing his readers of his writing up until this year. Boilot (1955) edited, translated and annotated this bibliography, and also updated it with additional works composed by al-Bīrūnī after 1036. This list reveals that al-Bīrūnī also translated Varāhamihira's *Laghujātaka* into Arabic, under the title *Translation of the Small [Book] of the Births by Varāhamihira* (براهيمهر),[15] as well as a medical treatise, which he entitled *Translation*

---

10 The Scheffer manuscript (BNF no. 6080), dated to Ǧumādā al-ūla 4, 554 A.H. (May 5, 1159 CE), bears a note indicating that al-Bīrūnī completed the autograph in Ghazna, Muḥarram 1, 423 A.H., i.e., December 19, 1031 CE. Sachau discusses this note and other pieces of evidence in order to accurately date the composition of the *Taḥqīq mā li-l-Hind*. He concludes that al-Bīrūnī composed his monograph on India between April 30 and October 30, 1030 CE. Sachau's reasons for such dating are convincing (Sachau 1887: ix–x; see also Mishra 1985: 9). Therefore, in this book I adopt his dating.

11 In addition, al-Bīrūnī referred to parts of the *Brāhmasphuṭasiddhānta* in a text entitled *Translation of the Calculation Methods found in the Brāhmasiddhānta* (Boilot 1955: 189, no. 40) and expressed his intention to translate the whole work in the *Taḥqīq mā li-l-Hind*, a task, however, which he could not perform (*Taḥqīq* [1958], p. 119.8–9; Sachau 1910: I/154; Yano 2013). On al-Bīrūnī's *Brāhmasiddhānta* and his account of astronomical literature, see Verdon & Yano 2020.

12 N.A. Baloch (1973) edited the Arabic translation entitled *Ġurra al-zīǧ*.

13 Sachau provides us with a complete list of Sanskrit works quoted and/or referred to in the *Taḥqīq mā li-l-Hind* (1910: I/xxxix–xl). See also Baloch 1973: 11–12 and Shastri 1975.

14 Al-Bīrūnī states that he only had access to a poor translation of the original Sanskrit medical work prepared for the house of the Barmakids (*Taḥqīq* [1958], pp. 123.3–9, 126.4–7and 321.16–17; Sachau 1910: I/159, 162 and 382; see also Verdon & Yano 2020: 66).

15 *Taḥqīq* (1958), p. 122.5–6; Sachau 1910: I/158. Boilot 1955: 202, no. 79.

of the 'Kalab Yārah,' Indian Treatise on the Disease Which Behaves like Putrefaction.[16] In addition, al-Bīrūnī mentions in his bibliography several works which he translated from Arabic "into the Indian language" (إلى لغة الهند).[17]

In the domain of Indian philosophy, al-Bīrūnī produced two works, the *Kitāb Sānk* and the *Kitāb Pātanğal,* which are free translations based on classical Sāṅkhya and Yoga texts. Of these two, only the text of the *Kitāb Pātanğal* (*The Book by Pātanğal the Indian, on the Liberation from the Burdens,* [*being*] *a Translation into Arabic by Abū l-Rayḥān Muḥammad bin Aḥmad al-Bīrūnī*; كتاب باتنجل الهندى فى الخلاص من الاثقال نقل ابى الريحان محمد بن احمد البيرونى الى العربى) is extant. The *Taḥqīq mā li-l-Hind* includes scattered passages of the *Kitāb Sānk*.[18] These quotations, together with the extant *Kitāb Pātanğal,* constitute the earliest known instance of Indian philosophical texts rendered into Arabic. The many references to al-Bīrūnī in modern secondary literature attest to the significance of this figure for the history of South Asia and history of sciences. The following review focuses on a few key authors who have discussed al-Bīrūnī in their work. Numerous researchers of Indian and Islamic history or culture refer to him, including Alain Daniélou (1983), Hermann Kulke and Dietmar Rothermund (1986), Wilhelm Halbfass (1988), André Wink (1990 and 1997), Mohammed Hassan Syed (2003), Akhilesh K. Dubey (2005) and Brajadulal Chattopadhyaya (2006). In addition, three commemorative volumes, gathering contributions by experts in various fields, were published in English in 1951, 1976 and 1979 subsequent to international conferences.[19]

Most publications on al-Bīrūnī, however, pertain to the natural and exact sciences. Important authors who examined his input in the field of mathematics and astronomy are S.E. Kennedy, David Pingree and Michio Yano. Several scholars edited and/or translated several of al-Bīrūnī's writings—or parts of them, namely Carl Edward Sachau (1878, 1879, 1887 and 1910), Hellmut Ritter (1956), Jamil Ali (1967), Shlomo Pines and Tuvia Gelblum (1966, 1977, 1983 and 1989), Mohammed Hakim Said (1973 and 2001), N.A. Baloch (1973), Gotthard Stroh-

---

16 Boilot 1955: 206, no. 92.

17 Three non-extant books are listed in Boilot 1955: 238–239: nos. 175–177.

18 *Taḥqīq* (1958), p. 6.2; Sachau 1910: 1/8. The *Kitāb Sānk* and the *Kitāb Pātanğal* respectively correspond to numbers 97 and 98 in Boilot 1955: 208. Al-Bīrūnī entitled number 97 *Translation of a General Book on the Sensitive and Rational Existents* (ترجمة كتاب شامل فى الموجودات المحسوسة والمعقولة). The title of this book and its place in the list immediately before the *Kitāb Pātanğal* strongly suggest that it corresponds to the *Kitāb Sānk,* as Boilot hypothesised, despite his addition in the bibliography of the *Kitāb Sānk* under number 174 (1955: 238).

19 The present book refers to these edited volumes by the individual authors' names. In 2022 and 2023, al-Bīrūnī's 1050th birthday was celebrated in different countries of Central Asia, notably in Iran and Uzbekistan.

maier (1991), and Seyyed Hossein Nasr and Mehdi Mohaghegh (2005). The valuable works of Jacques-Dominique Boilot (1955) and of Jan Hogendijk (http://www.jphogendijk.nl/biruni.html [accessed October 2023]) provide information regarding editions and translations of al-Bīrūnī's books. Al-Bīrūnī's significant treatise on mathematics, *Al-qānūn al-Masʿūdī* (1030), has not yet been translated in its entirety into a modern Western language.[20]

Several well-grounded and useful biographies include those by Kennedy (1970), F.A. Shamsi (1979), Mohammed Hakim Said and Ansar Zahid Khan (1981), and Michio Yano (EI). Studies on the Ghaznavid rulers, the patrons of al-Bīrūnī, comprise works by Muhammad Nazim (1931), Clifford Edmund Bosworth (1963 and 1977), Minoru Inaba (2013) and Sarah Cappelletti (2015), while the Late Shahi kings whom the Ghaznavids encountered in their military campaigns to the East are dealt with by Yogendra Mishra (1972), Dinabandhu Pandey (1973), Abdur Rehman (1979) and Michel Alram (2016: 151–153).A forthcoming volume edited in the context of the international project *Cultural Formation and Transformation: Shahi Buddhist Art and Architecture* shall include new research and outcome about the history of the Shahi kingdoms.[21]

Several academic works examine al-Bīrūnī's methods of investigation and highlight his scientific objectivity when interacting with Indian society and culture. Their authors are M.S. Khan (1976), Bruce B. Lawrence (1978), G. Kaur (1982), Akbar S. Ahmed (1984), Vincent-Mansour Monteil (1996), M.A. Saleem Khan (2001), Floréal Sanagustin (2003), Kemal Ataman (2005) and Mario Kozah (2016). Only a few surveys, however, explore how al-Bīrūnī dealt with Sanskrit literature. Jan Gonda (1951) analyses passages from the *Taḥqīq mā li-l-Hind* ascribed to the Purāṇas by al-Bīrūnī. Arvind Sharma (1983) provides a study comparing al-Bīrūnī's quotations from the *Kitāb Gītā* found in the *Taḥqīq mā li-l-Hind* to the Sanskrit *Bhagavadgītā*. Pingree (1969 and 1983) examines al-Bīrūnī's quotations from Sanskrit astronomical works. Judith Stareček (2003) compares quotations from the *Kitāb Pātanǧal* and *Kitāb Gītā* found in al-Bīrūnī's *Taḥqīq mā li-l-Hind*.

Between the late nineteenth and late twentieth centuries, there have been various attempts to identify al-Bīrūnī's Sanskrit sources of his translations of Sāṅkhya and Yoga philosophical texts, respectively the *Kitāb Sānk* and the *Kitāb Pātanǧal*. Sachau (1910), Richard Garbe (1894, 1896 and 1917), Junjiro Takakusu (1904a) and Surendranath Dasgupta (1922, 1930), as well as Pines and Gelblum (1966 to 1989), are among those who examined the relationship between al-

20 See Boilot 1955: 210–212 and Hogendjik's website.

21 See fn. 56 of Chapter 1.

Bīrūnī's Arabic works and Sanskrit literature. However, they were unable to find conclusive answers concerning the Sanskrit sources he may have used. Recently, a book chapter by Maas & Verdon (2018) examined the question of al-al-Bīrūnī's sources for composing the *Kitāb Pātanğal* in light of his hermeneutics, while Kozah (2020) provides us with a new edition and English translation of this work.These are the sole studies of al-Bīrūnī's renderings of Sāṅkhya and Yoga texts into Arabic.

In addition, in-depth research on the context in which al-Bīrūnī encountered the South Asian subcontinent is still missing, despite a considerable number of academic publications dedicated to him and his work. Historical sources associated with al-Bīrūnī's time, as well as with the area he visited in South Asia, namely Gandhāra and northern Panjab, are missing or hardly documented.[22] Perhaps this is the reason why his life is relatively unknown and often mixed with legendary elements. Moreover, information regarding the larger historical context of al-Bīrūnī's translation project is fragmentary.

Al-Bīrūnī's studies of Indian culture and science, as well as his interpretations of Sanskrit literature, however, occurred in specific geographical, political, social and intellectual contexts. In order to use his work as a source for the study of South Asian history, these contexts need to be understood. Therefore, in the present book, I investigate several aspects of al-Bīrūnī's life: the geographical context, specifically the places he visited in early medieval India and the cultural boundaries between the Islamic and Indian spheres at the time; the political situation, or how al-Bīrūnī's work was connected to the rulers' interests; the social and intellectual environments, addressing the questions of whom he met when pursuing his research on India, what were his sources of information, how he learned about Indian sciences, literature and philosophy, and what types of intellectual exchange took place at the rulers' courts.

In order to tackle these questions, this book focuses on his *Kitāb Sānk* and *Kitāb Pātanğal* and on how al-Bīrūnī was able to produce such works. Based on various pieces of evidence, archaeological and textual, this research situates al-Bīrūnī's global work on al-Hind in its geographical and socio-cultural contexts. More specifically, it results in an updated biography of al-Bīrūnī. It localizes places where al-Bīrūnī travelled in the north-western subcontinent and determines the nature of his various sources of information alongside that of his interactions with Indian locals. The book further identifies and categorizes reasons for the many adaptations al-Bīrūnī made in his translations; this

22 To the extent possible, this book adopts the geographical terminology used in early medieval Arabic and Sanskrit literature.

makes it possible to analyse at a deep level the relationship between his Arabic interpretations and their possible Sanskrit originals. Finally, it enables to point to his possible Sanskrit sources with some confidence.

The present research takes two main approaches: historical and textual. The first two chapters of this book serve to fill in some gaps in our knowledge of al-Bīrūnī's life and to lay the foundation for the textual analysis which follows. Chapter 1 discusses geographical and cultural boundaries in relationship to al-Bīrūnī's descriptions of Indian culture. During his life, the scholar moved between modern Uzbekistan (Kāṯ), Turkmenistan (Ǧūrǧānīya), Iran (Ǧūrǧān and Ray), Afghanistan (Ghazna and Kābul) which belonged to the Islamic territory, and to Pakistan (Gandhāra and Panjab) in al-Hind, which had just been penetrated by the Ghaznavid rulers at the time.[23] The historical contexts of Kāṯ, Ǧūrǧānīya, Ǧūrǧān, Ray, Ghazna and Kābul are dealt with together in Section 1.1 which also presents preliminary remarks on the Late Shahi kings, also known as the Hindu Śāhis.[24]

Section 1.2 first delimits the geographical boundaries between Islamic and non-Islamic eastern lands as conceptualised by al-Bīrūnī who refers to the latter as al-Hind. Based on an analysis of the evidence, I then argue that his travels to al-Hind were chiefly confined to Gandhāra and Panjab. From the year 1017 onward, when al-Bīrūnī became a scholar at the Ghaznavid court, his freedom of movement was owed to the rulers, and notably to Maḥmūd of Ghazna, who had military and mercantile interest in al-Hind. With his role as a scholar appointed to work for this sultan, al-Bīrūnī was given the opportunity to travel eastward. In addition, while his sources of information were various, they consisted more in oral reports and written documents, and less in his own direct observations.[25] Section 1.3 discusses archaeological data and primary literary sources related to five locales which he visited in Gandhāra and Panjab and

23 This analytical method was inspired by a discussion with Prof. Najaf Haider (JNU, New Delhi).

24 The several designations commonly in use for the two lineages of the Shahi rulers are in my view not satisfactory solutions. Turki Śāhis and Kābulšāh for the Early Shahis (from the mid-7th to the early 9th c.) and Hindu Śāhis for the Late Shahis (from the early 9th to early 11th c.) can be all misleading in terms of religious and political history of the region. Therefore, in the present book, I designate them as Early and Late Shahis respectively, a terminology that is more inclusive and neutral, and at the same time that shows the distinction between the two ruling dynasties. See also Filigenzi 2015: 36.

25 Touati 2000: 13–14. Similarly, early Persian and Arabic geographical accounts of lands lying beyond the boundaries of the Islamic world were often based on oral and written sources of information rather than direct observation. See Bosworth 1970: xlviii and 26, Touati 2000: 154–156, and Zadeh 2011: 131, 154–155 and 172.

to the Late Shahis. Sections 1.2 and 1.3 also highlight the overlap between the regions al-Bīrūnī visited in al-Hind and the territory of the Late Shahi kingdom.

Chapter 2 is dedicated to the social and intellectual contexts which fashioned al-Bīrūnī's intellectual horizon before and during the composition of his book on India. From his birth in Khwarezm (Uzbekistan) up to his travels to the East, al-Bīrūnī had several opportunities to encounter and study al-Hind. As underlined above, Arabic translations of Indian works were available to him before he came into contact with Indian culture when working at the Ghaznavid court.[26] Section 2.1 treats al-Bīrūnī's preliminary knowledge of the Sanskrit language and Indian sciences based on these pre-existing translations. By the time he wrote the *Taḥqīq mā li-l-Hind*, his understanding of Sanskrit, Indian religion and sciences had grown significantly, enabling him to translate, among others, two works related to Sāṅkhya and Yoga.

Taking into account al-Bīrūnī's reliance on the Ghaznavids for coming in close contact with Sanskrit literature and Indian sciences, Section 2.2 explores the role of this royal court in providing favourable conditions for intellectual exchange, and Section 2.3 examines the nature of this exchange as well as the identity of al-Bīrūnī's informants. Lastly, Section 2.4 discusses the possibility that the *Kitāb Sānk* and *Kitāb Pātanǧal* were popular teachings among al-Bīrūnī's informants and points out the significance of his visits to Gandhāra and Panjab for his Arabic renderings of Sāṅkhya-Yoga works. More generally, Chapters 1 and 2 highlight the fertility of the period in terms of intellectual exchanges which accompanied martial and commercial interests of the rulers.

In the first part of this book (Chapters 1 and 2), I chiefly draw from al-Bīrūnī's own works, namely, *Al-āṯār al-bāqiya* (1000), the *Taḥdīd al-amākin* (1025), the *Tafhīm* (1029) and the *Taḥqīq mā li-l-Hind* (ca. 1030), in order to determine which places he visited in al-Hind, the circumstances of his exchanges with Indians, and his knowledge of Indian culture. Al-ʿUtbī was an official secretary at the Ghaznavid court. His account of the Ghaznavid dynasty and its military conquests is, however, not completely reliable because the available English translation by Reynolds (1858) used in this research is not based on the original Arabic text, entitled here *al-Yamīnī* (ca. 1021 CE), but on a later Persian translation of it, known as the *Kitāb-i-Yamīnī*.[27] Nevertheless, I occasionally make use of this source in the present book for recounting information about Maḥmūd's conquests, the life at the royal courts and the social contexts of regions in al-Hind visited by al-Bīrūnī. The accounts by Gardīzī and Bayhaqī, two historians

26 In the *Taḥqīq mā li-l-Hind*, al-Bīrūnī mentions the *Kalīla wa Dimna* and its versions in several languages (*Taḥqīq* [1958], p. 123.10–13; Sachau 1910: I/159).

27 See Anooshahr 2005 for an historiographical study on al-ʿUtbī's *Kitāb-i-Yamīnī*.

at the Ghaznavid court, help reconstructing the structure of the administration and general policies of the Ghaznavids. Other sources providing information related to al-Hind are the anonymous *Ḥudūd al-ʿālam* (982/983) and the historical chronicle *Rājataraṅgiṇī* (mid-12th c.) by Kalhaṇa.

The second approach taken in this book consists in an in-depth textual examination of al-Bīrūnī's *Kitāb Sānk* and *Kitāb Pātanǧal* in relationship to philological studies in Indology. Chapter 3 contextualizes these two Arabic works within the history of Indian philosophy. Sections 3.1 and 3.2 provide outlines of the Sāṅkhya-Yoga literature composed prior to al-Bīrūnī's time and of the philosophical tenets developed in two of their fundamental texts, the *Sāṅkhyakārikā* and the *Pātañjalayogaśāstra*. Section 3.3 discusses philological data, such as the authorships and titles of the Arabic texts in relation to their possible Sanskrit sources. Section 3.4 analyses how al-Bīrūnī regarded his two translations and made connections between the subjects developed in each book.

On the whole, Chapter 3 shows the intellectual and philosophical contexts to which al-Bīrūnī's interpretations belonged. They thus provide the background for the analysis of the subsequent chapters by addressing the question of their connection to the textual tradition in Sanskrit.

In Chapter 3, I also demonstrate that al-Bīrūnī's interpretations reflect the philosophical systems of the *Sāṅkhyakārikā* and *Pātañjalayogaśāstra* traditions. In this way, this outcome anticipates and corroborates the results of the further examinations conducted in Chapters 5 and 6, which investigate into the contents of the Arabic translations and their possible Sanskrit originals. As pointed out above, earlier attempts to identify al-Bīrūnī's sources were mostly unsuccessful in finding final answers. Several reasons may explain these difficulties. First, Louis Massignon discovered the manuscript of the *Kitāb Pātanǧal* in 1922,[28] while Hellmut Ritter critically edited in 1956. Prior to these years, the academic world only benefited from extracts of the *Kitāb Pātanǧal* scattered throughout the *Taḥqīq mā li-l-Hind*. The discovery of this manuscript, now kept in the Koprülü Library of Istanbul, enables a comprehensive analysis of its text. Research in Indology has also considerably developed since the time of Sachau, Garbe and Takakusu, who first undertook to identify the Sanskrit sources of al-Bīrūnī's Arabic *Kitāb Sānk* and *Kitāb Pātanǧal*. Sanskrit manuscripts containing new Sāṅkhya commentaries have been discovered and edited. Recent philological research on the *Pātañjalayogaśāstra* and its commentaries also have produced fresh perspectives on its dating and authorship.

28 Ritter 1956: 165–166.

This development in Indology makes it possible to deliver a more refined comparative analysis of the relation between the Sanskrit texts and the Arabic translations.

Furthermore, modern researchers generally noticed that al-Bīrūnī's translations and the Sanskrit works to which they compared them presented both important parallels and crucial differences. The problem, however, naturally disappears if al-Bīrūnī's hermeneutics are appropriately addressed. Many discrepancies can indeed be accounted for by al-Bīrūnī's interpretative choices. I thus argue that the actual implications of al-Bīrūnī's transformations for determining his sources have been generally overlooked so far.

Therefore, Chapter 4 suggests a new approach to examining this issue and resolves some problems that earlier scholarship faced. Section 4.2 demonstrates that al-Bīrūnī consciously transformed the original Sanskrit texts when he prepared his Arabic translations. Section 4.3 discusses his intentions behind these transformations and Section 4.4 considers his choices of interpretation in the light of the findings of Translation Studies. This method made it possible to identify several reasons for al-Bīrūnī's adaptations: his desire to transmit a message; his idiosyncratic understanding of Indian philosophical terms and concepts; his religious and intellectual backgrounds; his pre-existing knowledge of India; and his interactions with Indian thinkers.

In Chapter 4, I posit that investigating al-Bīrūnī's hermeneutics is a necessary step in the quest to determine the Sanskrit sources he may have used. This approach allows us to move beyond a pure philological and literal comparison between al-Bīrūnī's translations and their possible Sanskrit originals, while it offers interesting analytical tools for further research. Chapters 5 and 6, thus, building upon these observations, examine passages of the *Kitāb Pātanǧal* and the *Kitāb Sānk* in relation to Sanskrit works related to Yoga and Sāṅkhya, respectively.

The main sources of this textual study are al-Bīrūnī's *Kitāb Sānk*, *Kitāb Pātanǧal*, *Taḥqīq mā li-l-Hind*, the Sanskrit commentaries on the *Pātañjalayogaśāstra* and on the *Sāṅkhyakārikā*, as well as secondary literature on Sāṅkhya and Yoga philosophies. Rather than presenting comparisons between the Arabic translations and their sources as a whole, the analysis focuses on specific passages of the *Kitāb Pātanǧal* and the *Kitāb Sānk*.[29]

---

29 For a comprehensive understanding of the *Kitāb Pātanǧal*, readers may consult Ritter's edition (1956), as well as its English translation by Pines and Gelblum (1966, 1977, 1983 and 1989). Extracts of the *Kitāb Sānk*, or references to it, drawn from the *Taḥqīq mā li-l-Hind* are gathered in the Appendix to the present book.

Whereas in the present book I do not examine the question of the recipients of al-Bīrūnī's translations and works on al-Hind,[30] I thoroughly explore the contexts from which his works originated, with the aim to bring out the intimate connections between the translator's life, his intellectual career and his understanding of al-Hind and Indian philosophy.

Furthermore, the present book stands as a contribution to the discussion on the intercultural dialogue between two complex and lively cultural and intellectual spheres. There is no perfect coherent way of designating these two civilisations, considering that much of the terminology is external to them and created by Western thinkers. As such, commonly accepted designations are often too equivocal and general, while they also involve categorising elements of cultures in theory, whose limits are however permeable in practice.

In this book, the term "Indian" stands for ideas or traits relating to early medieval India and to its people or its culture in general. As is known, the Arabic terms *al-hind* (الهند) and *al-hindiyya* (الهندية) primarily referred to a territory and to the populations living in that territory. Accordingly, I generally deviate from Sachau's translation of *al-hindiyya* as "Hindu" and render it as "Indian." I make use of the terms "Brahminical" and "Brahminism" to qualify a society that follows the precepts of the caste system and acknowledges the Brahmins as supreme authorities, whereas I employ the words "Hindu" and "Hinduism" to specifically refer to religious activities and to a system of beliefs in which the cult of Hindu deities is dominant.

As for the terminology belonging to the second culture dealt with here, I resort to "Islamic" and "Muslim" in their sense of including different ethnic groups and political entities, referring to a whole characterised by its adherence to Islam. It may thus look inconsistent, at times, to see the adjective as "Indian" alongside the religious designation of "Islamic". I, however, understand them both as two wholes constituted by their respective—yet not homogeneous—sets of cultural traits. In this way, I chiefly employ this terminology as functional tools and I do hope that my usages shall not entail more misunderstanding or misinterpretation of this terminology.

Lastly, out of the main spotlight of this book, I bring forward several general topics of interest to historians of South Asia. First, this study reveals the vitality of the past in which two cultures, that is, the Indian and Islamic ones, meet as two permeable and moving spheres in terms of territory, cultures and ideology. Second, it shows the territorial conquests of Islam as a process, in

30 Sachau broached this question in the preface to his edition of the *Taḥqīq mā li-l-Hind* (Sachau 1887: xxxi).

which warring concerns intermingled with commercial, cultural and intellectual exchanges. Lastly, al-Hind, and in this case parts of Gandhāra and Panjab, appears much connected to its outside world. Thanks to this research, overall, I foreground aspects of the dynamic dialogue of these two cultural spheres and how this dialogue may have taken place.[31]

---

31 In this context, observations made in this book converge with the perspectives taken in Eaton 2003 and 2020, as well as Flood 2009.

CHAPTER 1

Cultural Contexts of al-Bīrūnī's Work and Writings

1.1 Persian and Islamic Spheres of Influence

1.1.1 *Kāṯ, Ǧūrǧān, Ray and Ǧūrǧānīya*

Al-Bīrūnī spent his youth in the region known as Khwarezm. He was born in the capital city of the region, Kāṯ,[1] also referred to as Kāṯ-Kala, and lived there from 973 to 995.[2] Khwarezm (a region extending between the territories of present-day Uzbekistan and Turkmenistan) was at the time governed by the Afrighid dynasty (ca. from the early 4th to the late 10th c. CE).[3] Qutayba b. Muslim al-Bāhilī had conquered the region in 712.[4] The Afrighids most probably only became dependent on the Islamic Samanid dynasty (819–1005)[5] by the end of the ninth century CE, that is, nearly two hundred years after Qutayba b. Muslim had penetrated the region.[6] By the tenth century, the Arabic language, as well as Islamic culture, was relatively well-established in Khwarezm.[7]

Situated on important trade roads, Kāṯ was an emporium in the tenth century. Arab sources report that the region benefited from great prosperity in terms of commerce.[8] For instance, the *Ḥudūd al-ʿālam* (982/983) describes Kāṯ as "the capital of Khwārazm and the Gate of the Ghūz Turkistān [… and as] the emporium of the Turks, Turkistān, Transoxiana and the Khazar" (Bosworth 1970: 121).[9] Khwarezm, due to its location on the south-eastern side of the Aral Sea, moreover constituted a fertile oasis in the middle of arid steppes and deserts. The region was also considered an important intellectual centre.

1 Well-grounded studies situate al-Bīrūnī's native town in Kāṯ (Kennedy 1970: 147–148; Shamsi 1979: 261–265; Yano 2013).

2 His observations recorded in the *Taḥdīd al-amākin* indicate that he stayed in the region of Khwarezm until the year 995 (Ali 1967: 77 and 211; Kennedy 1970: 148; Shamsi 1979: 268–269; Said & Khan 1981: 125). See the map picturing al-Bīrūnī's places of residence on Figure 2 of the present book.

3 Bosworth 2011a and b.

4 Le Strange 1930: 447; Bosworth 2012a.

5 Bosworth & Crowe 2012.

6 Bosworth 1976: 25 and 2011a.

7 Bosworth 1976: 21–22.

8 Bosworth 2012a.

9 See also Bosworth 2012b.

DOI:10.1163/9789004680302_003

In al-Bīrūnī's time, two dynasties, the Afrighids in Kāṯ and the Ma'mūnids in Ǧūrǧānīya (Kunya-Urgench, located in present-day Turkmenistan), were competing to rule Khwarezm,[10] a situation that culminated in a war in 995 between the two dynasties. These conflicts eventually caused al-Bīrūnī to flee from Khwarezm. The exact length of his sojourn outside Khwarezm is unknown, but it is known that he lived in Ray (now a south-eastern suburb of Tehran) probably some time between the years 995 and 997.[11] In *Al-āṯār al-bāqiya*, al-Bīrūnī mentions his visit to Ray, where he met other scholars and led several research projects.[12]

Followers of Islam first conquered the city between the years 639 and 644, more than three centuries prior to al-Bīrūnī's time. In the tenth century, the Buyids (r. from the mid-10th to the mid-11th c.) incorporated the city, which served as the seat of governing bodies, into their kingdom.[13] In addition to the role of the city as an administrative and trade centre,[14] Ray's reputation as a centre of knowledge made it an essential destination for scholars. For instance, the physician and philosopher Ibn Sīnā (980–1037) visited Ray around the years 1014 and 1015.[15]

From approximately 1000 to 1004, al-Bīrūnī dwelt in ancient Gorgan, referred to as Ǧūrǧān in Arabic and located at the south-eastern corner of the Caspian Sea. In 1000, al-Bīrūnī dedicated *Al-āṯār al-bāqiya* to Prince Qābūs bin Wušmagīr bin Ziyār (r. 977 to 981 and 998 to 1012/1013) who governed the region at the time. Prince Qābūs was known to be redoubtable and cruel, and at the same time an important patron of science and art.[16] The Arab Muslims came to the region in 650/651, but it appears that Islam only established there in the early eighth century.[17] In the ninth and tenth centuries, the town was wealthy and comfortable,[18] known for its silk, and strategically

---

10 Ali 1967: 78; Bosworth 2012a. On the rulers of Ǧūrǧānīya, see Debarnot 1985: 67–70 and Bosworth 2011c. Ǧūrǧānīya should not be mistaken for Ǧūrǧān, which is the ancient name of modern Gorgan, in present-day Iran.

11 Al-Bīrūnī was in the region of Kāṯ in the year 997, as he observed there a lunar eclipse conjointly with the mathematician Abū l-Wafā', who was based in Baghdad. However, during the two the periods extending from 995 to 997 and from 997 to 1000, the exact events of his life remain obscure (Ali 1967: 214–215; Kennedy 1970: 148–149).

12 *Al-āṯār* (2001), p. 433.18–19; Sachau 1879: 338.

13 Nagel 1990. See Guy Le Strange (1930: 186), referring to the tenth-century geographer Ibn Ḥawqal.

14 Le Strange 1930: 227.

15 Gutas 2011.

16 Bosworth 2012d.

17 Bosworth 2012c.

18 Referred to by Le Strange (1930: 377).

positioned for commerce.[19] Although only a few main roads passed through the city, Ǧūrǧān was an important station on the axis between the North and the South. Southward, the road led to Ray, and to the North the route reached Khwarezm.

In 1004, al-Bīrūnī returned to Khwarezm. The Ma'mūnid dynasty had won the war against the Afrighids, and a new capital was established at Ǧūrǧānīya (modern Kunya-Urgench). Al-Bīrūnī lived there until the year 1017.[20] Before becoming the capital city of Khwarezm, Ǧūrǧānīya was an emporium, linking the regions of Ghūz and Khurasan, in the same manner as Kāṯ had been.[21] Further, during the eighth century, several institutions known as *Bayt al-Ḥikma* or *Dār al-Ḥikma* (House of Wisdom) were flourishing in the Islamic territory.[22] These institutions generally housed large libraries and welcomed thinkers. One such establishment, the Ma'mūn Academy, was founded in Ǧūrǧānīya.[23] It was an important centre of knowledge, which hosted numerous scholars. In addition to al-Bīrūnī, other renowned scholars worked there, including the mathematician and astronomer Abū Naṣr 'Irāq,[24] the Christian physician Abū Sahl al-Masīḥī al-Ǧurǧānī[25] and Ibn Sīnā.[26]

1.1.2 *Ghazna and Kābul, the Gateways to Early Medieval India*

In the mid-seventh century, Arab Muslims made inroads on Sistan, from which they reached Ghazna and Kābul. Two centuries later, Alptigīn, a commander of the Samanid dynasty, took over Kābul probably around year 961, prior to seize Ghazna in 962. From the first incursions in the seventh century up to the tenth century, the political situation of these regions fluctuated between attempts of the Muslim governors to establish their authority in the area and upheavals of local rulers.[27] The process through which Islamic caliphate annexed these

---

19 Hartmann & Boyle 2012.

20 Al-Bīrūnī calculated the latitude of Ǧūrǧānīya and made other astronomical observations there up to the year 1016. See Ali 1967: 46–49, 50, 87, 96 and 113.

21 Le Strange 1930: 448.

22 Balty-Guesdon: 1992; Sourdel 2012a; 2012b.

23 Today, an institution of the same name is located in modern Khiva (Uzbekistan), approximately 170 km south-east of Kunya-Urgench.

24 Goldstein 2012.

25 Said & Khan 1981: 66–69; Dietrich 2012.

26 Gutas 2011.

27 Gibb 2012. See also Kuwayama (2002: 181–182) for a summarized table of Islamic incursions in the region in the seventh century. See Inaba (2015: 112–114 and 118–121) who provides a detailed account of the Muslims' attacks on the two cities, their interactions with regional dynasties and the foundation of the Ghaznavid dynasty.

eastern regions to its territory was thus relatively protracted, lasting from the mid-seventh to latter half of the tenth century.[28]

In 977, Sebüktigīn founded the Ghaznavid Empire with Ghazna (in present-day eastern Afghanistan) as its capital.[29] His son Maḥmūd (r. 997–1030) considerably expanded the empire and in 1017 annexed to it the kingdom of Khwarezm.[30] During his reign, Ghazna was also the administrative centre of his vast empire. From this year onward, al-Bīrūnī accompanied Maḥmūd at his court. Several of al-Bīrūnī's astronomical calculations point to his sojourns in Kābul and Ghazna.[31] He also composed numerous works in Ghazna, which include the *Taḥdīd al-amākin* and the *Taḥqīq mā li-l-Hind*, during Maḥmūd's rule. Al-Bīrūnī may have passed away a few years later than 1048, perhaps around 1050.[32] He spent thirteen years of his life, from 1017 to 1030, working for sultan Maḥmūd, and the remaining years for al-Masʿūd his son.

Kābul and Ghazna were positioned at the crossroads of two different cultural spheres. As Minoru Inaba points out, both cities were located in a network of roads leading to various cities and regions of Central Asia, by way of the Oxus River, and of the South Asian subcontinent,[33] through the Khyber Pass.[34] Geographically close to India, Kābul and Ghazna were places of important cultural and economic exchanges. The role of eastern Afghanistan at this crossroad remained crucial into the tenth and eleventh centuries, if for commercial reasons at the very least. According to Arabic sources, trade with India was prosperous.[35] Bust, a city located to the south-west of Ghazna, has been considered the "gateway to Hind."[36] The geographical location of these Ghaznavid sites conferred to them a crucial role in cross-cultural exchanges which were taking place with the South Asian subcontinent and through which goods, art and ideas travelled.[37]

---

28 On Islamization in Central Asia as a process, see, for instance, De La Vaissière 2008 and Arezou 2017.

29 Nazim 1931: 24–26; Inaba 2015a: 119.

30 Nazim 1931: 56–60. See Bosworth 2011e: 372–374 for an account of the relationship between the Khwarezm Šāh Abū' l-ʿAbbās Ma'mūn bin Ma'mūn and sultan Maḥmūd.

31 Ali 1967: 86 and 271; Shamsi 1979: 270–274.

32 Hermelink (1977) places the date of al-Bīrūnī's death in the year 1048 CE (440 A.H.), while Kennedy (1970: 151) argues that it occurred a few years later. See also Sami 1973: 27.

33 Inaba 2013: 81, figs. 2–3, and 85–87.

34 Dagens et al. 1964: 52.

35 For instance, Ibn Ḥawqal (10th c.), referred to in Bosworth 2012e. See also Le Strange 1930: 348–351.

36 As stated by Rehman (1979: 8) with reference to the *Ḥudūd al-ʿālam*. See also Inaba 2015a: 110–112.

37 Hallade 1968: 33; Schlumberger 1978; Sourdel-Thomine 1978; Elverskog 2010: 26; Dietz

Historically, eastern Afghanistan and Gandhāra were incorporated into a number of successive empires, including the Achaemenid Empire (ca. 8th–4th c. BC), that of Alexander the Great, the Greco-Bactrian kingdom and that of the Kuṣāṇas (ca. 1st–4th c. CE). After the Kuṣāṇas, during the period of Sasanian rule (224–651 CE) in the region, several dynasties were ruling over the whole territory. Numismatics tends to show that these kings had been ethnically and/or politically related to each other and that they governed side by side or successively some areas of Central and South Asia, such as the regions north of the Hindukush, Kābulistān, Zābulistān and the Gandhāra region. Broadly speaking, these dynasties include the Kidarites, the Alkhans, the Nezaks, the Rutbils, the Early Shahis,[38] the Rutbils and the Late Shahis.[39] While scholars have yet to comprehensively detail the socio-political and religious situations of pre-Islamic Gandhāra, Kāpiśī and Panjab, archaeological data reveals mixed influences in terms of art and architecture.[40] This is testimony to the specific location of this region, including Kābul and Ghazna, at the cross-road of different cultures.

Al-Bīrūnī mentions the Early Shahis (ca. mid-7th to early 9th c.) and the Late Shahis (early 9th to early 11th c.) in a well-known passage of the *Taḥqīq mā li-l-Hind*:

> And the [Indians] had kings in Kābul, Turks who were said to be of Tibetan origin. The first of them who came was Barhatigīn (برهتكين).[41] He entered a cave in Kābul, which it was only possible to enter by lying down. [...] He was dressed in Turkish clothes, such as the *qaba'*,[42] a hat, boots and weapons. [...] He took possession of these places, assuming the title of Šāhi of Kābul (شاهية كابل). The kingship remained in the hand of his descendants for around sixty generations.
>
> The Indians are careless about succession of things and negligent in the proper arrangement of the continuous chronicles of their kings. [When they] are at loss, they invariably resort to speculation. For this, we shall

2007: 49–50; Inaba 2015b; Alram 2016: 57–58. For the early connections between Kāpiśī-Gandhāra regions with Kashmir on the one hand, and with al-Hind on the other, see Grenet 2002: 212–214.

38 See fn. 24 of the introduction to this book.

39 Rehman 1979; Kuwayama 2002; Inaba 2004, 2005, 2010 and 2015a; Alram & Pfisterer 2010; Alram 2016: 45–151; Elverskog 2010: 27; Vondrovec 2010.

40 Rehman 1979: 289–292; Rowland & Rice 1971: 32–33. Other examples of mixed influences in the art and architecture of Central Asia are displayed in Dagens et al. 1964.

41 *-tigīn* is a usual ending of Turkish names, e.g., Alptigīn, Sebüktigīn.

42 On the *qaba'* see Flood 2009: 65–67.

> convey what some of their people mentioned. According to what I have heard, [the history of] such a lineage, [written] on a piece of silk, is found in the fortress of Nagarkot.[43] I desired to find it, but [I] was prevented from doing so for [different] reasons.
>
> In their group, there was Kaniṣka (كنك), who is at the origin of the *vihāra* (بهار) which is in Peshawar (برشاور) [...] called the Kaniṣka-Caitya (كنك جيت).[44]
>
> The last of them was Lagatūrmān (لگتورمان) and his minister, a Brahmin, was Kallara (كلّر). Times were auspicious to the latter, and, by accident, [Kallar] found hidden treasures, by which he gained the upper hand and became powerful. Then, the government turned away from his master because it had been with the members of his house for a long time.
>
> Then, the manners of Lagatūrmān became wicked, and his deeds were disgusting, in such a way that complaints to his minister [about that] increased. Therefore, [Kallar] tied him (i.e., Lagaturman) and imprisoned him as a punishment. He enjoyed being the sole master of the kingship. He had wealth [at his disposal], as resources, and he made himself master of [the kingdom]. The Brahmin kings succeeding him were Sāmanta (سامند), Kamalū (كملو), Bhīma (بهيم), Jayapāla (جيپال), Ānandapāla (اندپال), Trilocanapāla (تروجنپال), killed in the year 412 of Hegira (i.e., 1021/1022),[45] and his son, Bhīmapāla (بهيمپال), five years later. The [dynasty of the] Indian Šāhis (الشاهيّة الهنديّة) ended and not even a spark from the people of this House remains [today].[46]

These kings originally based in Kābul were among those whom the Muslims encountered in their successive waves of incursion into Kāpiśī, Gandhāra and Panjab. Therefore, it is worth discussing this passage. It is, however, likely that this excerpt only conveys few historical facts. The statement that the Early Shahis were of Tibetan origin is, for instance, probably not a historical reality. The generic term Turk (الترك; pl. الاتراك) in Arabic can refer to several different and distinct tribes or clans, originally coming from Western Eurasia, while

43 The ruins of Nagarkot, also referred to as Bhīmnagar (present-day Kangra Kot or Kangra Fort), are in today's Himachal Pradesh on the foothills of the Himalayan range. See also Dey 1927: 135; Nazim 1931: 89–91; Bhattacharyya 1991: 227. On Nagarkot, and its possible role in al-Bīrūnī's knowledge of India, see below pp. 77–79.

44 This is a transcription from the Arabic *kanika ǧit*. On the *vihāra* of Peshawar, see Dani 1969: 37–39 and Salomon 2018: 44 and 47.

45 The date of Trilocanapāla's death is corroborated by other literary accounts (Nazim 1931: 95).

46 *Taḥqīq* (1958), p. 348.10–351.3; Sachau 1910: II/10–13.

the term Tibetan (التبّت) generally designates the populations living beyond the Oxus River. Evidence indicates that Tibetan people never travelled beyond Gilgit in the upper Indus Valley, and that they were pushed from there in the beginning of the eighth century.[47] Thus, it is unlikely that the Early Shahis were of Tibetan origin. It is more probable that they belonged to a Turkish tribe, the Khalajs, as Inaba convincingly suggests. During the seventh century, a branch of the Khalajs migrated from the northern Hindukush to Kāpiśī and ruled the area until the ninth century.[48]

According to the above report, the founder of the Early Shahis lineage was named Barhatigīn.[49] Al-Bīrūnī connects this dynasty with that of the Kuṣāṇas and traces the lineage of the Early Shahis to a king named Kanik (كنك) who is to be identified with Kaniṣka (early 2nd c.).[50] This connection is, however, not supported by historical evidence.[51] Al-Bīrūnī's report about this genealogy either indicates that the dynastic account located in Nagarkot traced back the lineage of the Early Shahis to the Kuṣāṇas or reflects the complexity of this history as revealed by recent research. It may be added that evidence from several archaeological sites of eastern Afghanistan, such as Mes Aynak, Tepe Naranj, Tepe Sardar or Khair Khana, indicates a long occupation, that is, from the Kuṣāṇa period up to that of the Late Shahis.[52] The long period during which these sites were inhabited by different political groups may account for the narrative which connects the Early Shahis to the Kuṣāṇas and which al-Bīrūnī retransmitted in his account.

Nevertheless, the Early Shahis have long been considered as patrons of Buddhism,[53] current scholarship based on archaeological research also shows that some sites associated with them also hosted Hindu cults. Whoever

47 On references to Turkish people, see the study by P.B. Golden in Bazin et al. 2012. On references to Tibet in early medieval Arabic and Persian geographical accounts, see Akasoy 2011.

48 Inaba 2004: 107–108, 2005 and 2010: 443–449. See also Rehman 1979: 37–47 and 294–297, Wink 1990: 114–128 and Kuwayama 2002: 262–265.

49 Rehman 1979: 45–47.

50 On the date of Kaniṣka, see Dietz (2007: 57, n. 51).

51 See studies by De La Vaissière (2003), Grenet (2002), Inaba (2005 and 2010), Vondrovec 2010, Alram & Pfisterer (2010). Louis de La Vallée Poussin (1935: 17–18) and Dinabandhu Pandey (1973: 63) have previously cast doubt on the historical value of this section of al-Bīrūnī's account.

52 Verardi 2011: 308–309; Alram 2016: 106, 125 and 143. See the discussion on the relative chronologies of some of these sites, as well as on their religious affiliations, in Kuwayama 2002: 162–193, 200–207 and 222–259.

53 Rehman 1979: 285.

the Early Shahis might have been, the reign of these kings ended in the early ninth century, when the Late Shahis succeeded them.[54]

The second part of the above passage dealing with the Late Shahis appears to be relatively reliable historically, although the list of kings provided by al-Bīrūnī also diverges from the information provided by other sources, such as numismatics, epigraphy and the *Rājataraṅgiṇī*.[55] Nevertheless, the Late Shahis are of particular interest for the present investigation: being the first opponents of the Ghaznavids in Gandhāra and Panjab, they were at the forefront as subjects of al-Bīrūnī's research on India.

Thus, in Section 1.2, I put forth positive evidence of al-Bīrūnī's visits to al-Hind. The material presented below show that the territory visited by him mostly belonged to the Late Shahi kingdom. Further, in order to connect al-Bīrūnī's fieldwork with its socio-cultural context, I present, in Section 1.3, a few elements of this context and of the Late Shahi religion, based on a preliminary survey of archaeological and literary sources.[56]

Lastly, the towns where al-Bīrūnī resided and travelled in the Islamic cultural sphere were all prosperous in terms of material wealth and of intellectual developments, likely due to their locations on important roads. This situation, together with him travelling through several regions in the early part of his life, provided him with favourable conditions to have his mind open to novelty and to meet a variety of scholars and sciences. These conditions somehow prepared al-Bīrūnī's work on al-Hind and his intellectual exchanges with local thinkers.

---

54 La Vallée Poussin 1935: 19; Pandey 1973: 67; Rehman 1979: 88 and 298–3002; Mishra 1983: 31–32.

55 Rehman 1979: 89.

56 The society of the Late Shahis remains relatively unknown despite their important role in the early encounters between Islamic and Hindu cultures between the mid-ninth and the early eleventh century CE. The outline in Section 1.3 only superficially broaches the subject. Interest in the Shahi kings and their material culture has lately increased among scholars. See for instance the international project *Cultural Formation and Transformation: Shahi Buddhist Art and Architecture* (https://www.univie.ac.at/cirdis/research/cultural-formation-and-transformation [accessed October 2023]). This project is led by Prof. Em. Deborah Klimburg-Salter (Dept. of Art History, Univ. of Vienna; Harvard Univ.) and is part of the research centre CIRDIS (Center for Interdisciplinary Research and Documentation of Inner South Asia). An edited volume presenting the most recent research on the two Shahi kingdoms is one of the outcome of this project. Recent publications include Filigenzi 2015: 36–40, on the Shahi period, and Khan 2017 on the Late Shahis. On the Late Shahi coinage, see Thomas 1846, MacDowall 1968 and Alram 2016: 151–153. For earlier references, see also Cunningham 1875: 82–83, Dani 1969: 54–56, Mishra 1972, Pandey 1973: 77–132, Rehman 1979: 89–167 and Wink 1990: 125.

1.2 Al-Bīrūnī's Visits to al-Hind

1.2.1 *The Geographical Delimitation of al-Hind*

The elements considered so far have highlighted the strategic position of Zābulistān, Kābulistān and Gandhāra in terms of economy, politics and intercultural exchanges, and at the same time pointed to eastern Afghanistan as a border zone with al-Hind. Boundaries between cultural spheres, often represented by political groups, had fluctuated in the region. Before and at al-Bīrūnī's time, the frontier zones between the Islamic sphere and the Indian world in the north-western subcontinent were particularly subject to changes. Such changes are, for instance, apparent from the Arabic and Persian geographical accounts written between the eighth and the early eleventh centuries. Several of these early works, such as the *Šašnāma* (end of the 9th c.), the *Kitāb futūḥ al-buldān* (*The Book of the Conquest of the Countries*) by al-Balāḏurī, the *Kitāb al-masālik wa l-mamālik* (*The Book of the Roads and the Realms*, 9th c.) by Ibn Ḵurdāḏbah, the homonymous work by al-Iṣṭaḵrī (mid-10th c.) and the *Ṣūrat al-arḍ* (*The Shape of the Earth*) by Ibn Ḥawqal (mid-10th c.), do not describe at length the inland areas of al-Hind, but rather regions of Sind, Gujarat, or coastal areas of al-Hind, as well as their islands.[57]

The focus on these territories in the mentioned accounts reflects the progress of the Islamic military conquests to the East, as well as the state of exchanges between the Islamic world and early medieval India. For instance, while commercial contacts between Arab merchants and the subcontinent existed since an early time through maritime network, Arab Muslims first arrived in Sind in the early eighth century. At first a border zone with al-Hind, Sind, became then part of the Islamic world. Further, up to the tenth century, that is, up to the Ghaznavid incursions in Gandhāra, Panjab and the Gangetic valley, north-western and central India remained mostly unknown to Muslim geographers.[58] By the time of the composition of the *Taḥqīq mā li-l-*

57 On Islamic geography see Miquel 1967; for a global overview of the geographical accounts of al-Hind, see Wink 1990: 109–192. On the *Šašnāma*, see Elliot & Dowson 1867: 131–211, Ahmad 2005: 98, n. 1, and Friedmann 2012. On al-Balāḏurī, see Elliot & Dowson 1867: 113–130, Bosworth 1988 and Becker & Rosenthal 2012. On Ibn Ḵurdāḏbah, see Elliot & Dowson 1867: 12–17, Bosworth 2011d and Hadj-Sadok 2012. On al-Iṣṭaḵrī, see Elliot & Dowson 1867: 26–30, Bolshakov 2012 and Miquel 2012. On Ibn Ḥawqal, see Elliot & Dowson 1867: 31–40 and Khalidov 2011.

58 Verdon 2015: 34–46. An analogous example is that of the city of Ghazna which appears in geographical writings after being included in the Islamic territory (Inaba 2015a: 114–116 and 2015b: 105–107). See also observations by Rehman (1979: 1–3) on al-Hind and its evolving frontiers, and Brauer's work (1995) on the conceptualization of frontiers in medi-

Hind in approximately 1030,[59] these regions, however, were rather well known to Muslims. In addition, the Indus Valley, including Sind and Panjab, as the first place of contact between Muslims and non-Muslim locals via the land route, constituted a zone of economic and political interactions between the two cultural spheres at the time, witnessing important cultural and social transitions.[60]

In view of the above, the question of how these borders were regarded by al-Bīrūnī when he composed the *Taḥqīq mā li-l-Hind* must be addressed. It constitutes the first step in contextualizing his research on al-Hind in order to delineate which of his travels eastward were located in the actual territory of al-Hind. Answering this question also helps determine the places where he observed local customs and traditions. Al-Bīrūnī defines the frontiers of the territory of al-Hind as outlined by mountains and the sea, as shown in several passages of the *Taḥqīq mā li-l-Hind*:

> This sea (i.e., the Indian Ocean) is generally named after what it comprises (i.e., islands), or what is opposite to it (i.e., coastal areas). [Here] we [only] need [to discuss] the part [of this sea] which borders the land of the Indians (الهند) and which therefore is named after them. Furthermore, imagine in the inhabitable world a high and uninterrupted mountain [range], as if it had a vertebral spine extending in the middle of its width in its length from East to West.[61] [...] The [mountain range] has on its surface a wide flat land with a [large] stretch (i.e., a plateau?) and [has] curvings that surround inhabited plains. Rivers flow from it in both directions (i.e., North and South). The land of al-Hind is one of these plains, surrounded by the aforementioned sea in the South and by the high mountains on the other sides.[62]

---

eval Muslim geography. On the concept of frontiers in ancient times, see Thapar 2002: 47–48. Derryl N. Maclean (1989: 1–82) discusses at length the process by which Buddhist communities almost disappeared from Sind when Muslim settled in the region.

59 On the date of the compilation of the *Taḥqīq mā li-l-Hind*, see above fn. 10 of the introduction to this book.

60 See, for instance, Ralph W. Brauer (1995: 33–44) on the concept of boundaries in Arabo-Islamic geography.

61 The delimitation and description of the inhabitable world constituted a common topic among Arabic authors, much indebted to Ptolemy's views (Zadeh 2011: 88–91). Early medieval Muslim geographers generally conceptualized the division of the world into climes. See al-Bīrūnī's description of them in the *Tafhīm* (*Tafhīm*, pp. 143–145, no. 241). In the same work, al-Bīrūnī describes different regions of the world and provides a map of them (*Tafhīm*, pp. 121–125, nos. 211–212).

62 *Taḥqīq* (1958), p. 157.1–11; Sachau 1910: I/197–198.

In this excerpt, the delimitation is confined to a topographical description of the frontiers. Further passages provide the names of places located on these borders:

> Fort Rājagirī (راجكرى) lies to the south of the [mountain Kulārǧak][63] and Fort Lahūr (لهور) to its west.[64] I never saw stronger [forts] than these two; and at three *farsaḵ*s from the [mountain Kulārǧak],[65] is the town of Rājāwūri (راجاورى).[66] Our merchants trade with it, but do not go beyond it. This is the frontier of al-Hind from the northern side. In the mountains to the west of the [land of al-Hind] are several groups of Afghan tribes, [whose settlement] ends near the land of Sind (السند). The southern side of the [land of al-Hind] is [delineated by] the sea.[67]
>
> As for the eastern islands in this sea, they are closer to the border of China [than of al-Hind]. They are the islands Zābaǧ (الزابج). The Indians call them *Suvarṇadvīpa* (سورن ديب)[68] i.e., the islands of gold.[69]
>
> You must imagine that the borders of the land of al-Hind are surrounded by mountains. To its North is the snowy Himavant (همنت), whose centre is the land of Kashmir (كشمير) and which is adjacent to the land of the Turks.[70]

As was common at the time, border zones were conceptualized as rather wide regions. The territory was also delineated by natural boundaries, and at al-Bīrūnī's time the western frontier of al-Hind corresponded to the mountains west of present-day Pakistan.

---

63 This mountain (كُلارْجك) is located south of the capital of Kashmir, i.e., Srinagar, according to al-Bīrūnī (*Taḥqīq* [1958], p. 167.1–2; Sachau 1910: I/207). It may be one of the mountains belonging to the Pir Panjal Range (Stein 1900: II/297–298).

64 Fort Lahūr (or Lawhūr) does not stand for present-day Lahore, in Pakistani Panjab. Al-Bīrūnī appears to refer to the latter city by the toponym Mandahūkūr. On the locations of Fort Lahūr and Fort Rājagirī, see below p. 30. On al-Bīrūnī's mention of these places, see below pp. 51–52.

65 The *farsaḵ* is a historical unit of distance of Persian origin. Traditionally, it represented "the distance that men could march in an hour" (Bivar 2010) and equated to approximately 5 to 6 km.

66 Rājāwūri probably corresponds to the modern Rajauri district situated to the south-east of Punch in present-day Jammu and Kashmir. See also Sachau 1910: II/320, Dey 1927: 165, and Bhattacharyya 1991: 258.

67 *Taḥqīq* (1958), p. 167.5–9; Sachau 1910: I/208.

68 In Sanskrit, the compound *suvarṇadvīpa* means "golden island." It was probably used as a name for the Indonesian island of Sumatra.

69 *Taḥqīq* (1958), p. 169.3–5; Sachau 1910: I/210.

70 *Taḥqīq* (1958), p. 214.3–5; Sachau 1910: I/258.

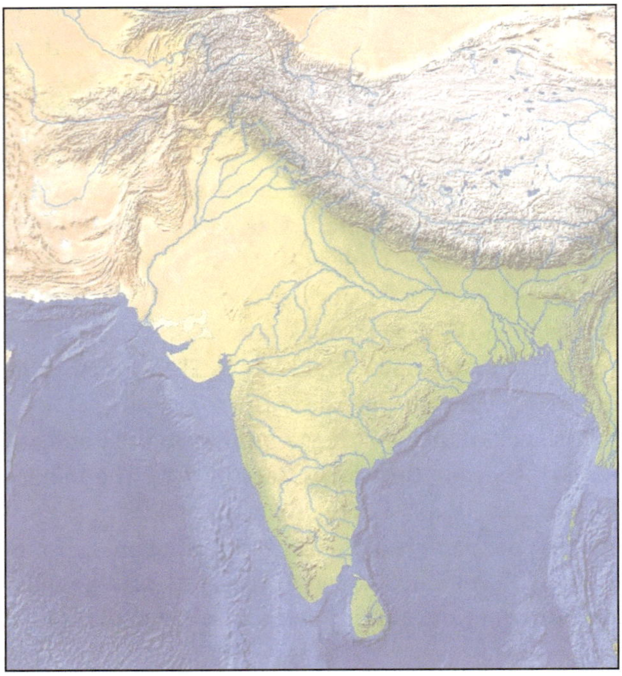

FIGURE 1
South Asian Subcontinent with visible natural borders
MAP PREPARED BY THE AUTHOR

1.2.2 *Evidence from His Writings*

Al-Bīrūnī's *Taḥqīq mā li-l-Hind* is in many respects a valuable source for studying Indian culture and history. Scholars consider his research methodology highly innovative for his time, and the data he provided is rich in historical information. Whereas the composition date of his work, namely around 1030, is known, his field of investigation, that is to say the territory covered by his research, is still uncertain, as are his actual sources of information. The following sections assess the question of al-Bīrūnī's visits to al-Hind and observations there.

Toward the end of the nineteenth century, Carl Edward Sachau edited and translated two of al-Bīrūnī's works, *Al-āṯār al-bāqiya* (Sachau 1878 and 1879) and the *Taḥqīq mā li-l-Hind* (Sachau 1887 and 1910). He also published a study on al-Bīrūnī's transliterations of Indic words into Arabic (Sachau 1888). Sachau's works significantly contributed to our knowledge of al-Bīrūnī's life, his research on India and his sources of information about India. In the preface to the translation of al-Bīrūnī's *Taḥqīq mā li-l-Hind*, Sachau writes that al-Bīrūnī stayed "at Multan, Peshawar, &c."[71] He also admits the "absence of

71 Sachau 1910: I/xv.

positive information" but infers "with a tolerable degree of certainty, that our author [...] stayed in different parts of India [...]."[72] Al-Bīrūnī's life can indeed only be reconstructed by compiling and analysing passing comments scattered throughout his works, and his astronomical observations conducted at different places that are found within his writings. The lack of direct evidence on al-Bīrūnī's travels in al-Hind thus constitutes a fundamental difficulty in answering the above question.

Further, Sachau only supports his statements about the places which al-Bīrūnī visited with evidence and arguments in two of his works, which have been insufficiently read and acknowledged in the scholarly world. They are the English preface to his Arabic edition of the *Taḥqīq mā li-l-Hind* (Sachau 1887)[73] and his thorough study on al-Bīrūnī's Arabic transliterations of Indic words (Sachau 1888). Based on al-Bīrūnī's calculation of latitudes of places in al-Hind and on linguistic observations, Sachau writes in these two works that al-Bīrūnī only stayed in the Kābul Valley and Panjab.[74] Later scholars, although generally coming to similar conclusions, merely refer to these two works. Furthermore, when *Alberuni's India* was published by Sachau, British India included present-day Pakistan, as well as Bangladesh. Thus, when Sachau used the term "India" to describe al-Bīrūnī's visits to al-Hind, this toponym included Pakistan. As I argue below, however, al-Bīrūnī travelled mostly to present-day Pakistan. The impacts of partition on place names, due to the division of the subcontinent into several nation states, have to be taken into account when considering al-Bīrūnī's visits to al-Hind.

Finally, so much has been written on al-Bīrūnī and his life that difficulties arise when one tries to distinguish history from legend. Numerous authors assumed that al-Bīrūnī visited many places in India, whereas only a few modern scholars have provided some details of their arguments when assessing al-Bīrūnī's travels in al-Hind. The objective of the paragraphs below is to summarize their statements about the geographical delimitation of al-Bīrūnī's visits to the territory of al-Hind.

Suniti Kumar Chatterji locates some of al-Bīrūnī's visits in western Panjab, adding that he must "have stayed for some time in Multan."[75] V. Courtois, for his part, maintains that "al-Biruni stayed in India several years and spent most of his time in the North West, within the limits of pre-partition

72 Sachau 1910: I/xvi.

73 Sachau's comments in the preface to his edition of the *Taḥqīq mā li-l-Hind* were brought to my attention by Prof. Emeritus Michio Yano.

74 Sachau 1887: xiii and 1888: 6.

75 Chatterji 1951: 86.

Panjab."[76] Bimala Churn Law (1955) discusses the fact that al-Bīrūnī observed the forts of Rājagirī and Lahūr situated south to the Kashmir Valley.[77] The well-documented biography of al-Bīrūnī by E.S. Kennedy briefly touches upon his travel in al-Hind. Kennedy notes that al-Bīrūnī's "travel and residences in various parts of India [...] were confined to the Punjab and the borders of Kashmir" and mentions al-Bīrūnī's visit to Nandana, as recorded in his *Taḥdīd al-amākin*.[78] M.S. Khan writes that "[i]t seems unlikely that al-Bīrūnī visited South India, but this question must remain open for investigation."[79] Baloch states that al-Bīrūnī must have visited Nandana in 1017, refers to places for which latitudes were calculated by him, and concludes that he "had visited parts of the Peshawar region, of Kashmir, Western Panjab, and of the Multan region of Sind."[80] Ahmad Hasan Dani casts doubt on the common view that al-Bīrūnī stayed in the city of Lahore, the capital of modern Pakistani Panjab.[81]

The view that he visited Lahore originates from the fact that Lahore became the second capital of the Ghaznavids under al-Masʿūd, Maḥmūd's son and successor,[82] and thus hosted scholars of the dynasty's court. Dani, however, stresses that it "would not be unreasonable to say that al-Biruni's account is more pertinent to the areas that fall within the Indus region, i.e., within the present territorial limits of Pakistan" and that al-Bīrūnī's observations made in al-Hind "can hardly be perfectly true of the Ganges Valley much less of South India."[83] More recently, M.A. Saleem Khan notes that "[a]l-Biruni [stayed] in India—and present modern Afghanistan was [...] part of India—and [visited] other places in the rest of India, [learned] its most important and difficult language i.e., Sanskrit, meeting with the learned pundits, [and] studying books."[84] Mohammed Hassan Syed (2003) argues that al-Bīrūnī stayed for a short period of time in today's Pakistan.[85] Most of the above statements are partially accurate and, as aforementioned, the above authors only partly substantiate their claims. Furthermore, only two scholars, Courtois and Saleem Khan, under-

76 Courtois 1952: 35.
77 Law 1955: 9–10.
78 Kennedy 1970: 150.
79 Khan 1976: 91, n. 24.
80 Baloch 1973: 39.
81 Dani 1979: 186–187.
82 Bosworth 1977: 64. On Lahore in the context of Maḥmūd's rule, see below pp. 32–38.
83 Dani 1979: 187.
84 Khan 2001: 21.
85 Syed 2003: 36.

line the fact that the boundaries of eleventh-century India were different from today's, yet at the same time they continue to use the concept of India in a rather vague manner.

Mohammed Hakim Said and Ansar Zahid Khan (1981) provide a relatively detailed account of al-Bīrūnī's life. According to these two authors, al-Bīrūnī travelled across some regions of al-Hind during three possible time periods, between the years 1020 and 1021, 1023 and 1024, or 1028 and 1029. These periods coincide with years during which the scholar's presence in Ghazna is not attested by his astronomical observations or writings. The two authors further conclude that the two time periods between the years 1020 and 1024 constitute the most likely periods of al-Bīrūnī's journeys to the East.

They also write that al-Bīrūnī visited Multan, Sialkot, near today's Lahore, Nandana, Fort Rājagirī and Fort Lahūr. Furthermore, they refute the assumption that al-Bīrūnī accompanied Maḥmūd on all of his military expeditions. Ultimately, they state that "al-Bīrūnī seems to have travelled along Kābul and the Panjab's routes."[86]

Jai Shankar Mishra's account similarly constitutes one of the most detailed analyses of al-Bīrūnī's travels, including their possible duration and geographical limits. Mishra refutes "the view that he travelled in many provinces of India,"[87] asserting that al-Bīrūnī only visited western Panjab. He bases his argument mainly on the study of the *Taḥqīq mā li-l-Hind*, mentioning the forts of Rājagirī and Lahūr as places that al-Bīrūnī actually saw, as well as the locales for which the scholar calculated the latitudes.[88] Said, Khan and Mishra are among the rare modern authors who attempted to determine specific places of al-Bīrūnī's visits to al-Hind, as well as dates for them, based on historical evidence.

Thus, the existing scholarship on al-Bīrūnī's visits to al-Hind tends to show that he chiefly travelled in Gandhāra, western Panjab and to some extent Sind. These surveys, however, remain largely speculative and tentative. In addition, and more importantly, despite these conclusions, the common view that al-Bīrūnī visited many regions of al-Hind seems to have persisted in numerous studies that mention his life and work on Indian culture. If his work is to be used in order to draw information on the historical, religious or intellectual context of the early eleventh century, it is, however, necessary to determine what could have been his field of investigation before he composed the *Taḥqīq mā li-l-Hind*. The difficulty of the task is explained by the lack of positive evidence, as Sachau

86 Said & Khan 1981: 82–86.

87 Mishra 1985: 11.

88 Mishra 1985: 11–13. On the latitudes, see below pp. 28–29.

already noted. Al-Bīrūnī did not indeed inform his readers about the places he himself had visited, nor did, for instance, his contemporaries at the Ghaznavid court al-ʿUtbī and Bayhaqī.[89]

Limited evidence is available and thus arises the need to examine the past through several types of evidence, philological, archaeological and circumstantial. Works by Said and Khan (1981) and Mishra (1985) stand among the rare studies that consider al-Bīrūnī's visits to al-Hind from a multidisciplinary perspective. The present book takes into account these two studies with the aim of supplementing them. Although absolute certitude might never be reached, several factors help clarify the question of where al-Bīrūnī's travelled in al-Hind. They are: the historical circumstances related to his life, such as Maḥmūd's military interests, the social and political contexts at the sultan's court, and the various ways al-Bīrūnī collected data on India—be it through his direct observations, his readings of Sanskrit literature or his interactions with Indians. In addition, observations about al-Bīrūnī's sources of information help contextualize his interpretations of Indian philosophy, which are the subjects of Chapters 4 through 6 of this book.

Al-Bīrūnī's works particularly conducive to understanding the extent of his actual travels in al-Hind are the *Taḥdīd al-amākin*, composed in 1025,[90] and the *Taḥqīq mā li-l-Hind*, written in approximately 1030.[91] In the *Taḥqīq mā li-l-Hind*, a few passages refer to al-Bīrūnī's direct observations.[92] Yet, they are of no help here, as they do not specify where they did take place. The only positive evidence naming places that al-Bīrūnī visited in the territory of al-Hind lies in a passage where he provides the latitudes of a few locales and in four portions of text pointing to five toponyms.

In the *Taḥqīq mā li-l-Hind*, al-Bīrūnī states that he calculated the latitudes of the following sites: Fort Lahūr (34°10′), Ghazna (33°35′), Kābul (33°47′), Kindī, the stronghold of the Prince (كندى رباط الأمير; 33°55′),[93] Dunpūr (34°20′), Lagh-

89 On Bayhaqī, see Bosworth 2004: 20.

90 *Taḥdīd* [1992]; Ali 1967.

91 *Taḥqīq* (1958); Sachau 1910. Al-Bīrūnī's mathematical treatise, *Al-qānūn al-Masʿūdī*, is only sparsely referred to in the present book.

92 Al-Bīrūnī witnessed the way in which the Indians catch gazelles (Sachau 1910: I/195) and a struggle between an elephant and an animal he calls *gaṇḍa* (Id. I/204). He informs the reader that he had seen Brahmins (Id. II/134.; on al-Bīrūnī's meeting with Brahmins and astronomers, see below Section 2.3) and enumerates more than thirty Indian religious festivals, some of which he may have observed (Id. II/178–185. See Verdon 2019a: 68–71).

93 According to Sachau, the reading Kirī is also found. This stronghold was most probably located between Kābul and Peshawar near Jalalabad (Sachau 1910: II/341–342; Bivar 1979: 169–170 and 172–173).

mān (34°43′), Peshawar (34°44′), Wayhind (34°30′), that is, Hund or Udabhāṇḍa, Jhelum (33°20′), Fort Nandana (32°0′), Sialkot (32°58′), Mandahūkūr (31°50′) or modern Lahore,[94] and Multan (29°40′).[95] Longitudes and latitudes of some of these places are also provided by al-Bīrūnī in his mathematical treatise *Al-qānūn al-Masʿūdī*.[96] These sites are located in Kāpiśī, Gandhāra and Panjab with the exception of Multan located in Upper Sind. Following this list of latitudes in the *Taḥqīq mā li-l-Hind*, al-Bīrūnī states:

> We have not passed beyond these aforementioned places in the land of the [Indians], nor have we learned about [other] longitudes and latitudes in their books.[97]

Thus, according to al-Bīrūnī's own words he did not travel beyond these regions. Another passage in the *Taḥqīq mā li-l-Hind* in which al-Bīrūnī writes that he could not go to faraway places such as the Valley of Kashmir and Varanasi[98] lends support to the validity of this statement. Furthermore, in the *Taḥdīd al-amākin* al-Bīrūnī explains that he visited the area of Laghmān and Fort Nandana, while in the *Taḥqīq mā li-l-Hind* he says that he saw Peshawar, Fort Rājagirī and Fort Lahūr. These are the only explicit references by al-Bīrūnī to places he himself visited in al-Hind in these two works.

Laghmān[99] was situated to the north of the Kābul river, between modern Kābul and Jalalabad.[100] It was located on one of the roads possibly taken by Maḥmūd, which leads from Ghazna to Peshawar via Kābul.[101] Al-Bīrūnī must

---

94 In the *Taḥqīq mā li-l-Hind*, al-Bīrūnī most probably refers to Lahore as Mandahūkūr (مندهوكور), described as the capital of the region of Lawhāwūr (لوهاور; also Lawhāwar) and located to the east of the Ravi River (*Īrāwah*; ايراوه; *Taḥqīq* (1958), p. 165.7; Sachau 1910: I/206). Mandahūkūr may be a corrupted form of Maḥmūdpur (see also below pp. 34–38).

95 *Taḥqīq* (1958), p. 270.5–11; Sachau 1910: I/317. Al-Bīrūnī also provides the latitudes of Ujjain, Taneshwar and Kanauj based on Arabic works, by Yaʿqūb Ibn Ṭāriq and Abū Aḥmad Ibn Ǧīlaġtakīn, and on Sanskrit sources, such as Balabhadra's work and the *Karaṇasāra* by Vaṭeśvara (*Taḥqīq* (1958), pp. 269.10–270.5; Sachau 1910: I/316–317). On Vaṭeśvara, see Pingree 1991: 555–556.

96 Sachau 1887: xii–xiii and 1910: II/317. See also *Al-qānūn* (1955), pp. 561–562 and 573–574. The reader may refer to A.D.H. Bivar's study (1979) on the coordinates of places located between Ghazna and Peshawar as found in the *Taḥqīq mā li-l-Hind* and in *Al-qānūn al-Masʿūdī*. Bivar, for instance, points out the discrepancies between the coordinates found in both works, as well as between al-Bīrūnī's calculations and actual coordinates.

97 *Taḥqīq* (1958), p. 270.13–15; Sachau 1910: I/318.

98 *Taḥqīq* (1958), p. 16.17–18; Sachau 1910: I/22. The passage is translated below p. 43.

99 This place is also known as Muraṇḍa (Dey 1927: 113).

100 Rehman 1979: 13.

101 *Taḥqīq* (1958), p. 165.5–11; Sachau 1910: I/206.

have visited the region of Laghmān some time after he had followed Maḥmūd as a member of his court and prior to the composition of the *Taḥdīd al-amākin* in which he mentions his observation, namely between the years 1017 and 1025. More specifically, he states that he was there during a solar eclipse.[102] An almost total (97%) solar eclipse occurred on April 8, 1019, in this region. Therefore, it is possible to infer that al-Bīrūnī was in Laghmān on this specific date, as already suggested by Kennedy.[103]

The exact locations of Fort Lahūr and Fort Rājagirī are problematic. The two toponyms refer to several places on the subcontinent, and scholars made various hypotheses regarding their locations. A passage of the *Taḥqīq mā li-l-Hind*, however, describes an itinerary starting from Kanauj to Kashmir, via Rājagirī.[104] Although not all places located on this route have been identified, it is clear that the itinerary goes north-north-west from Kanauj and then appears to run alongside the southern foothills of the Himalayan range from east to west. At Rājagirī, the road turns to the north and directly leads to the Kashmir Valley. Accordingly, Rājagirī was most probably located to the south of the Pir Panjal Range, at the entrance of one of its passes, such as the Pir Panjal pass, or alongside the Jhelum Valley.[105] The localisation by al-Bīrūnī of Rājagirī to the south the mountain Kulārğak in the Pir Panjal Range also points to the same location. Fort Lahūr, which is often mistaken for present-day Lahore, the capital of Pakistani Panjab, was probably located to the south-west of the Pir Panjal Range, and may well correspond to Loharakoṭṭa referred to in the *Rājataraṅgiṇī*.[106]

As aforementioned, al-Bīrūnī calculated the longitudes and latitudes of the two forts in his mathematical treatise *Al-qānūn al-Mas'ūdī*,[107] while he provides

102 Bivar 1979: 169. See below p. 49, for al-Bīrūnī's description of this eclipse.

103 Kennedy 1970: 150.

104 *Taḥqīq* (1958), pp. 164.15–165.2; Sachau 1910: I/205.

105 Abdur Rehman (1979: 275–276 and 2003: 9) argued that Rājagirī should be identified with the so-called Rāja Gīrā's castle, a site located on the Mount Rāja Gīrā near modern Uḍegrām in the Swāt Valley. The ruins of another fort, known as Rāja, are located around 8 km north-east of the modern city of Taxila. Considering al-Bīrūnī's description of Rājagirī Fort's location, these two other options are, however, unlikely.

106 According to Arabic and Persian literary sources, Maḥmūd of Ghazna attempted twice to capture a place situated on the foothills of Kashmir, known as Lohkot or Loharin (Nazim 1931: 104–105). Marc Aurel Stein identified Lohkot with the Castle of Lohara, referred to as Loharakoṭṭa in the *Rājataraṅgiṇī*, and located south-west of the Pir Panjal Range (RT IV.177, p. 50; Stein 1900: I/138 and II/293–300). This identification is more probable than with Chota Lahore (small Lahore) lying to the east of the Swāt Valley which has been at times identified with the Fort Lahūr of al-Bīrūnī. See also Khan 1979: 223–224.

107 Sachau 1887: xii–xiii; *Al-qānūn* (1955), p. 562; 574.

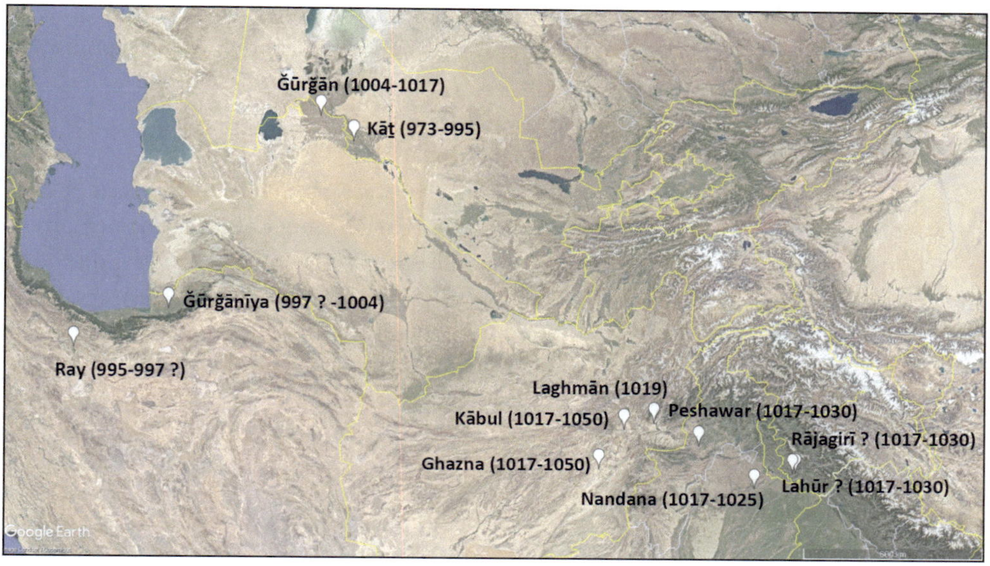

FIGURE 2 Al-Bīrūnī's places of residence
MAP PREPARED BY THE AUTHOR

the latitude of Fort Lahūr in the *Taḥqīq mā li-l-Hind*.[108] In general, however, some problems arise to determine places based on the coordinates provided in these two works, as the figures are not always exact, need some revisions and diverge in the two works.[109] Nevertheless, a study by the present author shows that one possible localisation based on these longitudes and latitudes concur with the localisation of the two places south to the Pir Panjal as proposed above.[110] Fort Nandana is located in modern Pakistani Panjab, on a hilltop belonging to a series of mountains called the Salt Range.[111] Based on the above, Figure 2 is a map depicting places where al-Bīrūnī lived and travelled, with known dates or time range.

Thus, it can be established that al-Bīrūnī surely travelled in Gandhāra, northern Panjab and to a lesser extent in Multan in Upper Sind.[112] Although the possibility that he visited other provinces of al-Hind after the composition of the *Taḥqīq mā li-l-Hind* cannot be completely discarded, evidence for such a hypothesis may never come forth. These observations concur with the assump-

108 See above p. 28.
109 On al-Bīrūnī's calculations of these coordinates, see Bivar 1979.
110 Verdon forthcoming.
111 Bhattacharyya 1991: 229.
112 Although al-Bīrūnī gives the latitude of Multan, the research of this book suggests that he did not conduct his research there. See below p. 78

tions made in earlier studies discussed above.[113] Hereafter, I offer circumstantial evidence, discussed in a systematic manner, that corroborate these preliminary remarks.

1.2.3 *Maḥmūd's Interests and His Court*

Beyond al-Bīrūnī's list of latitudes and observations made on five places of al-Hind, other pieces of evidence indicate that his direct observations chiefly applied to Gandhāra and Panjab. First, al-Bīrūnī's descriptions correspond to the interests of the Ghaznavids, particularly that of Maḥmūd. Maḥmūd focused much of his political and military attention to Gandhāra and Panjab. Thus, I argue that this interest, also motivated by economic concerns, is one reason for why al-Bīrūnī's direct observations and research occurred in these two regions.

Prior to the Ghaznavids' conquests, Gandhāra and Panjab chiefly remained a *terra incognita* to Muslim authors and rulers.[114] Sebüktigīn, Maḥmūd's predecessor, was the first known Muslim ruler to attack the cities of Laghmān and Peshawar. As for Maḥmūd, he concentrated many of his raids on the area by launching assaults on Laghmān in 1000, on Peshawar in 1001, on Wayhind in 1001 and 1008/1009, on Bhātinda (southern Panjab) in 1004/1005, on Fort Nandana in 1014, on Taneshwar in 1014, and on Fort Lahūr in 1015/1016 and 1021/1022, while in Sind he led expeditions to Multan several times in 1006, 1008 and in 1010/1011. In further southern and eastern regions, he also assailed Nārāyaṇapura (modern Rajasthan), also designated as Nārīn, in 1009, Kanauj and Mathura in 1018/1019, the forts of Gwalior and Kalinjar (in central India) in 1022, as well as the Somnāth temple in 1025/26.[115] During his military excursions, he confronted several political and religious groups, the Late Shahis ruling over Gandhāra and Panjab, the Ismāʿīlīs in Multan, the Pratihāras occupying Somnāth, some parts of Rajasthan and of Mālava, and the Chandelas established in Kanauj and Mathura. Among those, the sultan's conquests caused the Late Shahis' dominion over Gandhāra and Panjab to end.[116]

Maḥmūd aimed to access the fertile land of Panjab, as well as to control the important routes leading to the Gangetic Plain.[117] With the possession of Sind,

---

113 See above pp. 24–28.

114 Grover 2006: 44; Verdon 2015: 38–40.

115 Nazim 1931: 86–122. Inaba (2013: 77–79, table 1), provides a table listing Maḥmūd's conquests in Central Asia and India, based on several primary sources.

116 Inaba 2016. See also Thapar 2002: 508: map 12. Several Indian dynasties that ruled over other parts of al-Hind, which correspond to present-day north-western India, and were attacked by Maḥmūd are enumerated in Mishra 1983: 69–70. On the encounter between the Late Shahis and the Ghaznavids, see Rehman 1979: 130–167.

117 Nazim 1931: 88–89.

FIGURE 3 Example of Maḥmūd's bilingual coins
CAPPELLETTI 2015: 150, ILLS 4B

the sultan could reach the seaport at Debal, near modern Karachi, enabling him to benefit from an important trade network through the Indian Ocean. As Inaba has shown on a map representing the territory of the Ghaznavids at Maḥmūd's death, the sultan borrowed two main roads leading to the East, a southern one through the Sind and a northern one through Gandhāra, illustrating these two complementary advantages.[118] Maḥmūd also proceeded on the northern route leading from Ghazna to Kanauj, via Peshawar and Lahore. Incidentally, al-Bīrūnī describes this road, when he deals with a network of routes starting from Kanauj and leading into various directions of the subcontinent.[119] A parallel can thus be drawn between Maḥmūd's territorial conquests and al-Bīrūnī's intellectual exploration of Gandhāra and Panjab.

Furthermore, the Ghaznavids attempted to establish their authority in these two regions while they chiefly conducted raids into other provinces of al-Hind. The minting of bilingual silver dirhams bearing their marginal and central legends in both Arabic (in Kufic script) and Sanskrit (in Śāradā script) (Figure 3) in Lahore located in modern Pakistani Panjab, reflects this attempt. A series of these coins was minted between the years 1027 and 1028 (418/419 AH).[120]

118 Inaba 2013: 76, fig. 1. I am grateful to Professor Minoru Inaba for having drawn my attention to more specific military interests of Maḥmūd.

119 *Taḥqīq* (1958), pp. 155–170; Sachau 1910: 1/196–212. See map 2.1 in Verdon 2015: 42. Kanauj being the starting point of al-Bīrūnī's itineraries has to be seen in the light of the prestige this city enjoyed in the post-Gupta period. See Thapar 2002: 405–407 and Elverskog 2010: 45. Similar itineraries have been reconstructed in Schwartzberg 1978: 33 and Deloche 1968: planche VII. Grover (2006: 46–48) also describes routes passing through Panjab. For an account of the routes described by al-Bīrūnī, see Verdon 2015: 40–43. Indian cities are also mentioned by al-Bīrūnī in the *Tafhīm* (*Tafhīm*, pp. 143–144, no. 241).

120 Numbers 13–14 in Thomas 1859: 23–24. V.S. Agrawala (1943) and Sara Cappelletti (2015: 69–

The legends in the margins of the coins in Arabic and Sanskrit provide dates of the coins as well as the place where they were minted, that is, Maḥmūdpur, possibly a name given after Maḥmūd to the capital of the region of Lahore (Lawhāwūr [لوهاور] according to al-Bīrūnī).[121] The Sanskrit legend in the margin is found in two variants. Variant 2 of this marginal legend reads: *ayaṃ ṭaṅkaṃ mahamūdapura ghaṭita tājikīyena saṃvatī 418/419*.[122] As V.S. Agrawala and Sara Cappelletti have shown, this reading stands as a translation of the Arabic original "In the name of Allah, this dirham was struck at Maḥmūdpur in the year 418/419." The Sanskrit term *ṭaṅka* in the masculine gender means "stamped coin" and renders the Arabic dirham (درهم). The Sanskrit past participle passive *ghaṭita*, meaning "made" or "produced," translates the Arabic passive verbal form *ḍuriba* (ضرب) translated as "was struck."[123] The meanings of the terms correspond in both languages.[124]

As for the central legend in Arabic, one finds the Islamic declaration of faith (*šahāda*), as well as a formula expressing the legitimization of Maḥmūd's power on the obverse, as follows:

> There is no God but Allah, Muhammed is the messenger of Allah. Maḥmūd, the right hand of the state, the custodian of religion.[125]

Before considering Maḥmūd's political intentions behind the striking of the bilingual coins, it is worth discussing the underlying process of translation, because it resonates with the reflections in Chapter 4, 5 and 6 of the present

---

102) examine these coins in detail. Two other coins minted in 1021/1022 CE (412 AH) have been studied by Thomas under numbers 11 and 12 (1859: 22–23). Readers further interested in these bilingual dirhams may consult the studies by Thomas (1847 and 1859), Agrawala (1943), Bhattacharyya (1964) and Cappelletti (2015). See also Chatterji 1951: 96–100, Said & Khan 1981: 88, Khan 2001: 62–63 and Flood 2009: 41–42.

121 According to Thomas's reading: *maḥmūdsar* (Thomas 1859: 23–24). See Khan 1979: 221–226 and fn. 94 of Chapter 1 of the present book. Bosworth (2007: 299) doubts the identification of Lahore as Maḥmūdpur.

122 Agrawala 1943: 155 and 157–158; Bhattacharyya 1964: 54–56.

123 For a clear and synthetic exposition of the two variants, see Cappelletti 2015: 72 and 82. See also Agrawala 1943: 155–161 and Bhattacharyya 1954: 114.

124 The Sanskrit legend is, however, grammatically incorrect, as the agreement of the grammatical gender is not accurately observed. The correct form in the nominative case should be: *ayaṃ ṭaṅkaḥ ghaṭitaḥ* (or *ayaṃ ṭaṅko ghaṭitaḥ*), not *ṭaṅkaṃ*, since this word in the sense of "a stamped coin" is a masculine noun (I am grateful to Dr. Maas for having brought this to my attention). The Sanskrit *mahamūdapura* refers to the minting place. The word is, however, not found in the expected locative case. These features may indicate that the person(s) behind this translation lacked grammatical knowledge of Sanskrit.

125 Agrawala 1943: 156; Bhattacharyya 1964: 53; Cappelletti 2015: 87.

book.[126] The central Sanskrit legend on the reverse is puzzling. Several scholars have attempted to decipher it.[127] The direct transliteration of the legend runs as follows: *abyaktameka* (1) *muhammada avatāra* (2) *nṛpati mahamūda* (3).[128] The legend is written in an approximate Sanskrit and shall be emended in order to be grammatically correct, and thus interpreted. One possible emendation is the following: *avyaktam ekam muhammado 'vatāro nṛpatir mahmūdaḥ*. If one accepts this reading, the meaning of the legend is: "the unmanifest is one, Muhammed is the incarnation, Maḥmūd is the king." This emendation facilitates drawing parallels between the Arabic and the Sanskrit legends in their communicative aim. The Sanskrit text here constitutes an attempt to translate the Arabic *šahāda* on the obverse, as is the case for the marginal legend discussed above.

The message communicated by the three Sanskrit sentences of this legend (see below Table 1) indeed matches that of the Arabic *šahāda*. In this interpretation, the Sanskrit term *avyakta* (unmanifest) was chosen by the translator(s) to render "Allah." As shown in Section 2.4 below, al-Bīrūnī and his informants who possibly lived in Gandhāra and Panjab, namely in the same geographical context from which these legends originate, knew Yoga and Sāṅkhya philosophies. Further, al-Bīrūnī defined the concept of *avyakta* in the *Taḥqīq mā li-l-Hind* in a way similar to the Sāṅkhya philosophy as exposed in the *Sāṅkhyakārikā* and its commentaries.[129] In this philosophy, *avyakta* is a key concept. It refers to the original cause (*prakṛti*), also known as the primary matter (*pradhāna*), which is the unique active origin of the phenomenal world, whereas God (*īśvara*) does not play an active role in the creation of existents.[130]

The Sanskrit word *deva* (deity) that one could have expected here is in general not used to refer to Allah by Persian and Arabic medieval authors, who rather relate *deva* to the concept of angels (ملك).[131] Further, while the *deva*s and the angels are several, *avyakta* and Allah are unique. The translator(s) of the legend had considered this feature when using the Sanskrit

---

126 Two variants (A and B) of the Arabic central legend also exist (Cappelletti 2015: 72, 87 and 93). The differences between these variants are, however, relatively minor, as far as the present discussion is concerned.

127 See the discussion by Cappelletti (2015: 69–72).

128 Thomas first described one of these coins in 1847 (no. XLII) (1847: 269–270 and 323–324). After the finding of new exemplars of this type of coin, Thomas revised his reading (nos. 11 to 14) (1859: 22–24). I follow here Agrawala's reconstruction (1943: 156, lines 1–3).

129 See below pp. 86, 104 and 148.

130 See chapter 3, section 2 and chapter 6, section 3.3 below.

131 See Ernst 2003: 177.

word *eka* (one) to transmit the idea of God's uniqueness.[132] The use of the Sanskrit term *avyakta* to translate "Allah" is also supported by variant 1 of the Sanskrit marginal legend, which has the expression *avyaktīya nāme* as a probable rendering of the Arabic *bi-smi-llāh* (بسم الله), meaning "in the name of Allah."[133]

Thus, the concept of *avyakta* is the Sāṅkhya notion that best renders the concept of the Islamic God, as the unique creator of the phenomenal world. Another parallel between *avyakta*, that is, *prakṛti*, and Allah lies in the fact that they are conceived to be invisible to the common senses of perception. The term *avyakta* appears in the *Bhagavadgītā*, and its use in the Sanskrit legend of the bilingual coins could also be drawn from this work.[134] Al-Bīrūnī abundantly quoted both the *Kitāb Sānk* and the *Kitāb Gītā* in the *Taḥqīq mā li-l-Hind*. These two books were thus presumably scriptures well known among his informants. Therefore, it is possible that principles of Sāṅkhya or of the *Bhagavadgītā* lie behind the use of the term *avyakta* in the Sanskrit legend of this coin.

Furthermore, the Sanskrit term *avatāra* (descent of a deity upon earth) appears to translate the Arabic *rasūl* (رسول; messenger), referring to Muhammed. The word *avatāra* does not constitute a literal translation of *rasūl*. Each of the terms bears distinct meanings according to the two religious contexts to which they belong, namely Hinduism and Islam, and at the same time shares the common point of referring to a figure who connects the divine sphere to the human world.[135] Table 1 below shows the parallels between the two legends and illustrates the above discussion.

This method of interpretation that consists of substituting the source-concept with a target-concept whose meanings only partly overlap, such as in the cases of *avyakta* translating Allah and *avatāra* rendering *rasūl*, reminds one of al-Bīrūnī's methods when he prepared his *Kitāb Sānk* and *Kitāb Pātanǧal*.[136] Did al-Bīrūnī himself translate—or help the translation of—the Arabic legend of these bilingual coins? The similarity between the way the *šahāda* was translated into Sanskrit on these bilingual coins and the way in which al-Bīrūnī translated Sanskrit literature into Arabic may suggest so, although this cannot be demonstrated. At any rate, if he had, he must have been helped by

132 See also Cappelletti 2015: 95–96 on the interpretation of this expression.

133 Cappelletti 2015: 72 and 82.

134 Cappelletti 2015: 94–95.

135 The term *avatāra* was also used in a Kālacakra text, the *Vimalaprabhā*, to qualify Muhammed (Cappelletti 2015: 98–102). See also Chatterji 1951: 96–97.

136 For a definition of the translational strategy of substitution, see below Section 4.4.2.

TABLE 1 Synthesis of the text of the bilingual dirhams of Lahore, based on Cappelletti[a]

Arabic	Sanskrit	Message
There is no God but Allah (لا اله الا الله).	The unmanifest is one (*avyaktam ekaṃ*).	God / the creative cause is unique.
Muhammed is the messenger of Allah (محمد رسول الله).	Muhammed is an incarnation [of God] (*muhammado 'vatāraḥ*).	Muhammed is an intermediary figure between the human and divine worlds.
Maḥmūd, the right hand of the state, the custodian of religion (يمين الدولة و أمين الملة محمود).	Maḥmūd is the king (*nṛpatir-mahmūdaḥ*).	Maḥmūd possesses the authority associated with his status.

a Variant B of the Arabic legend is represented in this table. See Cappelletti 2015: 87 and 93–102.

Indian scholars well acquainted with a form of Sāṅkhya philosophy and/or the *Bhagavadgītā* who conveyed to him the original concepts and were able to connect them with Islamic concepts and culture. Lastly, the above observations indicate that the method of cultural translation was common at the time.[137]

Nevertheless, to come back the main discussion of this section, the bilingual coins were intended to legitimize the Ghaznavid power in the region.[138] The transmission of the common Islamic *šahāda* in the form of a Sanskrit legend likely served as a means to enhance Maḥmūd's authority through the appropriation of Indian concepts referred to by *avyakta* and *avatāra*. The Sanskrit legend addresses a non-Muslim Indian audience. Maḥmūd, expecting that illiterate Muslims would at least recognize the *šahāda* as a symbol, may have assumed the same for an Indian audience. However, as there exists no such thing as the *šahāda* in the Sanskrit tradition, local inhabitants of the region of Lahore would not have been able to recognize the Sanskrit legend as a symbol. Thus, despite the ruler's attempt to integrate local traditions by way of the text of these coins, the Sanskrit legend probably had less impact on the popu-

137 Bhattacharyya 1964: 56.

138 Bhattacharyya (1954: 115–116) provides other examples of attempts of legitimization by using pre-existing symbols on coinage.

lation than desired by those who minted the coins.[139] Even so, the example of these coins clearly illustrates the ruler's efforts to integrate local concepts for the sake of establishing his authority in the region and reveals that intercultural exchange took place between Maḥmūd's administration and local communities in Panjab at the beginning of the eleventh century.

After Maḥmūd's death, the city of Lahore, Mandahūkūr in al-Bīrūnī's writings, became the eastern capital of the Ghaznavid Empire and the outpost for the administration of the subjugated provinces. Governors were appointed in Lahore by the Ghaznavids and a Muslim community established there at an early date.[140] In addition to the Ghaznavid attempt to control Lahore, literary sources reveal that the dynasty posted governors in Nagarkot[141] and Nandana,[142] and that the principles of Islam were being practiced in Bhātinda.[143] Arabic and Persian primary sources also record that Maḥmūd appointed a certain Sukhpāl to administer Multan.[144] No extant account suggest similar endeavours in cities such as Taneshwar, Kanauj or Somnāth, located further east and south in present-day India.

Further, the relative proximity of Ghazna to Gandhāra, to Panjab and to Sind suggests that these areas were more accessible to al-Bīrūnī than other provinces of al-Hind. These three regions were rather close to the centre of the Ghaznavid Empire compared to other places further east. Stabilization following Muslim incursions was often a long process, especially in regions distant from the Islamic centre. For instance, an entire century was necessary for the official establishment of Islam in Khwarezm (from the early 8th to early 9th c.) and approximately three centuries were required in Kābul (between the end of the 7th and 10th c.).[145] The Ghaznavids had to repeat attacks on territory in al-Hind, including Laghmān, Wayhind, Multan, Nandana, Kanauj and Fort Lahūr, in order to establish and maintain control.[146]

If, in the areas of Panjab and Sind that were geographically close to Ghazna, political trouble between the Ghaznavids and the local rulers existed, it is likely that additional tension also occurred in the regions farther east. The region of Khwarezm and some parts of al-Hind were particularly far from Ghazna. Such

139 This observation has been inspired by a discussion with Ms. Sara Cappelletti.

140 In 1163, the Ghaznavids lost Ghazna and established their government in Lahore (Bosworth 2007: 147; Jackson & Andrews 2012).

141 Nazim 1931: 90.

142 Id. 93.

143 Id. 101.

144 Rehman 1979: 149–150 and 326–328.

145 See above pp. 13–15.

146 Nazim 1931: 29, 86–99 and 104–113.

remoteness prevented the Ghaznavids from holding them under their rule.[147] Thus, I posit that due to the greater distance between Ghazna and some territories conquered by Maḥmūd, the sultan only conducted intermittent raids there, rather than establishing his authority through a governor or other officials. Al-Bīrūnī, however, needed a long-term cooperation with Indian scholars in order to gather his material on India and to pursue his translations of Sanskrit literature into Arabic. Consequently, I suggest that he did not visit faraway places in central and southern India, let alone stay for an extended period of time there.

Lastly, scholars largely presumed that al-Bīrūnī always accompanied Maḥmūd in his conquests of the East, and thus visited every place attacked by the sultan. This led some to conclude that al-Bīrūnī could observe the culture and customs of many regions of al-Hind. The subsequent paragraphs pose the question of his position at the Ghaznavid court and the extent to which he accompanied the sultan in his military campaigns in the East.

Al-Bīrūnī's contemporary, Bayhaqī, reveals that conditions of the officials at the Ghaznavid court changed in accordance with the plots being orchestrated at the court, the sultan's dispositions and other officials' behaviours.[148] To start with, the following anecdote related to the poet Firdawsī exemplifies the insecurity and volatility of official positions at Maḥmūd's court. Having presented his epics to the sultan, the poet was not satisfied with his reward. After expressing his discontent, he was forced to go into exile in order to survive.[149] While the details of this story vary from author to author and may not all be historically accurate, it offers a portrayal of Maḥmūd's reputation and attitudes toward members of his court.

Al-Bīrūnī's status at the court was also likely precarious. In the following passage drawn from the *Taḥqīq mā li-l-Hind*, al-Bīrūnī states:

> I have found it very hard to work my way into this [subject] (i.e., al-Hind), despite my desire which I alone possess in my time, and although I sacrificed myself generously and as much as possible in collecting their books from places where they were likely to be found and in gathering those who were on the right way to [find] them from places where they were hidden. Who else besides me has the same [opportunity to learn this subject],[150]

147 Bosworth 1963: 73.

148 See for instance Bosworth 1963: 60–61 and 64.

149 Huart & Massé 2012.

150 Sachau's Arabic edition (*Taḥqīq* [1887]) here offers a better reading of the text than the Hyderabad edition (*Taḥqīq* [1958]).

> unless he were endowed with Allah's help, which I was deprived of in my capacity of movement, in which I was unable to come and go completely freely and independently. Thank Allah for what he granted me.[151]

Al-Bīrūnī was clearly aware of his reliance on the sultan's benevolence, and, simultaneously, his precarious position at Maḥmūd's court. On the one hand, he benefited from some support for his research, and, on the other, he had to subordinate himself to the ruler's will. This passage does not, however, reveal the extent to which al-Bīrūnī was dependent upon Maḥmūd and his court. In addition, in the postface to the *Kitāb Pātanǧal*, he explains:

> As for the impossible [things] which are [referred to] in this book (i.e., the *Kitāb Pātanǧal*), they can be accounted for in two ways. [...] The second way is that the Indians have a greater propensity [for recounting absurd things] and a lesser one for reflection and study to such an extent that I could only compare their books on astronomical calculations—with respect to the meaning and with respect to the order and the arrangement. When pearls are mixed with dung, and jewels with clay, the [Indians] are not rightly guided to distinguish between these and they make no effort to study them and refine them [...].[152]

In this passage, al-Bīrūnī explicitly rejects the very doctrines that he had detailed in the *Kitāb Pātanǧal*. It is possible that, as a Muslim scholar, he disagreed with some philosophical principles presented in this book. He may also have included these comments to guard himself against a censorship by Maḥmūd, who would have considered the *Kitāb Pātanǧal* unorthodox.

---

151 *Taḥqīq* (1958), p. 18.5–10. Sachau 1910: 1/24. The last portion of the quotation chiefly consists of idiomatic Arabic expressions. Therefore, I also provide Sachau's translation of the full passage: "I have found it very hard to work my way into the subject, although I have a great liking for it, in which respect I stand quite alone in my time, and although I do not spare either trouble or money in collecting Sanskrit books from places where I supposed they were likely to be found, and in procuring for myself, even from very remote places, Hindu scholars who understand them and are able to teach me. What scholar, however, has the same favourable opportunities of studying this subject as I have? That would be only the case with one to whom the grace of God accords, what it did not accord to me, a perfectly free disposal of his own doings and goings; for it has never fallen to my lot in my own doings and goings to be perfectly independent, nor to be invested with sufficient power to dispose and to order as I thought best. However, I thank God for that which He has bestowed upon me, and which must be considered as sufficient for the purpose."

152 KP, pp. 199.7–200.4; Pines & Gelblum 1989: 272.

The Persian version of al-ʿUtbī's work, the *Kitāb-i-Yamīnī*, records that Maḥmūd kept many captives from his military campaign in Khwarezm at the court, without specifying their identity or social rank. This work further comments that the captives were held in Ghazna and later sent to various regions throughout al-Hind.[153] Al-Bīrūnī may have been among these men who were held captive in Ghazna or al-Hind, although no definitive evidence supports this hypothesis.

The opinion of modern scholars on the issue of al-Bīrūnī's freedom and position during Maḥmūd's reign is divided.[154] The above remark by al-Bīrūnī does not necessarily indicate that he was a prisoner, but simply dependent on the royal court. However, in my view, he benefited at least from the space, resources and time necessary to pursue his work. In any case, al-Bīrūnī stayed for approximately thirty years (from 1017 to ca. 1050) at the Ghaznavid court, thirteen of which (from 1017 to 1030) were under Maḥmūd's patronage. Therefore, whatever problems occurred between the scholar and the sultan, the two did collaborate for quite some time.

Evidence exists that some members of Maḥmūd's court accompanied the sultan when he travelled. For instance, Farruḵī, a poet at the Ghaznavid court, states that he accompanied Maḥmūd on some of his conquests in al-Hind, notably to Somnāth, Kathiawar, Bulandshar, Kanauj and Taneshwar, as well as at the time of the sultan's battle with the king Trilocanapāla.[155] Bosworth notes that Bayhaqī and Gardīzī accompanied Maḥmūd during some of his campaigns[156] and explains that the *Dīwāns*, that is, governmental bodies, generally followed the royal courts.[157]

Bayhaqī explains how a court official was required to organize and equip the sultan's quarters, which included providing herds of sheep to allow the sultan to welcome guests wherever he was.[158] Furthermore, Maḥmūd's army required the contributions of engineers, prospectors, blacksmiths and others in order to enable the army to proceed, for instance by building roads and strongholds on the way to the foreign lands or by providing the facilities required by Maḥmūd's military campaigns. In addition, numerous soldiers belonging to the army, elephants and other military equipment were part of Maḥmūd's expeditions.[159]

153 Reynolds 1858: 448.
154 Sachau 1910: I/ix–xvi; Shamsi 1979: 270; Said & Khan 1981: 70–82.
155 Bosworth 1991: 43.
156 Bosworth 1963: 127.
157 Bosworth 2004: 18.
158 Quoted in Bosworth 1963: 65.
159 Bosworth 1963: 118.

These were large-scale excursions, and their organization necessitated the participation of a variety of specialists. Experts in various domains thus escorted Maḥmūd during his travels.

While al-Bīrūnī may have accompanied Maḥmūd during some of his military campaigns to the East, at times the scholar also certainly resided in places secured by the Ghaznavids. In addition, the sultan consulted with al-Bīrūnī as an astronomer and adviser to the court. Since early times, astronomers and astrologers assisted Muslim rulers in planning their military campaigns or other political matters. During the Delhi Sultanate, historians also counselled the rulers on political matters.[160] At least two known examples indicate that al-Bīrūnī may have held such a position under Maḥmūd. In 1024 in Ghazna, the scholar met with a delegation sent by the Volga Turks and in 1026 with another one attached to the dynasty K'itan which ruled over southern Manchuria and northern China at the time.[161]

Further, due to his knowledge of Indian science and language, al-Bīrūnī certainly assisted the sultan as an interpreter and mediator in his interactions with Indians.[162] One story narrates how Maḥmūd, returning from his raids against Mathura and Kanauj in 1018/1019, had met al-Bīrūnī somewhere between Kābul and Ghazna. Maḥmūd then showed him a precious stone stolen from a temple situated in Mathura.[163] While it remains uncertain whether this incident really occurred, this narrative suggests that al-Bīrūnī did not travel with Maḥmūd's army and court to Kanauj and Mathura. More convincingly perhaps, in *The History of Khwarazm* as handed down by Bayhaqī, a reference is made to al-Bīrūnī who narrates an episode during which Maḥmūd sent him an envoy from al-Hind while he was in Kābul. Thus, both pieces of evidence indicate that al-Bīrūnī did not always accompany the sultan on his military campaigns.[164]

If the scholar had accompanied Maḥmūd on some of his campaigns, it is, however, difficult to know exactly at which dates. Al-Bīrūnī visited India after 1017, more than fifteen years after Maḥmūd's first raids in Laghmān and Peshawar, in 1000 and 1001 respectively. Thus, based on the above, al-Bīrūnī may have travelled during the sultan's campaign of Fort Lahūr (1021/1022), Gwalior/Kalinjar (1022) and Somnāth (1025/1026). However, as shown in the present chapter, al-Bīrūnī did not venture eastward beyond Panjab during his

160 Auer 2012: 16.
161 Minorsky 1951: 234–235; Said & Khan 1981: 80–81; Tetley 2009: 65–66.
162 On intercultural exchanges between Indians and Muslims taking place at the Ghaznavid court, see below Sections 2.2 and 2.3.
163 Said & Khan 1981: 76–77.
164 Bosworth 2011e: 376–377.

visits to al-Hind. Thus, I conclude that the scholar accompanied Maḥmūd to the East during his journeys but did not necessarily travel as far as the actual battlefields. He indeed probably remained in regions where Maḥmūd had secured a certain level of political stability through his earlier raids.[165] It is also likely that al-Bīrūnī at times travelled with a military escort independent from Maḥmūd and thus spent time in places where the Ghaznavids had already established some authority, enabling him to interact with Indian Brahmins.

1.2.4 *The Various Sources of Information*

In the preceding sections, I presented evidence from al-Bīrūnī's writings and from his socio-political context showing that he chiefly travelled in Gandhāra and Panjab. Al-Bīrūnī's descriptions in the *Taḥqīq mā li-l-Hind* of numerous places in India, however, led scholars to believe that he personally visited many regions of al-Hind. Yet, as I argue below, many of his descriptions are not based on direct experience, but on oral and written sources.

For instance, in the following passage al-Bīrūnī suggests that he did not personally see the regions of Kashmir and Varanasi:

> This is the reason,[166] too, why their (i.e., the Indians) sciences have disappeared beyond the limits [of the world] conquered [by the Muslims] and have fled to places where our hands cannot reach, namely Kashmir [كشمير], Varanasi [بانارسى] and other similar [places].[167]

In all likelihood, the toponym Kashmir here stands for the Kashmir Valley, as al-Bīrūnī describes it as "a plain that high and inaccessible mountains surround."[168] This extract is unique in that it explicitly indicates places where al-Bīrūnī did not visit at least prior to his composition of the *Taḥqīq mā li-l-Hind*: the Valley of Kashmir and Varanasi.[169] This passage also suggests that the scholar was not able to cross the frontiers of the world conquered by the Ghaznavids, and rules out the possibility of him having travelled to South India. Another passage concurs with this latter observation:

165 This was suggested by Said and Khan (1981: 84–86).

166 Al-Bīrūnī refers here to the animosity of the Indians towards Muslims due to Maḥmūd's invasions.

167 *Taḥqīq* (1958), p. 16.17–18; Sachau 1910: I/22.

168 *Taḥqīq* (1958), p. 165.11; Sachau 1910: I/206.

169 In his *Pharmacology* (*Kitāb al-ṣaydana fī l-ṭibb*), compiled at the end of his life in approximately 1050, al-Bīrūnī asserts that he had seen apples in Kashmir (Said 1973: 91, under the entry *tuffaḥ*, no. 20).

> Before, one or two foreigners could enter [Kashmir], especially Jews. Now, they do not let any Indians whom they do not know [enter it], let alone the others.[170]

Even so, al-Bīrūnī abundantly refers to Kashmir in the *Taḥqīq mā li-l-Hind*. He describes geographical, ethnic and social features at length, names cities and mountains, adduces itineraries leading to the Kashmir Valley, mentions Kashmiri customs,[171] reports on the alphabets and scripts in use there,[172] and presents detailed accounts of religious and astronomical practices.[173] He portrays the Kashmir Valley in more detail than any other region discussed in the *Taḥqīq mā li-l-Hind*. Since al-Bīrūnī did not visit it, he must have based his account of Kashmir on sources of information other than his direct observation.

A few passages in al-Bīrūnī's writings confirm this statement. In 1036, the scholar compiled a catalogue (فهرس) of the works of the physician and philosopher Muḥammad Ibn Zakariyyā' Rāzī (ca. 854–925/935). This catalogue also includes a list of his own works.[174] This auto-bibliography notably provides the following two titles: *Answers to the Questions of the Astronomers of al-Hind* (الجوابات عن المسائل الواردة من منجمى الهند)[175] and *Answers to the Ten Kashmiri Questions* (الجوابات عن المسائل العشر الكشميرية).[176] These works are no longer extant, but their titles indicate that al-Bīrūnī interacted in some manner with Indian astronomers and with residents of Kashmir.

Several passages found in the *Taḥqīq mā li-l-Hind* also indicate that al-Bīrūnī had some contact with Kashmiri thinkers. For instance, regarding a festival celebrated in Kashmir al-Bīrūnī draws his information from the account of a certain Jīvaśarman. On this festival, he states: "The people of Kashmir whom I have seen do not agree with this [account] regarding the place and the time [of the festival]."[177] In another passage, al-Bīrūnī mentions calendars of the year 951 of Śakakāla[178] which had been brought from Kashmir.[179] In addition, a fur-

170 *Taḥqīq* (1958), pp. 165.19–166.2; Sachau 1910: I/206.
171 Sachau 1910: I/206–208.
172 Sachau 1910: I/173–174.
173 Sachau 1910: I/393, I/116–117 and II/178.
174 Boilot 1955: 165–166.
175 Boilot 1955: 199, no. 71.
176 Boilot 1955 200, no. 72.
177 *Taḥqīq* (1958), p. 489.10–15; Sachau 1910: II/181.
178 The year 951 of Śakakāla corresponds to the year 1029 CE. See the online converter at http://www.cc.kyoto-su.ac.jp/~yanom/pancanga/index.html [accessed October 2023].
179 *Taḥqīq* (1958), p. 328.9–10; Sachau 1910: I/391.

ther statement by the scholar testifies to intellectual interaction between him and certain Kashmiris. He declares: "We have verified these [methods] in the *Zīǧ* (i.e., astronomical handbook) which we have composed for Syāvapala (?) (سياوپل) the Kashmiri."[180]

These passages show that al-Bīrūnī met Kashmiris and exchanged books with them. M.S. Khan, noting the comprehensiveness and accuracy of al-Bīrūnī's account of Kashmir, declares that Kashmiri scholars probably helped him in gathering information;[181] this remark finds support in the above observations. Thus, although the Kashmir Valley was unreachable for al-Bīrūnī and his peers, intellectual exchange were taking place between the territory of Gandhāra and Panjab on the one hand, and that of Kashmir on the other. These interactions enabled al-Bīrūnī to describe at length a region that he had not visited himself at the time of the composition of the *Taḥqīq mā li-l-Hind*.

As for other regions of India, the *Taḥqīq mā li-l-Hind* also reveals that al-Bīrūnī communicated with people from Kanauj, Multan and Somnāth.[182] As a result, the scholar could convey pieces of information about Kanauj, the alphabet and the calendar in use there, and its history.[183] As for Somnāth, al-Bīrūnī informs his readers about the year when Maḥmūd attacked its temple, namely 416 AH (1025/1026),[184] provides a detailed account of its idol and reports some myths associated with the temple.

Further, in Chapter 7 of the *Taḥqīq mā li-l-Hind*, al-Bīrūnī discusses the views on cosmography in the *Ādityapurāṇa*, the *Vāyupurāṇa*, the *Viṣṇupurāṇa* and the *Matsyapurāṇa*. He furnishes comparative tables with the different names of the regions of the world presented in these texts. In two tables, he provides the names of netherworlds (*pātāla*), oceans and islands (*dvīpa*) which he had heard (مسموع من الألسنة).[185] Not only had he recourse to an interlocutor for information related to subjects such as cosmography; he was also able to supplement his knowledge of such topics based on Sanskrit literature with information provided by this oral source. However, al-Bīrūnī did not identify his interlocutor or mentioned his place of origin.

180 *Taḥqīq* (1958), p. 512.18–19; Sachau 1910: II/208. The name has not yet been identified. Sachau makes some assumptions about this figure, but with little certainty (Sachau 1910: II/400).

181 Khan 1976: 92, n. 28.

182 *Taḥqīq* (1958), p. 125.5–6, 129.3–4, 170.4–5, 347.15–18 and 451.4–5; Sachau 1910: I/161, 165, 211, II/9 and 129. The case of Multan, as a place where al-Bīrūnī may have been, is discussed below p. 77–78.

183 Sachau 1910: I/173, 199, II/5, 9 and130.

184 *Taḥqīq* (1958), pp. 347.20–348.2; Sachau 1910: II/9.

185 *Taḥqīq* (1958), pp. 187 and 193; Sachau 1910: I/230 and 235.

The *Taḥqīq mā li-l-Hind* contains other elements pointing to the importance of al-Bīrūnī's interactions with Indians. Al-Bīrūnī refers to a traveller who communicated with him on the area located to the north-east of Varanasi and the realm of Nepal.[186] It is furthermore possible to deduce from the *Taḥqīq mā li-l-Hind* that pilgrims also transmitted him information.[187] In addition, al-Bīrūnī's account of the Early and Late Shahis originates from an oral report. He states: "According to what I have heard, [the history of] such a lineage, [written] on a piece of silk, is found in the fortress of Nagarkot."[188] In the same passage, al-Bīrūnī explains that he could not find this silk document. The person(s) who informed him of its existence may have been the very narrator(s) of this royal chronicle. The scholar mentions the titles of several grammar books about which he came to know through an oral account.[189] Although he did not visit many provinces of al-Hind himself, al-Bīrūnī met people, such as merchants, ascetics and pilgrims from various parts of India.[190]

Furthermore, al-Bīrūnī interacted with Indian scholars—Brahmins in general, and specifically astronomers and philosophers—some of whom must have belonged to the court of the Late Shahis, as I show below.[191] Other passages indicate that al-Bīrūnī drew on oral sources concerning the custom of eating beef and the status of low-caste people vis-à-vis Brahmins.[192] Al-Bīrūnī also devoted a complete chapter of the *Taḥqīq mā li-l-Hind* to the description of the lifestyle and duties of the Brahmins.[193]

Al-Bīrūnī provided references and descriptions of certain places located in al-Hind. These include Mandahūkūr, namely modern Lahore in Panjab, Mathura and Taneshwar. The scholar sparsely refers to Gujarat, Prayāga (Allahabad), the Kannara region, Varanasi and some places in present-day north-eastern India. His description of various itineraries starting from Kanauj also suggests that this information was orally transmitted to him, because these routes link

186 *Taḥqīq* (1958), p. 160.5–6; Sachau 1910: I/201.

187 *Taḥqīq* (1958), p. 466.5–6; Sachau 1910: II/148.

188 See the full excerpt from the *Taḥqīq mā li-l-Hind* presented above pp. 17–18.

189 *Taḥqīq* (1958), pp. 104.14–105.1; Sachau 1910: I/135.

190 In the *Taḥdīd al-amākin*, al-Bīrūnī collects pieces of information about distances between cities from travellers' accounts (Ali 1967: 14). On different sources of information related to the territory of al-Hind described in the *Taḥqīq mā li-l-Hind*, see Verdon 2015: 43–45.

191 *Taḥqīq* (1958), pp. 17.16–18.5, 456.12 and 475.14; Sachau 1910: I/23–24, II/134 and 163. On his interactions with Indian scholars, see further Section 2.3 below.

192 *Taḥqīq* (1958), p. 458.2–7; Sachau 1910: II/152–153.

193 *Taḥqīq* (1958), pp. 452.5–457.7; Sachau 1910: II/130–135.

many cities or regions of India he could not possibly have experienced first-hand. His account includes places on the eastern coast of present-day India (modern West Bengal), in the North (modern Nepal, Kashmir), in the North-East (modern Assam), in the centre (modern Bihar, Madhya Pradesh and Uttar Pradesh), in the South-West (modern Sind and Gujarat) and in the South (Kannara). It cannot be concluded, however, that al-Bīrūnī travelled to all of these places. Even though he demonstrates extensive knowledge of al-Hind, this knowledge was not necessarily based on personal visits. Instead, it wasoften accounted for by his interactions with Indians.

In addition, al-Bīrūnī accessed a large number of written sources, which built up his knowledge on the culture and sciences of al-Hind.[194] He was acquainted with the Vedas, the *Smṛti* of Manu, the *Viṣṇudharmottarapurāṇa*, which he entitles *Viṣṇudharma*, and the *Viṣṇupurāṇa*, through oral accounts of Brahmins or quotations found in other books he read.[195] He also consulted the *Ādityapurāṇa*, the *Matsyapurāṇa*, the *Vāyupurāṇa*,[196] the *Bhagavadgītā*—referred to as the *Kitāb Gītā* by him—and the *Mahābhārata*. Al-Bīrūnī rendered into Arabic the two foundational texts related to the Sāṅkhya and Yoga schools of thought, respectively, the *Kitāb Sānk* and *Kitāb Pātanǧal*. He also translated—or began translations of—astronomical works, such as Brahmagupta's *Brāhmasphuṭasiddhānta*, referred to by him as the *Brāhmasiddhānta*, the *Pauliśasiddhānta* by Puliśa, and the *Bṛhatsaṃhitā* and the *Laghujātaka* by Varāhamihira.[197] Lastly, he quotes works whose authors were known to him, but have yet to be identified: the *Srūdhava* by Utpala from Kashmir, the *Karaṇatilaka* by Vijayanandin from Varanasi, and certain works by Vaṭeśvara, who hailed from Nāgarapura, by Durlabha, a native of Multan, Śrīpāla and Jīvaśarman.[198]

Thus, thanks to informants coming from many regions of al-Hind and to accessing a large amount of Sanskrit literature, al-Bīrūnī gathered data that

194 See the list of al-Bīrūnī's literary sources in Sachau 1910: I/xxxix–xl and Shastri 1975. See also Mishra 1985: 35–43.

195 Although the *Viṣṇupurāṇa* is quoted at length in the *Taḥqīq mā li-l-Hind*, al-Bīrūnī admits that he could not read it himself. See *Taḥqīq* (1958), p. 101.5; Sachau 1910: I/130–131.

196 Jan Gonda (1951) scrutinizes the way in which al-Bīrūnī transmits information drawn from the Purāṇas, solves several of Sachau's doubts about the identification of Arabic transliterations of Sanskrit proper names, and states that some of al-Bīrūnī's readings might be valuable for scholars interested in purāṇic studies.

197 *Taḥqīq* (1958), pp. 6.2, 119.8–9, 122.5–6 and 327.2; Sachau 1910: I/8, 154, 158 and 389.

198 *Taḥqīq* (1958), pp. 121.6–13, 122.4, 128.17, 198.5, 250.2, 281.19, 304.15, 309.2, 348.6, 388.11, 489.10 and 490.1; Sachau 1910: I/156–157, 164, 240, 298, 334, 361, 367, II/9, 54 and 181–182.

exceeded the scope of his travels in al-Hind. In contrast, I have presented evidence above that delineates the territory of al-Hind visited by al-Bīrūnī, a territory mostly confined to Gandhāra and Panjab. As a result, I hypothesise that some of his interlocuters belonged to this region and to the society living there, and that they constituted the primary layer of people he interacted with. Therefore, in the subsequent section, I discuss some preliminary socio-cultural traits of this area and, in Chapter 2, I further investigate this hypothesis.

1.3 Elements of Culture of Gandhāra and Panjab

1.3.1 *Laghmān, Peshawar, Fort Rājagirī, Fort Lahūr and Fort Nandana*

The present section focuses on the historical and social contexts of the five locales that al-Bīrūnī certainly visited, namely Laghmān, Peshawar, Fort Rājagirī, Fort Lahūr and Fort Nandana, all located in today's eastern Afghanistan and Pakistani Gandhāra and Panjab.[199] These five sites belonged to the kingdom of the Late Shahi dynasty, immediately prior to al-Bīrūnī's journeys there. Arabic and Persian sources detail the encounters between Alptigīn, Sebüktigīn and Maḥmūd, and these local rulers.[200] Sebüktigīn (977–997)[201] launched several raids against the Late Shahis in Kābul as well as in the regions of Laghmān[202] and Peshawar. Maḥmūd continued the attacks against the Late Shahis, mainly in regions stretching from present-day eastern Afghanistan to Pakistan. He defeated four kings of this dynasty: Jayapāla (ca. r. 964–1002),[203] Ānandapāla (ca. 1002–1010),[204] Trilocanapāla (ca. 1010–1021/1022)[205] and Bhīmapāla (ca. 1021–1026/1027).[206] These kings, originally established in Kābul,

199 As for other places located in al-Hind and whose latitudes al-Bīrūnī calculated, such as Kindī, Dunpūr, Jhelum and Sialkot, they are not dealt with here because the *Taḥqīq mā li-l-Hind* contains too little information on them, despite their possible significance for al-Bīrūnī's encounter with Indians.

200 See for instance Nazim 1931 and Rehman 1979: 125–167.

201 Nazim 1931: 28–33.

202 Reynolds 1858: 38.40. Laghmān, Laghman or Lamghan was situated in eastern Afghanistan, lying on the northern side of the Kābul River; see Rehman 1979: 13.

203 Id. 469. The dates of the reigns of Late Shahi kings are drawn from Rehman 1979: 89–167. For a comparative table of the different datings proposed in Pandey 1973: 80–114 and Mishra 1972: 9–223, see Khan 2017: 48.

204 Rehman 1979: 4, n. 17 and 2003: 3–4; See also Reynolds 1858: 327–328. For more references on the Late Shahis, see above fn. 56 in Chapter 1 of the present book.

205 The *Rājataraṅgiṇī* describes a battle between Maḥmūd and Trilocanapāla (Majumdar 1957: 67). See also Rehman 1979: 4, n. 18.

206 Nazim 1931: 86–121.

shifted their capital city most probably at first to Wayhind/Udabhāṇḍa, and later on to Nandana in the Salt Range and to Lahore, ultimately taking shelter in Kashmir.[207]

A preliminary examination of archaeological data and literary sources allows for a reconstruction of some details regarding these five locales. Laghmān is the first place which will be dealt with here, as it is the city closest to Ghazna. When al-Bīrūnī visited this region, he observed a solar eclipse, which he describes in the *Taḥdīd al-amākin*:

> Again, though they (i.e., the Khurasanian calculators) had not discussed the solar eclipse that took place in Dhū al-Qaʿda, year four hundred nine of the Hijra, the reserved amongst them said that it would occur below the horizon of Ghazna, and that it would not be seen there. However, it happened that we were near Lamghān, between Qandahār (i.e., Gandhāra)[208] and Kābul, in a valley surrounded by mountains, where the sun could not be seen unless it was at an appreciable altitude above the horizon. At sunrise, we saw that approximately one third of the sun was eclipsed and that the eclipse was waning.[209]

Archaeological excavations have not yet been conducted in Laghmān, which makes it difficult to reach conclusions on the site or on the type of society that lived there. However, in 1960 the head of a statue, probably dating to the second half of the first millennium, was found by accident in the region on a mound named Qal'a Amir Muhammad (Tagao). According to Klaus Fischer who examined it, the head may be associated with the Early or Late Shahi dynasties.[210] It represents a female goddess, Durgā Mahiṣāsuramardinī or Pārvatī.[211] In addition, eight (Proto-)Śāradā inscriptions are found in caves at

---

207 Al-Bīrūnī describes Wayhind/Udabhāṇḍa as the capital of Gandhāra (*Taḥqīq* [1958], pp. 165.8–9 and 215.7–216.1; Sachau 1910: I/206 and 259; *Qānūn* (1955), p. 562; *Ǧamāhir*, p. 236.8; Said 2001: 293), while al-Muqaddasī (ca. 945/46–991) relates that it is a provincial capital (referred to in Bosworth 1970: 254 and Rehman 1979: 17). On Udabhāṇḍa, as an administrative and political centre for the Late Shahis, see Verdon 2021, and for a discussion on the status of Lahore at the time of Jayapāladeva and his son Ānandapāla, see Rehman 1979: 139–141. See also Cunningham 1871: 52–54; Stein 1893: 198–200 and 1900: II/337; Rehman 1979: 4.

208 Arabic textual sources used the name Qandahār for two different places: a place located in south-eastern Afghanistan and a second one corresponding to Gandhāra in Peshawar region.

209 Translation by Jamil Ali (1967: 261). See also *Taḥdīd* [1992], p. 291.21–292.3.

210 Fischer 1964: 38. See also the brief description in Kuwayama 2002: 225–226.

211 Fischer 1964: 37–38. Whereas Durgā is honoured by herself, Pārvatī is almost exclusively worshipped as the spouse of Śiva.

Laghmān.[212] They, however, deserve examination to either decipher them or to update their existing decipherment in order to provide any historical information.

Literary sources indicate that the city was an important site during the last centuries of the first millennium CE. The account by Xuanzang, who visited Laghmān in the early seventh century, bears witness to the importance and prosperity of the region located on a trade road. Xuanzang also reports that Laghmān belongs to a country of Brahmins.[213] In 982/983, the *Ḥudūd al-ʿālam* describes Laghmān as "an emporium of Hindūstān and a residence of merchants […] [which] possesses idol-temples."[214] Similarly, the Persian translation of al-ʿUtbī's *Kitāb-i-Yamīnī* portrays the region of Laghmān as one of the most prosperous of the time and as belonging to the land of the Late Shahi king Jayapāla.[215]

Al-Bīrūnī also mentions the city of Laghmān in the *Taḥqīq mā li-l-Hind*, in which he gives its latitude[216] and locates it across the stream of the River Sāwa (today's Alishing River ?).[217] In addition, when he discusses different calendars of al-Hind, he states that the people of Laghmān start the year with the month Mārgaśīrṣa (November–December).[218] He also provides an alternative name for the city of Laghmān: *Lanbaga* (لنبگا).[219] Since al-Bīrūnī most probably visited the region during the year 1019,[220] namely nineteen years after Maḥmūd's takeover of Laghmān, his descriptions of this region show that local people kept their calendrical systems, based on ancient Hindu traditions, even after Muslims reached the region.

Further, in the *Taḥqīq mā li-l-Hind*, al-Bīrūnī conveys his observations made in Peshawar. There, he witnessed local practices:

> After seven and a half *gaṭī* (گهری ; pl. گهریات) have elapsed, they beat the drum and blow a winding conch, named *śaṅkha* (شنگ) [in the Indian language] and *spīd-muhra* (سپید مهره) in Persian. I have seen this in the land of Peshawar (بلد پرشور).[221]

212 See Foucher 1947: 386–387; Pl. 37–38 and Humbach 1986.
213 Watters 1904: I/181–182; Kuwayama 2002: 204.
214 Bosworth 1970: 92.
215 Reynolds 1858: 35–40. See also Pandey 1973: 35.
216 *Taḥqīq* (1958), p. 270.9; Sachau 1910: I/317.
217 *Taḥqīq* (1958), p. 215.3–4; Sachau 1910: I/259; Rehman 1979: 13.
218 *Taḥqīq* (1958), p. 347.12–15; Sachau 1910: II/8–9.
219 The inhabitants of Laghmān are referred to as *lampāka* in the Purāṇas (Bhattacharyya 1991: 202).
220 See above pp. 29–30.
221 *Taḥqīq* (1958), p. 285.2–4; Sachau 1910: I/338.

The modern city of Peshawar lies in present-day northern Pakistan, east of Laghmān.[222] According to Xuanzang, the population and the wealth of the city, designated then as Puruṣapura, were declining in his time, as was the practice of Buddhism.[223] Al-Bīrūnī describes a timekeeping ritual involving the use of a conch (*śaṅkha*), a symbol of the Hindu god Viṣṇu. Nevertheless, in the absence of additional pieces of contextual information, it is difficult to determine which specific Indian religious group celebrated this ritual in the region of Peshawar. In addition to this passage, al-Bīrūnī mentions Peshawar at a few other places. He explains that it lies opposite of the Ghorvand River, to be identified with today's Kābul River,[224] and provides its latitude.[225] He also recalls that Kaniṣka had a *vihāra* built there.[226] His visit(s) to the region of Peshawar took place between the years 1017 and 1030. Lastly, al-Bīrūnī's observation of the above ritual shows that locals continued their religious practices several years after the Ghaznavids annexed their territory to their empire, as was observed above in the case of Laghmān.

The Persian *Kitāb-i-Yamīnī* based on the Arabic work *Al-Yamīnī* by al-ʿUtbī, in the description of Maḥmūd's attack on Peshawar, refers to the city as being located "in the midst of the land of Hindustan,"[227] thereby suggesting that this territory was outside the frontiers of the Islamic boundaries and possible inhabited by Hindus. According to a recent study, numerous archaeological sites of the Valley of Peshawar could be associated with the Late Shahi dynasty, most of which have not yet been studied.[228] Among those, the ancient site of Hund, that is, Wayhind/Udabhāṇḍa, was an important city for the Late Shahi kings, before the Ghaznavids definitively pushed them eastward, and most probably served as their main capital until the end of Jayapāla's rule in the year 1002. It is thus possible that the former eminence of Peshawar waned with Wayhind/Udabhāṇḍa emerging as a new centre for the Shahi rulers from the mid-seventh century onward.[229]

In another passage of the *Taḥqīq mā li-l-Hind*, al-Bīrūnī describes two forts as strong places situated to the south of the Kashmir Valley:

---

222 Dey 1927: 162; Bhattacharyya 1991: 256.
223 Wriggings 2004: 60. See also Kuwayama 2002: 211.
224 *Taḥqīq* (1958), p. 215.5–6; Sachau 1910: I/260.
225 *Taḥqīq* (1958), p. 270.9; Sachau 1910: I/317.
226 *Taḥqīq* (1958), p. 349.8–9; Sachau 1910: II/11. See above p. 18.
227 Reynolds 1858: 280.
228 Khan 2017: 57–59.
229 Dani 1969: 56; Rehman 1979: 16.

> Fort Rājagirī (راجکری) lies to the south of [the mountain Kulārğak], and Fort Lahūr (لهور) to its west. I never saw stronger [forts] than these two; and three *farsakh*s from it (i.e., the mountain Kulārğak), is the town of Rājāwūri (راجاوری). Our merchants trade with it, but do not go beyond it. This is the frontier of al-Hind from the northern side.[230]

Al-Bīrūnī mentions the two forts a few times in the *Taḥqīq mā li-l-Hind*. In one passage, he describes Fort Rājagirī as being situated on the road from Kanauj to the Kashmir Valley, via Taneshwar, to the south of the Pir Panjal Range, before bifurcating north to the Valley.[231] Fort Lahūr was most probably located to the south-west of the Pir Panjal Range[232] Maḥmūd attempted to seize the fortress of Lohkot (i.e., Lahūr) twice. This would have facilitated his access to Kashmir.[233] However, the sultan was never able to take it. When he attempted to attack Fort Lahūr, Trilocanapāla (ca. 1010–1021/1022), the Shahi ruler of the time, asked assistance to the Kashmiri king Saṅgrāmarāja (r. ca. 1003–1028).[234] Al-Bīrūnī visited these regions between the years 1017 and 1030. He does not, however, provide any details that may suggest which religious traditions were followed in this region.

Farther east lies Fort Nandana where al-Bīrūnī calculated the circumference of the earth. He states:

> When I happened to be living in the fort of Nandana in the land of India, I observed from an adjacent high mountain standing west of the fort, a large plain lying south of the mountain. It occurred to me that I should examine this method there. So, from the top of the mountain, I made an empirical measurement of the contact between the earth and the blue sky.[235]

The remains of two temples were found there in a rather impaired state which does not allow for thorough qualitative archaeological interpretations (see

---

230 *Taḥqīq* (1958), p. 167.5–7; Sachau 1910: I/208. This passage has already been quoted earlier in this book, see p. 23.

231 Quoting Jīvaśarman, al-Bīrūnī reports that the Swāt country is opposite the district of Girī, which may be the same district to which Fort Rājagirī belonged (*Taḥqīq* [1958], p. 390.1–2; Sachau 1910: II/182).

232 *Taḥqīq* (1958), pp. 164.15–165.2; Sachau 1910: I/205. See above pp. 30–31 for a discussion on the location of these two forts.

233 Nazim 1931: 104–105.

234 RT VII.47–53, pp. 23–24; Stein 1900: I/270–272. On other relationships between the Kashmiri kings and the Late Shahis, see below p. 82.

235 Translation by Ali (1967: 188). See also *Taḥdīd* [1992], p. 222.10–223.1.

FIGURE 4
Temple A, Nandana
MEISTER 2010: FIG. 52

Temple A in Figure 4). However, these two edifices belong to a larger group of temples located in the Salt Range that have been associated with the Late Shahis. The discovery of various coins at the sites has made it possible to date this group of structures between the sixth/seventh and the eleventh centuries.[236] According to some scholars, Nandana became the capital of the Late Shahis after they had been defeated in Wayhind/Udabhāṇḍa and shortly before they were attacked by Maḥmūd.

To my knowledge, al-Bīrūnī is the only Arabic author from this period who mentions Nandana, most probably because this site located much farther east than the four others was not known to Arabic and Persian authors who preceded him. He may have spent some time in Fort Nandana between 1017 and 1025.

1.3.2 *The Late Shahis and Their Religion*

As shown above, all locales dealt in the preceding section were connected with the Late Shahi kingdom. Therefore, I set forth here pieces of archaeological material related to the Late Shahis, in order to paint a preliminary picture of their religion and society. The present section also illustrates the difficulties of dealing with the historical material associated with the Late Shahis due to the lack of systematic and comprehensive research ever since Rehman's study was published in 1979. Before the Late Shahis were pushed eastward by Alptigīn and

236 Rehman 1979: 266–267 and 273–274; Meister 1996 and 2010: 32–38; Khan 2017: 70–71.

subsequently by the Ghaznavids, their kingdom extended from the north-west in Kābul and Wayhind/Udabhāṇḍa to the north-east in some areas of northern Panjab.[237]

Xuanzang (early 7th c.) explains how the ruler of Kābul region followed a specific pattern of seasonal migrations, residing in Kābul in summer and in the territory of al-Hind in winter.[238] The succeeding rulers, the Early Shahis (ca. mid-7th to early 9th c.) and the Late Shahis (early 9th to early 11th c.), also probably originally maintained their capital in Kābul during summer and in al-Hind during winter. The Late Shahis established several administrative and political centres in the territory of al-Hind, namely Wayhind/Udabhāṇḍa, Fort Nandana and Lahore, successively, depending on the wars taking place between the Muslim rulers and these kings. When Alptigīn attacked Kābul, followed by the Ghaznavid army to Gandhāra, in the second half of the tenth century, the local rulers withdrew further east and expanded their kingdom to western Panjab.

According to Rehman, epigraphy and numismatics indicate that the Late Shahis were worshippers of the Hindu god Śiva.[239] A stone found at the site of Wayhind/Udabhāṇḍa bears a (Proto-)Śāradā votive inscription that can be roughly dated from the second half of the tenth century, that is, during the reign of Jayapāla (ca. r. 964–1002).[240] The inscription, mostly written in *śloka*s, is dedicated to Śiva. It begins with a formula to praise Śarva, which refers to one of the eight forms (*aṣṭamūrti*) of Śiva, known under the collective name of *mūrtyaṣṭaka*.[241] In this inscription, Śiva is referred to in his form of Pinākin (lit. the one armed with a bow), which figures in a myth in which Śiva destroys three cities of demons (*tripura*). The text of this inscription also praises Śaṅkara, another name of Śiva,[242] and Umā.[243] The inscription mentions Jayapāladeva and his predecessor Bhīmadeva, gives a description of the town of Udabhāṇḍa and indicates the occasion on which it was written, that is, the construction of a temple devoted to Śiva.

---

237 For further references on the Late Shahis, see above fn. 56 of Chapter 1.

238 Inaba 2013: 89–90.

239 Rehman 1979: 33–34. Pandey (1973: 187) is of the same opinion.

240 See Pandey 1973: 135–137 and Rehman 1978 and 1979: 246–247, 308–318. For a transliteration and translation of this inscription, see Rehman 1979: 310–313. See also Rodziadi Khaw 2015: 119–121.

241 Rao 1997: II.2/398–406.

242 Rao 1997: II.1/40.

243 On Umā's representation in sculptures see Rao (1997: 122, 124 and 137). Umā, Durgā and Devī can also be regarded as a feminine aspect of Viṣṇu (Rao 1997: 332).

There are many inscriptions associated with the territory and/or the period of the Late Shahis. However, their state of study and decipherment is poor, while in many cases their texts only consist of a few (partial) lines. Rodziadi Khaw, who recently catalogued and described numerous (Proto-)Śāradā inscriptions found in Gandhāra, classifies them into five categories, ranging from "unpublished" to "published and satisfactorily deciphered" (2015: 94). These labels in his categorization highlight the poor degree to which these inscriptions have been studied so far.[244] All of the inscriptions associated to the Late Shahis reflect a society involving Brahmins in its official activities and celebrating Hindu deities, including Śiva.[245] Some of these inscriptions bearing the mention of a date, with or without a year, show that the persons who had the texts engraved followed the traditional luni-solar system commonly in use in ancient India.

Coins associated with the Late Shahis are difficult to interpret. The following description of these coins is mostly based on Rehman's work and aims at offering an impression of their types.[246] Two common types of coins are associated with the Late Shahis. The first type portrays a bull and a horseman (gold, billon, and silver).[247] This type, which is already seen with coins minted by the Indo-Scythian and Indo-Greek kings, was also widespread later on in Gandhāra. The second common type of coin (made of copper) linked to the dynasty depicts an elephant and a lion. The motifs seen on both types of coins are recurrent not only in early Indian coinage, but also in Hindu iconography. Originally the lion figures as the vehicle (*vāhana*) of the goddess Durgā, and the bull often represents Nandin, Śiva's vehicle, both deities being commonly depicted together with these two animals in Indian art and coinage. Even so, the occurrence of these motifs does not necessarily point to a possible cult of these deities.[248]

The coins of king Sāmantadeva (late 9th c.), referred to as Sāmanta by al-Bīrūnī, display a trident (*triśūla*), Śiva's attribute, and a star-shaped pendant as a decorative feature of a horse. A coin, issued by Bhīmadeva (ca. r. 921–964), referred to as Bhīma in al-Bīrūnī's report, represents a king seated on a throne and a woman on its obverse. The two persons display the clothing and

244 See Rodziadi Khaw 2015 and 2016: 64–114 for a catalogue and description of the Śāradā inscriptions of Gandhāra. See also Rehman 1979: 218 and 241–248.

245 See for instance Rodziadi Khaw 2015: 97–98.

246 Rehman 1979: 194–217. See also Thomas 1846, MacDowall 1968 and Alram 2016: 151–153. New research is now conducted by Arturo Annucci on the Late Shahi coinage. See for instance Annucci 2023.

247 Billon is an alloy of silver and copper (Bhattacharyya 1954: 118).

248 Cappelletti (2015: 54–55 and 152, fig. 6B) also provides us with a generic description of the first type.

hairstyles of the time. Above their heads are a *triśūla* and a diamond-shaped object. On the reverse, a king, similar to the one shown on the obverse, is seen beside Lakṣmī, the consort of Viṣṇu and the goddess of wealth and prosperity. This representation on Indian coinage reveals a close connection between kingship and this goddess.[249] However, this representation cannot suggest that the Late Shahis were specifically devoted to Lakṣmī and Viṣṇu. This coin may have been minted for a special occasion on which celebrating Lakṣmī as a provider of wealth was deemed necessary, or for the celebration of Dīpāvali, the Hindu festival dedicated to her.

As for architecture, several temples belonging to the territory of the Late Shahis, including the aforementioned Fort Nandana in the Salt Range, display similar features.[250] These temples dating from the sixth to the eleventh century, have, for instance, conical *nāgara* roofs, a type of *śikhara* construction.[251] In Barikot in the Swāt Valley, an area also belonging to the Late Shahi territory before the arrival of the Ghaznavids, fragments of marble sculptures, perhaps of Viṣṇu and of figures representing his attribute and associated with the ruins of a monumental temple, has been uncovered. According to archaeological surveys, this temple was in use from the late seventh to the late tenth century.[252] Moreover, as Rehman discusses, a few sculptures representing Hindu deities, such as Viṣṇu, Śiva, Kārttikeya and Durgā, were found at different sites in the kingdom of the two Shahi dynasties. However, further information, such as on their exact dates and archaeological contexts, is often missing.[253] Lastly, a passage of the *Rājataraṅgiṇī* refers to a temple dedicated to Viṣṇu and built by king Bhīmadeva.[254]

This material demonstrates that populations living in the Swāt Valley, Gandhāra and Panjab celebrated various Hindu gods from approximately the seventh to the early eleventh century, a geographical area and chronological period which correspond to that of the kingdoms and rules of the Early and Late Shahis. Furthermore, this material suggests a certain continuity in ritual practices of each of the two ruling dynasties, rather than a fracture between the practices of both dynasties. Yet, a systematic and thorough investigation is necessary in order to reach conclusions about how these rulers lived their religion and how the dynastic change took place.

---

249 Singh 2017: 166 and 189.

250 A study by Ijaz Khan (2017) focuses on the political aspect of the Late Shahi society. As an up-to-date study of sites located north of the Valley of Peshawar, it highlights the importance of the defensive activities of the Late Shahi in the region.

251 Rehman 1979: 281–284; Meister 1996 and 2010.

252 Filigenzi 2015: 36–38.

253 Rehman 1979: 285. See also Pandey 1973: 233–236 and Kuwayama 2002: 222–248.

254 RT VI.178, p. 97; Stein 1900: I/249.

Furthermore, in the *Taḥqīq mā li-l-Hind*, al-Bīrūnī describes thirty-seven Hindu festivals, most of which were likely taking place in the regions he visited in al-Hind and celebrated thus within the society ruled over by the Late Shahis. Among these festivals, ten were dedicated to women or to female deities, one to Śiva, four to Kṛṣṇa, the *avatāra* of Viṣṇu, and six to Brahmins.[255] If the celebrations of these festivals are indeed to be connected to the society of the Late Shahis, then they indicate that this society did not worship one specific Hindu god at the exclusion of another one and rather venerated several Hindu deities concurrently. However, at the current state of research on the Late Shahis' material culture, and without an extensive and multidisciplinary examination of historical data related to them, any further conclusions would remain conjectural. Lastly, the above overview shows that these rulers used Sanskrit as their literary religious language and (Proto-)Śāradā as their script in inscriptions and coinage.[256]

1.4 Concluding Remarks

This chapter highlighted the importance of determining the historical and social contexts of al-Bīrūnī's life in order to better understand his journeys in al-Hind. It showed that he spent his life in three major geographical zones: 1) western Central Asia (Kāṯ, Ray, Ǧūrǧān and Ǧūrǧānīya), 2) eastern Central Asia (Kābul, Ghazna and Gandhāra), and 3) the western Panjab (Fort Nandana). The cultural and political contexts diverged in these three zones. In the western part of Central Asia, Islam became well-established by the tenth century. In its eastern areas, Islamic authority had been continuously challenged by the Early and Late Shahis, and the region could be subdued only at the end of the tenth century. Islam, with the Ghaznavids, entered western Panjab in the early eleventh century, that is, at al-Bīrūnī's time. The territory of the above three geographical zones was divided into various political entities: the Afrighids ruled in Kāṯ,

255 Verdon 2019a: 68–78.

256 Al-Bīrūnī does not mention Śāradā as one of the scripts used in al-Hind. His silence on this type of script confirms Slaje's remark that this term was not used before the eleventh century (Slaje 1993: 15–16.). Observations by Rehman on the Śāradā script lead to the same conclusion (1979: 237–241). Al-Bīrūnī explains that the script *siddhamātṛkā* (*siddamātrika*; سد ماترك) was in use in the regions between Kashmir and Kanauj (*Taḥqīq* [1958], p. 135.3–16; Sachau 1910: I/173). For al-Bīrūnī, Śāradā is the name of a Kashmiri idol (*Taḥqīq* [1958], p. 89.12–13; Sachau 1910: I/117), most probably referring to the Śāradā *pīṭha*, i.e., the ancient Hindu temple whose ruins are located to the east of the Kashmir Valley.

the Buyids in Ray, the Ziyārids in Ǧūrǧān, the Ma'mūnids in Ǧūrǧānīya, and the Ghaznavids in Kābul, Ghazna and western Panjab.

The present chapter further pointed out the economic, cultural and intellectual prosperity of each of these regions, which fostered communities of literates with whom al-Bīrūnī could interact. Ray and Ǧūrǧānīya in particular were influential and respected intellectual centres where he could access important libraries. In addition, the regional ruler of Ǧūrǧān, Qābūs, who supported him in his efforts, and Maḥmūd, whose court included many scholars, most probably facilitated al-Bīrūnī's research. Kābul, Ghazna and western Panjab, far from being isolated or sterile areas, were at the centre of various exchanges between the West and East. Located in a frontier zone, but connecting different cultural areas, these regions witnessed important cultural changes and exchanges.

As al-Bīrūnī crossed this cultural frontier, he discovered Indian religion, sciences and literature chiefly in Gandhāra and Panjab. Several elements suggest that he only visited a confined area of the South Asian subcontinent: al-Bīrūnī's direct and explicit observations made in Laghmān, Peshawar, Fort Rājagirī, Fort Lahūr and Fort Nandana located in Gandhāra and Panjab, Maḥmūd's interest in these areas and control over them, and the variety of al-Bīrūnī's sources of information for the *Taḥqīq mā li-l-Hind*. Thus, whereas his writings indicate sojourns in the mentioned regions, there is no positive evidence of him travelling to cities like Taneshwar, Kanauj, Somnāth or Mathura, which were, however, conquered by Maḥmūd. I therefore argue that al-Bīrūnī chiefly observed cultural traditions of the above five places that are Laghmān, Peshawar, Fort Rājagirī, Fort Lahūr and Fort Nandana.

The above discussion of the cultural history of the five places visited by al-Bīrūnī and the religion of the Late Shahis is a preliminary one. In order to grasp the complex history of the Late Shahis and the practices of their society, a thorough and deeper investigation into texts, as well as archaeology of their sites, epigraphy and numismatics, is necessary. The little material presented above merely suggests that the Late Shahi adopted a form of Hinduism as their religion and the structure of a Brahminical society. However, for the purpose of the present book it is sufficient to note that most of al-Bīrūnī's travels in al-Hind actually took place in the territory of the Late Shahis.

Lastly, al-Bīrūnī observed local and living traditions in Laghmān and Peshawar, two sites defeated by the Ghaznavid long before his visits there. These examples point to the survival of pre-Islamic practices in territories conquered by the Ghaznavids. In addition, these traditions are also to be connected to customs adopted by the society of the Late Shahis. These observations are supported by the analyses presented in Sections 2.3 and 2.4 that show that al-Bīrūnī observed Hindu traditions and mainly interacted with Brahmins.

CHAPTER 2

The Social and Intellectual Contexts

2.1 Building up Theoretical Knowledge on al-Hind

Our journey toward al-Bīrūnī's encounter with al-Hind necessarily leads us to examine the ways in which became acquainted with its language, culture, philosophies and sciences. In his writings available to me, al-Bīrūnī does not use the term "Sanskrit." He occasionally uses the expression "in the Indian language" (*fī l-luġa al-hindiyya*; في اللغة الهنديّة), and generally employs the word *al-hind* (الهند) as a collective term to designate "India" or to refer to "what is Indian." He also employs the adjective derived from it, *al-hindī* (الهندي) meaning "Indian." The accuracy of his Arabic transliterations in *Al-āṯār al-bāqiya* and *Taḥqīq mā li-l-Hind*, however, leave little doubt that he had knowledge of the classical Sanskrit lexicon related to astronomical science, literature, geography, philosophy and religion. In the *Taḥqīq mā li-l-Hind*, al-Bīrūnī also comments on the grammatical, semantic and phonological complexities of the literary Indian language, namely Sanskrit, the religious and scientific language at the time.[1]

In the year 1000, al-Bīrūnī dedicates *Al-āṯār al-bāqiya* (*The Chronology of Ancient Nations*) to Prince Qābūs of Ǧūrǧān.[2] This treatise describes the astronomical calendars of different civilizations, explaining various methods by which their societies calculate days and nights, months and years, as well as longer eras. It also presents festivals linked to different calendars and covers historical events. The main civilizations considered in *Al-āṯār al-bāqiya* are those of Persians, Sogdians, Khwarizmians, Greeks, Jews, Syrians, Christians (Nestorians and Melkites), Zoroastrians (or Magians), Sabians, Arabs before Islam and Muslims. This work further includes information regarding India, especially terminology related to astronomy and the calendrical system.

An examination of the examples drawn from *Al-āṯār al-bāqiya* below reveals that al-Bīrūnī's knowledge of Indian astronomical terminology was relatively accurate before he visited al-Hind. His proficiency in Sanskrit terminology is evident in how he provides transliterated Sanskrit names of the months, seven heavenly bodies and zodiacal signs in Arabic, as displayed in the following three tables:

1 Sachau 1910: I/18. See also Strohmaier 1991: 153.
2 *Al-āṯār* (1878); *Al-āṯār* (2001); Sachau 1879.

TABLE 2 Names of the months in Sanskrit, as transliterated into Arabic by al-Bīrūnī in *Al-āṯār al-bāqiya*[a]

Arabic		Sanskrit	
baišāk	(بيشاك)	*vaiśākha*	(April–May)
zyašt	(زيشت)	*jyaiṣṭha*	(May–June)
āsār	(ااسار)	*āṣāḍha*	(June–July)
srāwān	(سراوان)	*śrāvaṇa*	(July–August)
bhadrabad	(بهدربد)	*bhādrapada*	(August–September)
aswiǧ	(اسوج)	*āśvina*	(September–October)
kārṯ	(كارپ)	*kārttika*	(October–November)
mankis	(منكس)	*mārgaśīrṣa*, also *mārga*	(November–December)
bawš	(بوش)	*pauṣa*	(December–January)
māk	(ماك)	*māgha*	(January–February)
bākr	(باكر)	*phālguna*	(February–March)
ǧaitra	(جيتر)	*caitra*	(March–April)

a *Al-āṯār* (1878), p. 71; *Al-āṯār* (2001), p. 80; Sachau 1879: 83. The following comments concern variant readings of some of the names of the months given in this table: *bhadrabad* (بهدربد) is an emendation by Sachau; the manuscripts read *bharūnda* (بهروند). Azkaei's edition reproduces the manuscripts' readings as follows: *bawš* (بوش) reads *bawšn* (بوشن) in *Al-āṯār* (2001), and *bākr* (باكر) reads *yākn* (ياكن).

TABLE 3 Names of the seven planets in Sanskrit, as transliterated into Arabic in *Al-āṯār al-bāqiya*[a]

Arabic		Sanskrit	
sanasǧar	(سنسجر)	*śanaiścara*	(Saturn)
brhasbatī	(برهسبتي)	*bṛhaspati*	(Jupiter)
mankal	(منكل)	*maṅgala*	(Mars)
ādīda	(اديد)	*āditya*	(the Sun)
šurk	(شرك)	*śukra*	(Venus)
bud	(بد)	*budha*	(Mercury)
sūm	(سوم)	*soma*	(the Moon)

a *Al-āṯār* (1878), p. 192; *Al-āṯār* (2001), p. 221; Sachau 1879: 172. The sun and the moon were included in the concept of the planets (*graha*) during a specific phase in the history of this Indian concept (Yano 2003, 2004: 331–332 and 335–337). My comments on the variant readings are as follows: *brhasbatī* (برهسبتي) is an emendation by Sachau, as the manuscripts have various readings; *ādīda* (اديد) reads *adiṯah* (ادثه) in *Al-āṯār* (2001), and *šurk* (شرك) reads *šūk* (شوك).

TABLE 4 Names of the zodiacal signs in Sanskrit, as transliterated into Arabic by al-Bīrūnī in *Al-āṯār al-bāqiya*[a]

Arabic		Sanskrit	
miš	(ميش)	*meṣa*	(Aries)
brša	(برش)	*vṛṣa*	(Taurus)
maṯūn	(مثون)	*mithuna*	(Gemini)
karkar	(كركر)	*karkaṭa*	(Cancer)
sink	(سنك)	*siṃha*	(Leo)
kan	(كن)	*kanyā*	(Virgo)
tul	(تل)	*tulā*	(Libra)
wšǧika	(وشجك)	*vṛścika*	(Scorpion)
dhan	(دهن)	*dhanus*	(Sagittarius)
makar	(مكر)	*makara*	(Capricornus)
kum	(كم)	*kumbha*	(Aquarius)
mīn	(مين)	*mīna*	(Pisces)

a *Al-āṯār* (1878), p. 193; *Al-āṯār* (2001), p. 222; Sachau 1879: 173; Yano 2003: 384–385; *brša* (برش) reads *bršā* (برشى) in *Al-āṯār* (2001), and *makar* (مكر) reads *makad* (مكد).

Al-Bīrūnī provides Arabic transliterations of the names of months, planets and zodiacal signs that in general correspond to their Sanskrit counterparts. Only two transcriptions significantly differ from their Sanskrit originals: *mankis* for *mārgaśīrṣa* and *bākr* for *phālguna*.[3] Further, in two passages below, al-Bīrūnī discusses Indian astronomical methods to divide the celestial globe:

> We say that the Indians divide the celestial globe by the number of lunar mansions (منازل), that is, twenty-seven for them. Accordingly, the [celestial globe] is divided by this number. Each mansion approximately amounts to thirteen degrees and a fourth. They draw [their] rules (احكام) from the stars' entrances in their *ribāṭāt*s (رباطات);[4] this [process] is generally known as *ǧufūr* (جفور)[5] [and] it is applied [by them] to every single

3 In the *Taḥqīq mā li-l-Hind*, al-Bīrūnī equally enumerates the names of the Indian months transliterated from Sanskrit. The spellings there slightly differ from the same list found in *Al-āṯār al-bāqiya* (*Taḥqīq* [1958], p. 302; Sachau 1910: 1/358).

4 The Arabic term *ribāṭāt* is understood as an astronomical technical term referring to stations.

5 The expression *ʿilm al-ǧafr* can be translated as "divination," but the exact meaning of the plural term *ǧufūr* in the context of the above quotation is unclear.

situation and need. Its report requires a long discourse foreign to [our] purpose and found in the books of sciences known by this [name] (i.e., *ǧufūr*). [...] The [Arabs] used the [notion of the lunar mansions] in a different manner than the Indians, as they intended to learn the conditions of the atmosphere at [different] times and weather phenomena according to the seasons of the year by [using] them. They were illiterate people, unable to [have] knowledge, except for visible [things].[6]

[...] This is a testimony of Abū Maʿšar showing that this method leads to correct results.[7] Further, if examined by the Indian way of the *ribāṭāt* and the *ǧufūr*, the matter would approach the correct [result].[8]

These passages show the scope of al-Bīrūnī's knowledge on the topic and reflect his early interest in Indian astronomy. They also indicate the type of information which al-Bīrūnī had about Indian sciences before composing the *Taḥqīq mā li-l-Hind*. In the first extract, al-Bīrūnī explains—and shows appreciation for—the Indian lunar mansions and other related astronomical concepts. In the second quotation, he recognizes the use of the Indian concepts mentioned in the first quotation, that is, the *ribāṭāt* (stations) and the *ǧufūr*, in order to obtain relatively accurate results on the times of the rising and setting of the lunar mansions. In the second half of the eighth century CE, at the Abbasid court, astronomical and medical treatises had been translated from Sanskrit into Arabic, as aforementioned in the introduction of the present book. Thus, Indian astronomy, which was known to Muslim thinkers for two centuries before al-Bīrūnī composed his *Al-āṯār al-bāqiya*, enjoyed prestige, and al-Bīrūnī was certainly indebted to this tradition.

Furthermore, several Indian *siddhānta* texts were among the works that had been translated into Arabic during the Abbasid caliphate.[9] Some of these translations were known to al-Bīrūnī, as the following five excerpts show:

6 *Al-āṯār* (1878), p. 336.12–22; *Al-āṯār* (2001), p. 432.1–11; Sachau 1879: 335–336.

7 A quotation drawn from Abū Maʿšar precedes this passage. It exposes a method to calculate the influences of the rising and setting of a lunar mansion (Sachau 1879: 341–342). The astrologer Abū Maʿšar was a native of Balkh living in the eighth or ninth century CE. He played an important role in the transmission of Indo-Iranian astrology to the Islamic world (Sachau 1879: 375; Pingree 1963: 243–245).

8 *Al-āṯār* (1878), p. 341.6–7; *Al-āṯār* (2001), p. 437.13–14; Sachau 1879: 342.

9 These translations include the *Zīǧ al-arkand* and the *Zīǧ kandakātik*, both based on Brahmagupta's *Khaṇḍakhādyaka*, as well as the *Zīǧ karanatilaka* (Vijayanandin), *Zīǧ karanasara* (Vaṭeśvara) and the *Kitāb al-adwār wa l-qirānāt* (Ahmed 2001: 161–165). See also Pingree 1963, Baloch 1973: 24–33, Said & Khan 1981: 45 and Sarma 2009: 214–215.

According to Ptolemy, the revolutions [of the sun] are equal, as he did not find that the apogee of the sun has any movement. For others than him, I mean the authors of the *Sindhind* and modern thinkers, they are unequal, because their observations led them to [make conclusions] about the existence of movement [of the sun]. However, whether equal or unequal, [the revolutions] encompass the four seasons and control their nature.[10]

These [cycles of the stars determined by Abū Maʿšar] differ from the cycles determined by the observations of the Indians, known as the *Cycles of the Sindhind* (ادوار السندهند), [and likewise] differ from the *Days of Āryabhaṭa* (ايّام الارجبهز) and the *Days of al-Arkand* (ايّام الاركند).[11]

The discrepancy in the cycles [of the stars], not the discrepancy in the observations, is a sufficient argument and stronger evidence in refuting what Abū Maʿšar perpetrated. Stupid [people] rely upon him, discrediting religion and making the *Cycles of the Sindhind*, or others, a means to abuse those who warn about the approach of the [ultimate] hour and who inform them on the gathering [of the dead] for the reward and punishment in the everlasting abode.[12]

As for the day of [the vernal] equinox, the Indians calculate it with their astronomical handbook (*zīǧ*; زيج), which they say, with ignorance, is eternally ancient and that other astronomical handbooks make use of it. Their Nowrūz (نوروز) (i.e., the Persian New Year celebration occurring at the vernal equinox) is a great festival for them. During the first hour of this [day], they worship the sun and pray [to the] spirits for happiness and bliss. At noon, they worship the [sun] and pray for the life to come and the beyond. At the end of the day, they worship the [sun] and pray for their bodies and health. During that [day], they worship every object of value and [every] living creature. They say that during that [day], the winds blow great auspicious spiritual beings. The people of heaven and hell look at each other with affection. Light and darkness are in equilibrium. During the hour of [the equinox], fire is burnt in sacred places.[13]

10 *Al-āṯār* (1878), p. 9.15–18; *Al-āṯār* (2001), p. 13.6–9; Sachau 1879: 11.
11 *Al-āṯār* (1878), p. 25.12–13; *Al-āṯār* (2001), p. 31.11–12; Sachau 1879: 29.
12 *Al-āṯār* (1878), p. 26.18–20; *Al-āṯār* (2001), p. 32.15–18; Sachau 1879: 31.
13 *Al-āṯār* (1878), p. 259.2–8; *Al-āṯār* (2001), p. 323.1–7; Sachau 1879: 249–250.

> According to the astronomical handbook, the *Sindhind*, the second equinox is a great festival for the Indians, similar as Mihrǧān (مهرجان) for the Persians. During that [day], they exchange all important goods and delicate jewels. They gather in the temples and places of worship at noon. Then they go out in their parks, they gather in their public places and bow to their [god of] Time and show obedience to Allah[14]—respected and exalted be He.[15]

The above passages reveal that al-Bīrūnī's knowledge of India at the time of the *Al-āṯār al-bāqiya*'s composition, in the year 1000, was largely based on literary sources. He quotes and refers to writings on topics such as the astronomical revolution of the sun, cycles of the stars, the vernal and autumnal equinox or rituals performed on certain days. The Arabic term *zīǧ*, translated here as "astronomical handbook," was a generic appellation for a type of handbook which regrouped astronomical tables and explanatory material. The *Zīǧ al-Sindhind* refers to al-Ḵwarizmī's work on Indian astronomy, while the *Days of Āryabhaṭa* and of *al-Arkand* are Arabic works based on other Sanskrit astronomical treatises.[16] These books were available to al-Bīrūnī who drew upon them when discussing Indian astronomy.

Arabic sources that were no translations nor interpretations of Sanskrit works also played a part in al-Bīrūnī's account on India in *Al-āṯār al-bāqiya*, as the following three passages illustrate:

> I have heard that the Indians use the appearance of the new moon for determining the months. They intercalate one lunar month to every 976 days. [...] Abū Muḥammad al-Nā'ib al-Āmulī reported in the *Kitāb al-ġurra*, based on the work of Ya'qūb Ibn Ṭāriq, that the Indians consider four types of periods. One of them is the revolution of the sun [starting] from a point that consists in a star in the constellations, [and returning] to its source. This is a solar year. The second is the rising of [the sun] 360 times. It is called the middle year, because it is longer than the lunar year and shorter than the solar year. The third corresponds to

14 The original term *allāh* is kept here, as it is difficult to determine which Indian deity al-Bīrūnī is referring to.

15 *Al-āṯār* (1878), p. 274.13–16; *Al-āṯār* (2001), p. 339.15–18; Sachau 1879: 266.

16 Pingree 2012b. The mathematician and astronomer al-Fazārī also composed the *Zīǧ al-Sindhind al-kabīr* which is either based on the *Brāhmasphuṭasiddhānta* or on a hypothetical *Mahāsiddhānta* drawn on the former. See Pingree 2012a and fn. 2 of the introduction to the present book.

> twelve revolutions of the moon [starting] from the two [stars, called] al-Šaraṭāni (الشرطان), which are both at the top of the Aries [constellation] (الحمل), and [returning] to the two [same points]. This is their lunar year. It approximately amounts to 327 days, seven hours and two thirds. The fourth is the appearance [of the moon] twelve times. It is the lunar year [commonly] in use.[17]

> The author of the *Kitāb ma'ḵaḏ al-mawāqīt* (i.e., *Book on the Method for Determining Times*) claimed that those [who] follow [the calendar] with the intercalated fourth [day], namely the Greeks[18] and others, established the sun's entrance into the constellation of Aries in the beginning of April, which is Naysān (نيسان) for the Syrians, as the beginning of their era. […] Further, he [said], speaking about the Greeks, that after they understood that the beginning of their year had changed its place, they had recourse to the years of the Indians and intercalated in their year the additional [day] between two years. […] He (i.e., the author of the above-mentioned book) assimilated the differences between the Greek year and the solar year in the manner the Indians [did].[19]

> Al-Ǧayhānnī reported that, at the Indian Ocean, roots of a tree spread along the seacoast in the sand, that [its] leaf rolls up before separating from its root, and that [the leaf] changes into a male bee and flies away.[20]

These three extracts point to some of al-Bīrūnī's Arabic sources which informed him about India at an early date.[21] First, al-Bīrūnī quotes Abū Muḥammad al-Nā'ib al-Āmulī (*Kitāb al-ġurra*) who refers to Ya'qūb Ibn Ṭāriq in order to describe four different types of astronomical years in use among Indians.[22] In the next passage, he refers to the *Kitāb ma'ḵaḏ al-mawāqīt*[23] without naming the author of this book, in order to point out different manners of cal-

17 *Al-āṯār* (1878), pp. 12.19–13.10; *Al-āṯār* (2001), pp. 16.21–17.11; Sachau 1879: 15.

18 The Arabic term *al-rūm* (الروم) is employed to refer to the people of the Eastern Roman Empire, including the Greeks, in contrast to *al-yūnānī*, (اليوناني) which refers to the ancient Greeks.

19 *Al-āṯār* (1878), pp. 51.1–2, 5–6 and 9–10; *Al-āṯār* (2001), pp. 59.5–7, 9–11and 13–14; Sachau 1879: 60.

20 *Al-āṯār* (1878), p. 228.2–3; *Al-āṯār* (2001), p. 283.9–11; Sachau 1879: 214.

21 In the *Taḥqīq mā li-l-Hind*, al-Bīrūnī also makes mention of Arabic writers acquainted with India (*Taḥqīq* [1958], p. 351.3; Sachau 1910: II/18).

22 On Ya'qūb Ibn Ṭāriq see Pingree (1968).

23 This work is unknown to me.

culating days and years. In the last excerpt provided above, al-Bīrūnī quotes al-Ǧayhānnī[24] to depict a tree found on the coast of Indian Ocean endowed with fantastic characteristics. The use of the expression "I have heard that the Indians [...]" (سمعت أنّ الهند)[25] in the first of these excerpts also suggests that information was transmitted orally to him.

In conclusion, all these examples show that al-Bīrūnī had material regarding Indian astronomy at his disposal before the year 1000 CE and illustrate his respect for it.[26] Based on the above, I argue that al-Bīrūnī accessed oral and written sources for his account of India in *Al-āṯār al-bāqiya*, while he chiefly based his report on written documents. These writings were available to him in Khwarezm, Ray or Ǧūrǧān, that is, before he travelled eastward, came in contact with Indian thinkers, and veritably entered the territory of al-Hind.

Furthermore, as I aim to show below, by the time he had written the *Taḥqīq mā li-l-Hind*, al-Bīrūnī's knowledge of Sanskrit had increased considerably.[27] According to Sachau, al-Bīrūnī had gained good proficiency in the lexicon and grammar of Sanskrit when he composed this book. Sachau also considers that al-Bīrūnī's work is the result of both his endeavours to understand Sanskrit and his collaboration with Indian thinkers.[28] David Pingree, however, contends that al-Bīrūnī was not highly skilled in Sanskrit and that his translation of the Sanskrit astronomical treatise *Brāhmasphuṭasiddhānta* relied, for the most part, upon the Indian pandits whom he met and who misled him in his interpretation.[29]

Nevertheless, al-Bīrūnī's faithful transfer of Sanskrit terms into Arabic in his works provides evidence for his good command of the Sanskrit lexicon related to Indian astronomy, literature, philosophy, geography and religion. His trans-

---

24 Al-Ǧayhānnī was probably a vizier of the Samanid dynasty (ca. 10th c.); see Sachau 1879: 424 and Pellat 2012. Al-Bīrūnī also made reference to him in the *Taḥdīd al-amākin*, when he writes: "Once, I had the intention to glean the information provided by the method of Ptolemy, in his book, the Geography, and by the method of al-Jaihānī and others, in their books on al-Masālik, for the following purposes: the collection of data, the clarification of obscurities, and the perfection of the art" (translation Ali 1967: 14).

25 *Al-āṯār* (1878), p. 12.19; *Al-āṯār* (2001), p. 16.21.

26 It is worthy of note that more than the half of the *Taḥqīq mā li-l-Hind* deals with Indian astronomy.

27 In a passage of the *Taḥqīq mā li-l-Hind*, al-Bīrūnī explains how his knowledge of the language grew while staying with Indian astronomers. See below pp. 74–75.

28 Sachau 1887: xv–xix. Sachau (1887: xiv) also assumes that al-Bīrūnī used a grammar book and a dictionary. See also Chatterji 1951: 86–87 and 95 on al-Bīrūnī's knowledge of Sanskrit.

29 Pingree 1983: 353. Al-Bīrūnī titled his Arabic translation of the treatise *Brāhmasiddhānta* (see Verdon & Yano 2020: 60–62 and 68–71). See p. 135, below, on Pingree's assessment of al-Bīrūnī's knowledge of Sanskrit.

TABLE 5 Transliterations from Sanskrit to Arabic by al-Bīrūnī in the *Taḥqīq mā li-l-Hind*[a]

Arabic	Sanskrit	Arabic	Sanskrit
bīḏa (بيذ)	*veda*	*nārāyan* (ناراين)	*nārāyaṇa*
purānāt (پرانات)	*purāṇāḥ* (pl.)	*bāsudīwa* (باسديو)	*vāsudeva*
mīrū (ميرو)	*meru*	*bhārata* (بهارت), *bhārāṯa* (بهاراث)	*bhārata*
dībāt (ديبات)	*dvīpāḥ* (pl.)	*akšauhinī* (اكشوهنى)	*akṣauhiṇī*
lanka (لنك)	*laṅkā* (f.)	*adimāsah* (ادماسه)	*adhimāsa*
māna (مان)	*māna*	*ūnarātra* (اونرَاتر)	*ūnarātra*
brahmānda (برهماند)	*brahmāṇḍa*	*ahargana* (أهرگن)	*ahargaṇa*
kalpa (كلپ)	*kalpa*	*parba* (پرب)	*parvan*
catur jūga (چترجوك)	*caturyuga*	*sanbajjara* (سنبجر)	*saṃvatsara*

a Al-Bīrūnī appears to have transliterated the nominative plural masculine of the word purāṇa, even though the Sanskrit original, used with reference to the thus-designated works, should have read *purāṇāni*, i.e., the nominative plural neuter.

literations in *Al-āṯār al-bāqiya* and the *Taḥqīq mā li-l-Hind* are mostly true to the corresponding Sanskrit words. The above table displays a random sample of Sanskrit terms transliterated into Arabic drawn from the table of contents of the *Taḥqīq mā li-l-Hind* as printed in the Hyderabad edition.[30]

In this table, most of the long vowels in Sanskrit are rendered with long vowels in Arabic. The nasal *ṅ* (guttural) and *ṇ* (retroflex) are generally reproduced by the same letter *nun* in Arabic, as it is the sole (dental) nasal of the Arabic alphabet. The Arabic letters *bā*, *fā* and *wāw* are each employed at different times to transliterate the Sanskrit sound *v*, which does not exist in Arabic. In other cases where the Arabic language did not have sounds specific to Sanskrit, al-Bīrūnī inserted Persian characters, such as *ch*, *g* and *p*, into the Arabic script. He generally rendered the Indic phoneme *e* into *ī* when writing in Arabic.

Based on the renderings of Sanskrit terms into the Semitic alphabet in the *Taḥqīq mā li-l-Hind*, Sachau suggests that al-Bīrūnī knew several vernacular languages alongside Sanskrit and that his transliterations of Indic terms reflect this

30 *Taḥqīq* (1958), pp. 7–12. With regard to the Latin transliteration of short vowels when the Arabic script does not specify them, I attribute the length of the vowels of the corresponding Sanskrit term to them. Similarly, diphthongs have been inferred from the spelling of the original Sanskrit terms.

plurality.[31] A recent study of al-Bīrūnī's rendering of terms related to Indian religious festivals by the present author showed a relation between his various ways of transliterating Sanskrit terms into Arabic and the two types of sources, textual and oral, that he used to collect information on India. Al-Bīrūnī tended to transfer words that he had heard in an abbreviated form close to vernacular languages such as Panjabi and Sindhi, whereas he rendered words that he knew from literary sources in a way that is more faithful to the classical Sanskrit terms.[32]

Lastly, a comparison between al-Bīrūnī's use of Sanskrit terminology in *Al-āṯār al-bāqiya* and in the *Taḥqīq mā li-l-Hind* points to the evolution of his knowledge of this language. In the former, al-Bīrūnī's knowledge of Sanskrit terminology was largely confined to the astronomical field and based on literature. In the latter, his field of expertise, based then on literature and on oral transmission, expands to other areas of knowledge. Al-Bīrūnī indeed quotes Sanskrit works belonging to various literary genres and scientific domains, such as the Purāṇas, the *Kitāb Gītā*, two texts related to Sāṅkhya and Yoga, and to a considerable amount of astronomical literature.[33] Al-Bīrūnī's degree of proficiency in Sanskrit can also be appreciated by the two texts he enumerates in his auto-bibliography that are interpretations from Arabic into Sanskrit of Euclid's *Elements*, of Ptolemy's *Almagest* and of a book on astrolabes.[34]

Thus, I highlighted developments in al-Bīrūnī's knowledge of Sanskrit and of Indian sciences that occurred from the time of the composition of *Al-āṯār al-bāqiya* up to that of the *Taḥqīq mā li-l-Hind*, an evolution that necessitated a long-term collaborative work with Indian thinkers. Therefore, I next explore the role of intercultural and intellectual exchanges taking place in the Ghaznavid royal court to which al-Bīrūnī belonged.

2.2 Collaborations and Multiculturalism at Royal Courts

Socio-historical elements help retrace the manner in which al-Bīrūnī learned Sanskrit so that he could eventually acquire the ability to translate two works related to Sāṅkhya and Yoga philosophy into Arabic. Although al-Bīrūnī re-

31 Sachau 1887: xxii–xxvii and 1888.

32 Verdon 2019a: 71–75.

33 See above p. 47.

34 Boilot 1955: 238–239, nos. 175–177. These translations are not extant, but the *Elements* and *Almagest* are mentioned in the *Taḥqīq mā li-l-Hind* (*Taḥqīq* (1958), p. 102.5.7; Sachau 1910: I/137).

mains an isolated historical figure of his period known to have conducted in-depth research on al-Hind, the context of royal courts of the time, and that of the Ghaznavids in particular, played a crucial role in his work. The existence of various instances of intellectuals working together on translations demonstrates that this was common practice from an early date. The *Bayt al-Ḥikma* (House of Wisdom) of Baghdad, an academy where philosophers, translators, secretaries, clergymen, copyists, librarians and astronomers worked together, is just one example of these early collaborations.[35] Another instance centers around the work of Ḥunayn Ibn Isḥāq (b. 808) who explains how after he had translated Galen's *De motu muscolorum* into Syriac, another scholar translated it into Arabic.[36] As Travis Zadeh notes, Ḥunayn's explanation illustrates a "professional process of translation,"[37] as well as the necessity of teamwork in such processes.

The Marvels of India (عجايب الهند) is an example of a literary work that illustrates early intercultural exchanges and points to the existence of polyglotism. Authored by Buzurg Ibn Šahriyār in the mid-tenth century CE, it gathers 134 whimsical travellers' tales.[38] This book not only exemplifies how information was propagated throughout different regions connected by the Arabic Sea, the Indian Ocean and the Mediterranean Sea, but also demonstrates how sailors and other travellers used multiple languages to communicate. One report narrates how an Indian king in a region of Kashmir wanted to have the laws of Islam translated and requested a person from Iraq, who lived in India and knew several of its languages, to come to his court.

Incidentally, the same Indian king asked Buzurg Ibn Šahriyār to translate the *Quran* into his own Indian language.[39] Another story tells of a person from Siraf, in present-day Southern Iran, who travelled with an Indian guide. According to this story, the two travellers were able to converse, although no information regarding the language they used is provided.[40] While the report in *The Marvels of India* may not be completely historically reliable, it at least reveals the existence of polyglot persons able to speak different languages, most probably Arabic, Persian and some Indic languages, in the mid-tenth century.

In a context closer to al-Bīrūnī's, the Persian translation of al-ʿUtbī's *Al-Yamīnī* describes Maḥmūd's army as composed of many ethnic groups, includ-

35 Balty-Guesdon 1992: 141–146.
36 As quoted by Travis Zadeh (2011: 60).
37 Zadeh 2011: 60.
38 Fück 2012.
39 Devic 1878: 2–3.
40 Devic 1878: 90–91.

ing Indians, who may have spoken various languages.[41] In this period, it was common for the Muslim nobility to have foreign servants, referred to as *ġulām*s (slaves, servants, young man).[42] Indian *ġulām*s were regularly seized during military campaigns and some, after spending time serving their new masters, appear to have been assigned to relatively high-ranking positions at the Ghaznavid court. One case is that of Tilak, an Indian *ġulām* and son of a coppersmith, who became a military leader after having been an official interpreter for Maḥmūd's son in his administration. Bayhaqī reports that Tilak was known as having "a good hand for writing both the Indian (*hendavi*, *hendui*) and Persian languages."[43]

This example illustrates that the Ghaznavids included different ethnic and linguistic groups in their army and administration, that they appointed people of Indian origin to higher positions, and that they needed interpreters for Indic languages and Persian. Thus, as part of cultural and ethnic encounters, languages were exchanged at the Ghaznavid royal court. In addition, the practice of promoting former *ġulām*s to higher positions provides a valuable clue for the need of the Ghaznavids to employ Indian interpreters, some of whom may have been Brahmins proficient in Sanskrit, to help them govern their newly conquered territories and communicate with the population.[44]

Regarding bilingualism and polyglotism, further evidence points to the use of different languages in Islamic royal courts of early medieval times. Rehman calls attention to an epigraph dated to 1011 and inscribed on a foundation stone of a tomb found in Zalamkot in the lower Swāt. This epigraph, which dates to the reign of Maḥmūd, bears a bilingual inscription in Persian (seven lines) and Sanskrit (three lines in Śāradā script), indicating an early interest in writing records in these two languages.[45] In this official document, Persian was used next to Sanskrit, in contrast to the Arabic legend of the bilingual coins minted in the region of Lahore.[46]

41 Reynolds 1858: 335–336.

42 Sourdel et al. 2012.

43 Bosworth 2011e: 57. See also Bosworth 1963: 101 and 2011: 57–59; Flood 2009: 4 and 78. See further Cappelletti 2015: 110. On slavery under the Ghaznavid see Bosworth 1963: 99–106. Richard M. Eaton also highlights the significance of the inclusion of Indians in the Ghaznavid army by stating "despite the dynasty's rhetoric about defending Sunni Islam, religion posed no bar to military recruitment, as Indians had always been prominent in Ghaznavid armies" (2020: 34–35).

44 Said & Khan 1981: 89.

45 Rehman 1998, Rodziadi Khaw 2016: 142–144 and Shavarebi 2022.

46 See above pp. 34–38.

Further, intellectuals, alongside ideas and languages, travelled beyond cultural borders amid military and commercial interests of the rulers, thereby enabling early intercultural exchanges. As illustrations of intellectuals' moving through cultural and linguistic boundaries during this period, I provide a few additional examples. Al-Bīrūnī refers to an Indian physician who travelled in the region of Gardez, between Ghazna and the area of Panjab now located in Pakistan, in his introduction to the *Pharmacology* (*Kitāb al-ṣaydana fī l-ṭibb*) composed at the end of his life.[47] Not many years after al-Bīrūnī had composed the *Taḥqīq mā li-l-Hind*, the Persian Sufi Ali Huǧwīri, born in Ghazna, spent the latter part of his life in Lahore and died there in 1071/1072.[48] The poet Masʿūd-I Saʿd-I Salmān (1046/1049–1121/1122), who was of Persian origin and lived in Lahore, is said to have composed poems in Persian, Arabic and Indic languages. Only his Persian poems are however extant.[49]

Thus, the context of the Ghaznavid court was rich in intercultural and intellectual exchanges, while polyglotism was probably more common at the time than generally assumed. These above examples also support Finbarr Barry Flood's statement that "[b]ilingualism and/or polyglossia may in fact have been relatively common phenomena of the South Asian borderlands."[50]

Thus, this context favoured al-Bīrūnī's constant development of knowledge and, in this case, his learning of Sanskrit and of Indian sciences thanks to collaborations with Indian thinkers. The existence of intermediary languages known by the involved parties, including Sindhi and Panjabi, as Carl Edward Sachau and Suniti Kumar Chatterji have noted, but also Persian made these interactions possible.[51] As the available evidence suggests, al-Bīrūnī met people from Multan and surely spent some time in Panjab[52] which enabled him to become acquainted with some vernacular languages, in all likelihood a form of Sindhi and of Panjabi. Persian, which belongs to the Indo-Iranian linguistic family, became the official language of Islamic royal courts at the time and would also serve as an intermediary language in these intercultural exchanges.

47 Said 1973: 6.

48 Böwering 2012.

49 Grover 2006: 61; Clinton 2012.

50 Flood: 2009: 42.

51 Sachau 1887: xxiv and 1888: 37; Chatterji 1951: 93–94. Chatterji also observes various spellings in al-Bīrūnī's Arabic transliterations of Sanskrit words that do not reflect the pronunciation of northern Panjab or the Gangetic Plain. This linguistic observation leads Chatterji to think that al-Bīrūnī interacted with people from regions of India other than those two (Chatterji 1951: 89). See also Sachau 1888: 5–6 and 10–41. Fabrizio Speziale (2010: 419–420) observed that later Muslim authors studied Indian sciences under similar conditions.

52 See chapter 1, section 2 of the present book.

According to Houari Touati, Muslim travellers did not leave their native countries solely out of curiosity, but their journeys were often first motivated by governmental interests (embassies, conveyance of messages, administrative organisation in all regions of the empire).[53] The testimony of al-Balāḏurī (d. 892 CE) furnishes an example of how a ruler appointed somebody to visit India in order to gather information about this foreign land.[54] This report indicates that Muslim rulers showed interest in Indian countries at an early date for commercial, administrative or military reasons. Similarly, Maḥmūd benefited from al-Bīrūnī's skills in Indic languages in his conquests in the East. Further, al-Bīrūnī's wish to access Indian science, an interest that was part of an existing tradition among his peers is evident from a passage found in the preface to the *Taḥqīq mā li-l-Hind*, in which he explains that the composition of the work was requested by another learned man.[55]

Al-Bīrūnī's position at Maḥmūd's court was thus conducive for him to learn Sanskrit and study Indian sciences, religion and philosophies. He came in contact with Indians thanks to Maḥmūd's military and commercial interests. Maḥmūd is also known to have sought to gather scientific writings in Ghazna, for instance, from the cities of Ray and Isfahan in Iran,[56] and to have requested a considerable number of scholars and poets to come to his court.[57] As I suggested above in Section 1.2.3, numerous people accompanied the sultan during his campaigns: soldiers, workers, officials, poets, secretaries, interpreters, etc. Al-Bīrūnī records in his book on gemmology, *Al-ǧamāhir fī l-ǧawāhir* (*The Collection of Gemstones*), that the encounter with the ambassadors of the K'itan dynasty provided him with information on the Far East.[58] Farrukī's poems also offers information about the life of the sultan, who received delegates and military leaders from foreign states.[59] It is likely that there were members of the Indian elite among these delegates, such as royal advisors, astronomers or officials, who were educated Brahmins. Maḥmūd's court thus

53 Touati 2000: 12.

54 Elliot & Dowson 1867: 116.

55 In the *Taḥqīq mā li-l-Hind*, al-Bīrūnī writes: "[H]e incited me to write down what I know about the Hindus" (Sachau 1910: 1/7). He also states in *Al-āṯār al-bāqiya*: "[A] learned man once asked me regarding the eras used by different nations" (Sachau 1879: 2). See also Kozah 2016: 15 and 34.

56 Nazim 1931: 158.

57 Bosworth 1963: 132.

58 See above p. 42. This episode is referred to by Minorsky (1951: 233–234), Shamsi (1979: 271) and Said & Khan (1981: 80, 82 and 222, n. 178). For a complete English translation of *Al-ǧamāhir fī l-ǧawāhir*, see Said 2001. See also Boilot 1955: 230, no. 156.

59 Bosworth 1991: 47.

helped thinkers access various resources, whether written documents or scholarly interactions. This context enabled al-Bīrūnī to complement his—rather theoretical—knowledge of India preceding his actual travels in al-Hind, with a practical approach of Indian languages and sciences.

2.3 Al-Bīrūnī and Indian Scholars

After having highlighted how important were the socio-cultural context and the collaboration work in al-Bīrūnī's intellectual project, I examine his interactions with Indian thinkers with the aim of specifying the identity of his interlocutors. First, the *Taḥqīq mā li-l-Hind* reveals that Brahmins were essential informants to his project. The two following passages drawn from this work are direct evidence of al-Bīrūnī's interactions with Brahmins:

> I saw [some] Brahmins who allowed to their table companions, [who are also] relatives, [to eat] from one plate [with them], but the rest of them denied this.[60]

> I [repeatedly] heard that when Indian slaves escape [from another land] and return to their country and religion, they are forced to expiate by fasting, then they are soaked in cow's dung, urine and milk for a certain number of days, until they become mature there. Then, they go out of the dirt, and they eat additional similar things. I asked the Brahmins about this, but they denied it, pretended that there is no expiation for these [Indians], and that they are not allowed to return to their previous situation.[61]

As aforementioned in Section 1.2.4, al-Bīrūnī devotes an entire chapter to the life and practices of Brahmins, whereas he portrays the other classes of the society in one chapter.[62] Further, he describes the four *varṇas* (lit. colours) and the classes that are outside the caste system in a chapter entitled "On the classes called colours and on the [classes] which are lower [than them]."[63] In general,

60 *Taḥqīq* (1958), p. 456.12–13; Sachau 1910: II/134.

61 *Taḥqīq* (1958), p. 475.11–16; Sachau 1910: II/163. This passage shows that notion of impurity and pollution resulting from the contact with other castes or with foreigners (*mleccha*) was acknowledged by these Brahmins in the early eleventh century.

62 *Taḥqīq* (1958), pp. 452.5–459.5; Sachau 1910: II/130–139.

63 *Taḥqīq* (1958), pp. 75.11–80.1; Sachau 1910: I/99–104.

the *Taḥqīq mā li-l-Hind* stands as evidence of a Hindu society organized according to the so-called caste system and following a Brahminical model.[64] Another passage of the *Taḥqīq mā li-l-Hind* is instructive on the type of society al-Bīrūnī describes:

> Their [religion] chiefly [revolves] around the Brahmins, as the [latter] are trained to preserve and maintain it.[65]

This passage is found in a chapter entitled "On their belief in the existents, both intelligible and sensible."[66] It presents Brahmins as the custodians of the religion of the society encountered by al-Bīrūnī. Al-Bīrūnī's informants conveyed to him the picture of a society in which not only the caste system prevailed, but also the authority of the Brahmins. Further, the latter constituted the literate class of the population, as advisers, interpreters or astronomers/astrologers at the kings' courts. The Indian rulers, accompanied by their courts, formed the sector of the society that was the most likely to enter into contact with their Islamic counterparts. Therefore, it is no wonder that al-Bīrūnī, who held a position at the Ghaznavid court, collaborated with Brahmins, rather than with people at any other level of the society, such as soldiers or peasants.[67]

The question arises then of the field of expertise of these Brahmins: where they astronomers, priests, philosophers, or these several things at the same time? First elements of answer regarding the exact nature of al-Bīrūnī's intellectual encounters with Indians lie in his writings:

> [At first,] I stood among their astronomers in the position of a student with [his] master, because of my difficulty in speech (لعجمتى) [ignorant of their language] among them and my shortcomings about what they were [involved] in, such as their conventions.[68] When I had made some progress in these [matters], I began instructing them on the defects [of

64 Mishra 1983: 103.

65 *Taḥqīq* (1958), pp. 29.19–30.1; Sachau 1910: 1/39.

66 *Taḥqīq* (1958), pp. 24.4–34.4; Sachau 1910: 1/33–45.

67 This observation parallels Thapar's remark that early European thinkers chiefly based their research on India on their interaction with Brahmins and thus depended on the latter's view of the Indian society (Thapar 2002: 9).

68 Sachau paraphrases this passage, as follows: "being a stranger among them and not acquainted with their peculiar national and traditional methods of science" (Sachau 1910: 1/23).

> their conventions],[69] pointing out elements of demonstrations and correct methods of arithmetic. They swarmed around me, from all parts, being astonished [and eager] to learn [from me, and] while crowding in, they asked: "With whom among the Indians did you stay to acquire [this knowledge]?"[70]

Some of al-Bīrūnī's informants were thus astronomers. His particular interest in astronomical science and the need of the Indian rulers to be advised by astronomers and astrologers may account for this fact. As for the connection between religion and the science of the stars, the extract below is informative:

> The science of the stars is most well-known among them (i.e., the Indians), because matters of religion are dependent on it. Those among them who do not know how to make judgments [based on the stars] (i.e., astrology) see the [fundamental] characteristic of astrology as nothing but arithmetic.[71]

This connection is particularly relevant to the present discussion, as it opens the possibility that religious officials were astrologers/astronomers and suggests that al-Bīrūnī met such Indians expert in several domains.[72] In addition, al-Bīrūnī also met Indian philosophers who assisted him in his interpretations of the Sāṅkhya and Yoga texts, as shown in Chapters 4, 5 and 6. Interestingly, Gardīzī, a writer contemporaneous to al-Bīrūnī also attached to the Ghaznavid court, enumerates Indian communities in an account based on al-Ǧayhānnī's report. He describes one group of persons as including thinkers who combined the astronomical and medical sciences in their work and designates them as "philosophers" (Persian: *khudāvandān-i andīsha*; "masters of thought").[73]

Although the semantic field of this term is vast, Gardīzī's description suggests that Indian thinkers specialised in astronomy and medicine also cultivated the science of thought, in all likelihood including reasoning, logic and metaphysics. Gardīzī further describes these thinkers as eating "dates, plants and herbs so that it should be light for their senses" (Minorsky 1948: 633). This

69 I translate *ʿilal* (علل) as "defects," whereas Sachau renders it by "the elements on which this science rests" (Sachau 1910: 1/23).

70 *Taḥqīq* (1958), pp. 17.16–18.2; Sachau 1910: 1/23.

71 *Taḥqīq* (1958), p. 118.7–9; Sachau 1910: 1/152–153.

72 See below Section 2.3.

73 Minorsky 1948: 633.

diet and the reason for adopting it reminds one of the lifestyle promoted by ascetics (e.g., yogis) and orthodox Brahmins. Gardīzī's report may thus support the hypothesis that the Brahmins who were al-Bīrūnī's informants were versed in astronomy and philosophy.

Moreover, in the *Taḥqīq mā li-l-Hind*, al-Bīrūnī constantly distinguishes between the views of educated and uneducated people, referred by him as the elites (الخاصّ) and the masses (العامّ) respectively. He generally approves of the intellectual and religious attitudes of the elites, who, in his view, are able to consider abstract notions and whose conceptualization of the divine can be compared to the monotheism of Islam. He described the masses, on the contrary, as idolatrous people. Al-Bīrūnī certainly interacted with members of the elite, as his comments in the preface to the *Kitāb Pātanǧal* confirm:

> I had not ceased to translate books by arithmeticians and astronomers from the Indian [language into Arabic], until I turned to books on wisdom (في الحكمة) preserved by their elite (خواصّهم), and about which the ascetics compete [with each other] concerning the paths [leading] to devotion. When they were read to me letter by letter, and when I grasped their content, my conscience could not overlook [the occasion] to share [my knowledge] with those who wish to study these [books].[74]

In another excerpt drawn from the *Taḥqīq mā li-l-Hind* which presents the views of several Purāṇas concerning the names of the planets, al-Bīrūnī comments on those who assisted him in understanding these texts:

> Those who explained to me [these texts] by way of translation were well versed in the language and were not known to be treacherous.[75]

The above two passages illustrate how educated people, whom al-Bīrūnī considered reliable, helped him study philosophical and purāṇic Sanskrit literature. The elite and the people "well versed in the language", whom al-Bīrūnī refers to, likely belonged to the class of literate Brahmins. These two above passages also confirm two observations made earlier in this book: 1) al-Bīrūnī was originally interested in mathematics and astronomy, and then turned to other sciences; and 2) he did not translate his Sanskrit sources into Arabic all by himself, but in collaboration with Indians who orally conveyed to him their readings of these sources, probably using an intermediary language.

---

74 KP, p. 167.8–11; Pines & Gelblum 1966: 309.

75 *Taḥqīq* (1958), p. 186.11–12; Sachau 1910: I/229.

The prevalence of certain types of Indian literature, as opposed to others, reflects the intellectual and religious inclinations of al-Bīrūnī's informants. As seen earlier in Section 1.2.4, the Sanskrit works which al-Bīrūnī mentions or quotes in his book on India mostly belong to a Brahminical scientific and religious literature.[76]

The material presented above, in my view, strongly suggests that among the Brahmins al-Bīrūnī met, there were astronomers/astrologers, religious officers and philosophers. If so, Brahmins and advisers worked in several capacities for the rulers of royal courts of north-western early medieval India, including that of the Late Shahi kings.

Further, as I show in Chapters 4, 5 and 6, al-Bīrūnī reached an excellent proficiency of the Sāṅkhya and Yoga philosophies. In addition, I have already suggested in-passing that al-Bīrūnī had to engage in relatively long-term collaborations with Indian scholars in order to achieve such high level of knowledge about Sanskrit and Indian sciences, and to produce his numerous translations from Sanskrit to Arabic. I thus conclude this section with a discussion on the types of exchanges between the Ghaznavid court and that of local Indian rulers that occurred then, and on the possible geographical sites where a long-term collaboration would have taken place. In addition, the examination of these two questions offers an overview into some specifics of intercultural interactions of the time.

First, it is hard to fathom whether from the Indian side this collaboration occurred willingly or due to force. Abdur Rehman discusses how the relationship of the Late Shahi king Ānandapāla and his son Trilocanapāla with the Ghaznavid rulers fluctuated between being uncivil and amicable.[77] Other kings, such as the ruler of Nārāyaṇapura, surrendered and spontaneously offered to pay tribute to the Ghaznavids.[78] The existence of such peaceful tributes—discontinuous, however, in the case of the Late Shahis—point to the possibility of a dialogue between the two political spheres, which would have facilitated a dialogue between intellectual spheres.

Regarding the possibles places where al-Bīrūnī actually collaborated with Indians, evidence is scarce. However, both Kābul and Ghazna, located at the crossroad of the Islamic and Hindu worlds hosted books and prisoners brought from India. They may have thus served as centers of intellectual exchanges. Fort

76 See above p. 47.

77 Rehman 1979: 147–158.

78 Nazim 1931: 102. See also Cappelletti 2015: 109–110. Anooshahr (2021) discusses textual sources on the question of the political relationship between the Ghaznavids and local Indian rulers.

Nagarkot, situated in present-day north-western India, appears to have housed silk manuscripts and to have been a place where knowledge was preserved. Carl Ernst recalls that Muslim armies led by Sultan Firuz Ibn Tuqhluq looted some temples located near Fort Nagarkot in the mid-fourteenth century and that they collected a large quantity of books there.[79] In a similar way, Maḥmūd could have gathered books from his raids in temples and fortresses of al-Hind.

As for other sites located in the territory of al-Hind, the lack of available data concerning the sites of Laghmān, Peshawar, Fort Rājagirī and Fort Lahūr, locales visited by al-Bīrūnī in al-Hind, prevents us from determining their significance in his collaboration with Indians. Wayhind/Udabhāṇḍa was an important Late Shahi site located on the way of the Ghaznavid military conquests. Very little is known about this town, because much of the ancient site is at present submerged in the Indus River, or under the present-day town of Hund.[80] According to primary sources, Maḥmūd stayed in the region of Peshawar from September 1001 to April 1002.[81]

As for Lahore, it became the capital city of the Late Shahi kingdom, after Udabhāṇḍa and Nandana, and before they took shelter in Kashmir.[82] Indian scholars likely dwelt in Udabhāṇḍa and Lahore. The two cities were located on a road often used by Maḥmūd for his military excursions in al-Hind, which al-Bīrūnī describes, and which opened new territory to the Ghaznavids. However, al-Bīrūnī does not mention any observations or calculations he might have made in these two cities. It is thus possible, although not ascertainable, that al-Bīrūnī collaborated with Indians in both Udabhāṇḍa and Lahore.

Multan was an equally important city in al-Hind at the time, as al-Bīrūnī's many references to it suggest. He explains that different appellations were given to this city, describing it as a place of pilgrimage on account of its pond and Sun idol.[83] Al-Bīrūnī communicated with people from Multan and consulted books by authors from this city. Already in the mid-eighth century, scholars from Sind played a role in the translation of Sanskrit astronomical works into Arabic.[84] It is likely that 250 years later, when al-Bīrūnī wrote about India, the region still hosted Indian scholars who may have interacted with him. In addition, al-Bīrūnī says that he himself calculated the latitude of Multan.[85]

---

79 Ernst 2003: 175.

80 Verdon & Lončar 2016. Excavation reports are unpublished or inaccessible to me. On Wayhind, see Kimmet 2020 and Verdon 2021.

81 Rehman 1998: 472.

82 See Dar 1994 and 2001: 53–60.

83 Sachau 1910: I/116, 298, II/145 and 148.

84 Baloch 1973: 19–33.

85 See above pp. 28–29.

However, in contrast with the five locales discussed in Section 1.3.1, al-Bīrūnī does not explicitly express his presence at Multan. In addition, most of the other pieces of evidence point to al-Bīrūnī visiting some parts of Gandhāra and Panjab, not upper Sind. For instance, as I argued above, Multan was not located on the northern road that Maḥmūd chiefly utilized for his campaigns to the East and through which al-Bīrūnī travelled the most. Therefore, even if he visited the region, it is not certain whether he spent much time in this city. In my view, al-Bīrūnī did not visit Multan, or very briefly, despite him mentioning his calculation of its latitude. Lastly, Multan has seen Muslim Arabic speakers and local non-Arabic speakers coexisting since at least the eighth century CE. Maḥmūd was perhaps benefiting more from having an interpreter in newly conquered regions than in areas where bilingualism, or polyglotism, was already rooted.

With regard to Nandana in the Salt Range, the ruins of two important temples are found there. These temples most probably housed Indian Brahmins, along with Sanskrit texts. Traditional education and the opportunity to study may have been provided by the learned temple attendants.[86] After having plundered the temples of Nandana in 1014, the Ghaznavids appointed governors at the place, which, as argued above, implies a will to strengthen their local political authority.[87] In addition, al-Bīrūnī spent sufficient time in this fort to experiment with his method of calculating the circumference of the earth.[88] It is thus likely that Maḥmūd appointed the scholar to stay there for some time between the years 1017 and 1030. On this occasion, priests of the temples, who may have had proficiency in astronomy and/or philosophy, would have assisted al-Bīrūnī in learning Sanskrit and studying Indian culture.[89]

Lastly, Nagarkot, the modern Kangra in Himachal Pradesh, which is located farther east, may have been a significant site for al-Bīrūnī's study of Indian sciences. Maḥmūd attacked Nagarkot in Winter 1008. In the same way as in Fort Nandana and Lahore, he appointed a governor which indicates a strong desire to establish governmental authority in the place. As shown above, this fort was also a place where the chronicles of the Early and Late Shahis were stored, and thus where other texts might also have been found.

Although the above short outline may not suffice to ascertain where the collaborative work between Indians and al-Bīrūnī took place, the region of Ghazna

86 Scharfe 2002: 169.
87 Maḥmūd's political interest is discussed in Section 1.2.3.
88 Ali 1967: 188–189. Said & Khan 1981: 84.
89 On al-Bīrūnī's stay in Nandana, see Said & Khan 1981: 77–78. Baloch (1987) also wrote about al-Bīrūnī's sojourn at Nandana and his calculations there.

and Kābul outside al-Hind and Nandana Fort and Nagarkot within its territory are particularly good candidates. Other sites, such as Wayhind/Udabhāṇḍa, Lahore and Multan present, in my view, less indication for a possible long-term interaction between al-Bīrūnī and Indians.

2.4 The Transmission of Living Traditions

In the preceding sections, I argued that al-Bīrūnī described a society that adopted Brahminical precepts, literature and sciences, as well as Hindu cults. I have also highlighted that he relied on his interactions with Indian thinkers in order to study Indian literature, sciences and religion. In addition, elements coming from his direct observations pertains to a specific region of al-Hind, that is, Gandhāra and Panjab, which belonged to the kingdom of the Late Shahis not long before al-Bīrūnī's time. These rulers adhered to Brahminism and to a certain form of Hinduism. Thus, from this rather broad point of view, al-Bīrūnī's descriptions in the *Taḥqīq mā li-l-Hind* match the Late Shahi society. However, Brahminism and Hindu cult can be associated with almost all ruling dynasties of early medieval India. Therefore, I now explore the possibility that for more specific cultural features, notably the philosophy, al-Bīrūnī's descriptions also fit with the society that was based in Gandhāra and Panjab. In other words: were the teachings of Sāṅkhya-Yoga popular among the Brahmins of Gandhāra and Panjab in the early eleventh century CE?

This part of the investigation has two aims: 1) to consider the question of a geographical foyer of the Sāṅkhya-Yoga philosophies in the north-western subcontinent in the early eleventh century CE, an information which is often lacking when Indian philosophy is concerned; 2) to demonstrate that portions of the *Taḥqīq mā li-l-Hind* indeed constitute accurate testimony to the religious, literary and intellectual traditions of the Late Shahis, provided the information is well contextualized and analysed.

Addressing these questions presupposes that local traditions, for instance sciences and religion, continued to be practiced after the Ghaznavids entered the region. I have already observed that people continued to adhere to local customs, such as the timekeeping ritual and traditional calendrical system, in Laghmān and Peshawar areas, some years after Muslim conquests there.[90] Another example of such phenomenon is that of Somnāth. Richard M. Eaton notes that, after the expedition which Maḥmūd led there, extant local Sanskrit

90 See above Section 1.3.1.

inscriptions did not refer to the event, thus suggesting a relatively low impact on local life.[91] It is likely that in the domain of philosophy, too, locals continued to read their texts and to practice their teachings after the Ghaznavids' arrival in Gandhāra and Panjab. Al-Bīrūnī then most probably encountered such living traditions there.

Furthermore, if the philosophers who helped al-Bīrūnī understand the two philosophies were based in Gandhāra and Panjab, the intellectual context of the Late Shahis was thus conducive to the development of these two philosophies in particular. In other words, I posit that the Late Shahi rulers supported philosophers versed in these two philosophies, rather than any other one. Thus, the present section considers the question of how al-Bīrūnī accessed the Sanskrit Sāṅkhya and Yoga manuscripts. It explores the possibility that they were read among the Brahmins he met in Gandhāra and Panjab, and finally examines the reasons why he translated these two philosophical works in particular into Arabic.

There is however very little indication of the possible geographical provenance of the books related to Sāṅkhya-Yoga that al-Bīrūnī consulted for his translations. Therefore, I first discuss places that could most probably not constitute these geographical foyers for al-Bīrūnī, despite their importance as centre knowledge of the time. They are Kanauj, the Valley of Kashmir, Somnāth, Multan and Varanasi.

One passage of the *Kitāb Pātanğal* appears to rule out Kanauj as a place where al-Bīrūnī could have found his Sanskrit source for this translation. In the passage on the different means of knowledge, a simile is offered regarding *āgama*, that is, authoritative tradition:

> Just as our knowledge that the city of Kanauj (كنوج) is on the bank of the Gaṅgā (كنك) River, this [knowledge] results from [oral] report but stands for its (i.e., the knowledge's) apprehension by eyesight.[92]

This example is absent from any extant Yoga text predating al-Bīrūnī's time which could have thus inspired him, that is, the *bhāṣya*-part of the *Pātañjalayogaśāstra*, the *Vivaraṇa*, the *Tattvavaiśāradī* and the *Rājamārtaṇḍa*. It appears to be his own creation, or that of his informants. If it is so, the use of this illustration suggests either that al-Bīrūnī never went to Kanauj—which appears to be the case—or that his interlocutors, who read the *Kitāb Pātanğal* with him, were not from Kanauj.

91 Eaton 2020: 22. On the historiography of Maḥmūd's raid on Somnāth, see Thapar 2004.
92 KP 5, p. 171.4–5; Pines & Gelblum 1966: 315.

As for the Valley of Kashmir, as I have argued in Section 1.2.4, al-Bīrūnī was very well informed about this region, although he never visited it at least by the time of the composition of the *Taḥqīq mā li-l-Hind*.[93] At the time, important intellectual exchanges were indeed taking place between inhabitants of Gandhāra and Panjab on the one hand, and those based in the Kashmir Valley on the other. An extract drawn from the *Taḥqīq mā li-l-Hind* explicitly mentions such exchanges:

> I have been told that this man (i.e., Ugrabhūti, the grammarian) was the educator and teacher of the Šāhi Ānandapāla, son of Jayapāla (اندپال بن جيپال), [who ruled] in our time. After [Ugrabhūti] completed [his] book, he sent it to Kashmir (كشمير). People there did not accept it because of their arrogance in these [things] [...]. [Ānandapāla] ordered the sending of 200,000 dirhams and similar presents to Kashmir, in order to distribute [these gifts] among those who were occupied with the book of his master.[94]

This anecdote illustrates the vigour of intellectual exchanges between the two regions at the time. Further, as aforementioned, al-Bīrūnī's own bibliography also provides evidence suggesting that the scholar corresponded with Kashmiris, as one of his works is entitled *Answers to the Ten Kashmiri Questions* (الجوابات عن المسائل العشر الكشميرية).[95]

Interactions and relations between the kingdom of the Shahi rulers and the Valley of Kashmir were part of an earlier tradition. In the *Rājataraṅgiṇī*, composed in the mid-twelfth century, Kalhaṇa states that Lohara (Fort Lahūr) was dependent on the Kashmiri kings, at the time of Lalitāditya Muktāpīḍa (r. ca. 724–760).[96] According to the same author, the ruler of Lohara in the tenth century, Siṃharāja, was the son-in-law of Bhīma the Shahi (*śrībhīmaśāhi*), namely the king who precedes Jayapāla in the list of the kings provided by al-Bīrūnī. Siṃharāja had his daughter Diddā married to the Kashmiri king Kṣemagupta (r. ca. 950–958).[97]

Incidentally, Kashmir was flourishing at the turn of the first millennium, when Queen Diddā had a college (*maṭha*) built, which hosted young Brahmins from Madhyadeśa (Madhya Pradesh), Hāṭa (or Karahāṭa, in Uttar Pradesh) and

93 See above pp. 43–45.
94 *Taḥqīq* (1958), p. 105.1–6; Sachau 1910: I/135–136.
95 Boilot 1955: 200, no. 72.
96 RT IV.177, p. 50; Stein 1900: I/138.
97 RT VI.176–178, p. 97; Stein 1900: I/249; Majumdar 1957: 65; Pandey 1973: 94.

Saurāṣṭra (Surat, Gujarat).[98] The situation of Bilhaṇa, a Kashmiri minister and poet who lived in the eleventh century, also demonstrates this dynamism and mobility, as he travelled from Kashmir to Mathura, Kanauj, Prayāga, Anahilwada and Somnāth.[99] Al-Bīrūnī who describes Kashmir as a shelter for the Indian sciences and who had access to some of these sciences from Kashmir, may have thus received Sāṅkhya-Yoga manuscripts from Kashmiri travellers.

Moreover, Abhinavagupta, who lived in Kashmir during the second half of the tenth century, extensively elaborates upon the ideas referred to collectively as Kashmiri Śaivism. Both Kashmiri Śaivism and Śaiva Tantra make use of Sāṅkhya-Yoga concepts in their own philosophical constructions.[100] Thus, one may wonder whether the Sāṅkhya-Yoga ideas which are found in the *Taḥqīq mā li-l-Hind* were actually drawn from Kashmiri Śaiva theories or not. If so, then the *Kitāb Sānk* and the *Kitāb Pātanǧal* may be interpretations of Sanskrit works originally found in the Valley of Kashmir.

The content of the two books, that is, the *Kitāb Sānk* as it is transmitted to us in the *Taḥqīq mā li-l-Hind* and the *Kitāb Pātanǧal*, however, makes it clear that the ideas they developed relate to the traditions of the *Sāṅkhyakārikā* and the *Pātañjalayogaśāstra* respectively. This will become clear in the subsequent chapters of the present book. In addition, al-Bīrūnī's *Taḥqīq mā li-l-Hind* does not disclose explicit signs of influences of Kashmiri Śaivism. Thus, despite the favourable situation for intellectual exchanges with Kashmir, there is no evidence of such exchanges in the domain of philosophy between al-Bīrūnī and Kashmiri thinkers, nor is there any indication that his interpretations of Sāṅkhya and Yoga texts were based on manuscripts originating from this region.

Further, only in few cases, al-Bīrūnī provides the geographical provenance of the authors of the works he used: the *Brāhmasphuṭasiddhānta* is by Brahmagupta from Bhillamāla, a book by Durlabha from Multan, the *Srūdhava* by Utpala from Kashmir, and the *Karaṇatilaka* by Vijayanandin from Varanasi. In the same way, he very rarely specifies the geographic provenance of his informants. He only does so in the case of Kanauj, Multan and Somnāth. In addition, he never states in the *Taḥqīq mā li-l-Hind* if an information comes from the region he visited. I thus suggest that al-Bīrūnī may have only considered it important to inform his readers about the provenance of his information—oral and written—when it came from places he did not visit himself. In contrast, al-Bīrūnī may have not deemed it necessary to explicitly state the geo-

98 RT VI.300, p. 102; Stein 1900: I/260; Gopal 1989: 91.

99 Gopal 1989: 92.

100 Torella 1999: 555–557.

graphical provenance of books, sciences and customs when they were locally studied and practiced. If this is accepted and in view of the above discussion, the Sanskrit sources of the *Kitāb Sānk* and *Kitāb Pātanğal* are thus unlikely to have come from Kanauj, Multan, Somnāth or Varanasi. Lastly, some texts, like the Vedas, some great Purāṇas (Mahāpurāṇas), such as the *Viṣṇupurāṇa*, the *Ādityapurāṇa*, the *Matsyapurāṇa* and the *Vāyupurāṇa*, the *Bhagavadgītā* and the *Mahābhārata*, certainly belonged to a category of literature widespread among Brahmins of early medieval India. One may thus wonder whether the sources of the *Kitāb Sānk* and the *Kitāb Pātanğal* were also to be counted among this category of literature. This large diffusion would account for al-Bīrūnī's failure to specify their geographic provenance. A cross-examination of numismatic and textual data, however, suggests that the *Sāṅkhyakārikā* and the *Pātañjalayogaśāstra* traditions were popular teachings among people living in the north-western subcontinent at the beginning of the eleventh century.

First, the use of technical Sāṅkhya term *avyakta* in the legend of the bilingual coins of Lahore described in Section 1.2.3 of the present book may suggest that classical Sāṅkhya philosophy had adepts in Panjab.[101] The concept of *avyakta*, namely the original cause (*prakṛti*), is employed to refer to God (Allah) in the Arabic *šahāda*. This fits well with the fact that in classical Sāṅkhya there is no notion of a creator God: the original cause is the active origin of the phenomenal world.[102]

Thus, the notion of *avyakta* is the Sāṅkhya concept that best renders the concept of the Islamic creator God. The Sanskrit legend of Maḥmūd's bilingual coins likely suggests that the principles of Sāṅkhya metaphysics influenced the legend's composition, thereby serving as evidence that the literate population of the western Panjab, a part of the Late Shahi kingdom, most probably was familiar with classical Sāṅkhya.

Furthermore, there is a series of instances in the *Taḥqīq mā li-l-Hind* suggesting that al-Bīrūnī's informants considered the topics of the *Kitāb Sānk* and the *Kitāb Pātanğal* as essential teachings. In these instances, al-Bīrūnī does not attribute a specific geographical provenance to his statements, implying thus that their principles were well established practices among al-Bīrūnī's local informants. Al-Bīrūnī mentions the *Kitāb Sānk* and the *Kitāb Pātanğal* in the preface to the *Taḥqīq mā li-l-Hind*. These books are described as containing "most principles around which their (i.e., the Indians) faith revolves, without the subdivisions of their religious laws" (فيهما أكثر الاصول التى عليها مدار

101 See above pp. 34–38.

102 See below Section 6.3.3.

(اعتقادهم دون فروع شرائعهم).[103] In the *Taḥqīq mā li-l-Hind*, al-Bīrūnī also qualifies the *Kitāb Pātanǧal* as a "famous book" (الكتاب المشهور).[104] The *Kitāb Pātanǧal* and *Kitāb Sānk*—together with the *Kitāb Gītā* and the Purāṇas—are quoted throughout the *Taḥqīq mā li-l-Hind* when the latter does not deal with astronomy. The descriptions made by al-Bīrūnī of these two books, their mention at the very beginning of the *Taḥqīq mā li-l-Hind*, and the frequent references to them throughout this work indicate their central importance to the Indian thinkers he encountered.[105]

Another clue to the popularity of Sāṅkhya among the Indians with whom al-Bīrūnī interacted lies in the way he sometimes describes concepts related to this philosophy in the *Taḥqīq mā li-l-Hind*, as though they were common beliefs of these Indians and part of an oral local tradition. In a chapter entitled "On their belief in the existents, both intelligible and sensible" (فى ذكر اعتقادهم فى الموجودات العقليّة الحسّيّة), al-Bīrūnī presents the opinion of "those [among the Indians] who deviate from allusions [but direct themselves] to investigation" (الّذين يعدلون عن الرموز إلى التحقيق)[106] and enumerates twenty-five constitutive principles (*tattva*). For the most part, this exposition matches the definitions found in the *Sāṅkhyakārikā* and its commentaries.[107]

Al-Bīrūnī's enumeration begins with *pūruša* (پورش), the Arabic transliteration of the Sanskrit *puruṣa*, an essential Sāṅkhya principle that every living being possesses and which I translate here as "conscious self." Al-Bīrūnī defines it as the human soul or *nafs* (نفس). According to al-Bīrūnī's report, *puruṣa* is only characterized by life, and presents a succession of knowledge and ignorance, as it is ignorant in actuality and intelligent in potentiality, the cause of action being its ignorance.[108] This description of *puruṣa* reflects to some extent that of the *Sāṅkhyakārikā*. Indeed, according to this tradition, the conscious self is inactive, indifferent and defined as the knower (*jña*).[109]

---

103 *Taḥqīq* (1958), p. 6.3–4; Sachau 1910: I/8.

104 *Taḥqīq* (1958), p. 21.16–17; Sachau 1910: I/29.

105 On the relationship between the *Kitāb Sānk* and the *Kitāb Pātanǧal* according to al-Bīrūnī, see below Section 3.4.2.

106 *Taḥqīq* (1958), p. 30.10; Sachau 1910: I/40. See passage II in the Appendix (*Taḥqīq* [1958], pp. 30.10–34.4; Sachau 1910: I/40–44).

107 See chapter 3, section 2 of the present book for an overview of Sāṅkhya philosophy.

108 This definition echoes al-Bīrūnī's definition of the knower (العالم) in KP 36 and 37 of the *Kitāb Pātanǧal* (KP, p. 181.9–17; Pines & Gelblum 1977: 525). A similar description of the soul (النفس) is found in the following chapter of the *Taḥqīq mā li-l-Hind* (*Taḥqīq* [1958], p. 34.7–9; Sachau 1910: I/45).

109 GPBh, p. 4.4–5 on kā. 2.

The next element enumerated by al-Bīrūnī is *abyakta* (ابيكت), an Arabic transliteration of the Sanskrit *avyakta*, meaning "unmanifest," which al-Bīrūnī defines as absolute matter (المادّة المطلقة) or pure primary matter (الهيولى المجرّدة), a philosophical term drawn from Aristotle's works and known to his readership. It is inanimate and possesses the three forces (القوى الثلاث) in potentiality but not in actuality.

The three forces are *satta*, *raja* and *tama* (ست; رج; تم) and correspond to the three constituents (*guṇa*), *sattva*, *rajas* and *tamas*. Al-Bīrūnī describes them as: 1) quietude and goodness, from which existence and growth originate and which are ascribed to angels (الملائكة), namely the deities (*deva*), 2) exertion and labour, from which constancy and continuation originate and which are ascribed to man and 3) languor and indecisiveness, from which decay and annihilation originate and which are ascribed to animals.[110]

In his enumeration, al-Bīrūnī also describes the *byakta* (بيكت), a transliteration of the Sanskrit *vyakta*, meaning "manifest," and qualifies it as the shaped (متصوّرة) matter that possesses the three forces and moves outward in actuality (المادّة خارجة إلى الفعل). He reports that the term *prakriti* (پركرت) designates the whole of pure primary matter and of shaped matter. He then turns to *āhangāra* (آهنكار), which he identifies with the [innate] temperament (الطبيعة). The *mahābhūtas* (مهابوت), the five gross elements, are then described as constituting all existents of this world. He refers to them employing the Arabic expression commonly used to designate the four elements in Islamic tradition, namely the great natures (كبار الطبائع).

At this point in the passage, al-Bīrūnī quotes from the *Vāyupurāṇa*. After this quotation, he discusses the *panğ mātar* (پنج ماتر), a transliteration of Sanskrit *pañca tanmātra* which refers to the five subtle elements and interprets the expression as signifying "five mothers" (أمّهات خمسة) and "simple elements" (بسائط). According to him, these subtle elements precede the gross elements. In parallel with the *Sāṅkhyakārikā*, al-Bīrūnī connects each of the five *tanmātras* to one of the *mahābhūtas*: ether is sound, *šabda* (شبد); wind is what is touched, *sayiras* (سيرس);[111] fire is form, *rūpa* (روپ); water is what is tasted, *rasa* (رس); and earth is what is smelled, *ganda* (گند).[112] Attempting to explain the seemingly strange connection between sound and ether, he invokes quotations from Homer, Porphyrus, Diogenes and Pythagoras.[113]

---

110 Sachau 1910: I/40–41.

111 Here the reading should probably be *sapiras* (سپرس), as the corresponding Sanskrit word is *sparśa*, referring to the quality of tangibility.

112 For the related account in classical Sāṅkhya, see the commentaries on *kā*. 10.

113 Concerning the five gross elements (*mahābhūta*) and the five subtle elements (*tanmātra*),

Al-Bīrūnī also describes the five sense organs, *indriyān* (اندريان), corresponding to the *buddhīndriyas* of classical Sāṅkhya, which he defines as "hearing by the ear, sight by the eye, smelling by the nose, tasting by the tongue and touching by the skin." He further explains *mana* (من), namely the mind (*manas*), as "the will (إرادة) [that] directs [the senses] to [their] various locations (of action)" and "as [having] its residence in a [person] (منه) [... in] the heart (قلب)."[114] Al-Bīrūnī further explains the five senses by action (الحواسّ بلفعل), which he calls the *karma indriyān* (كرم اندريان)[115] and the five necessities (ضروريّات). He describes these principles as follows: production of a sound for [different] kinds of needs and wishes; strength by the hands for fetching and putting away; walking with the feet so as to seek [something] or to flee [from it]; and shaking off the excess of food through each of the two holes destined for it.

At the end of this explanation, al-Bīrūnī provides a summary in which he again lists all principles along with their generic designation as *tatwa* (تتو), that is, the Arabic transliteration of the Sanskrit term *tattva*.

Al-Bīrūnī appears to interpret and explain some of the above Indian concepts on the basis of his intellectual background, for instance, by using Aristotelian and Neo-Platonic terminology and concepts, such as "potentiality" and "actuality."[116] As I show in Chapters 4, 5 and 6 of this book, al-Bīrūnī often used a terminology that his Muslim readership knew so as to make his description of Indian thought and religion more palatable to them. This strategy of translation, however, results in transformed and unliteral translations when compared to their possible sources. As aforementioned, this strategy resembles that which was adopted to translate the Arabic *šahāda* into Sanskrit on the bilingual coins found in Lahore.

Sachau's translation is as follows: "As these elements are compound, they presuppose simple ones which are called *pañca mâtáras,* i.e. five mothers" (Sachau 1910: 1/42). This translation leaves space for some confusion about how al-Bīrūnī's understood the exact sequence of the subtle and gross elements. A literal translation of the above runs as follows: "These elements (i.e., the gross elements) are composite. Thus, they have simple [ones], which precede them and are called *pañca* [*tan*]*mātra* (پنج ماتر), meaning 'five mothers' (أمّهات خمسة)" (هذه العباصر مركّبة فلها بسائط تتقدّمها تسمّى "پنج ماتر" أى أمّهات خمسة) (*Taḥqīq* [1958], p. 32.3–4). This alternative translation renders the meaning of the fifth form of the verbal root *q-d-m* (تتقدّمها) as signifying "they precede them," an important specification in order to precisely understand al-Bīrūnī's description.

114 *Taḥqīq* (1958), p. 33.11–12; Sachau 1910: 1/43–44.

115 Interestingly, al-Bīrūnī here provides an Arabic transliteration of the inflected form *indriyāṇi* (nominative plural) of the neuter stem *indriya.*

116 See Verdon 2019b.

Al-Bīrūnī's account also diverges from the *Sāṅkhyakārikā* tradition on a few other points. Instead of using the terms *avyakta* and *vyakta* as generic designations for some of the twenty-five elements as the *Sāṅkhyakārikā* does, al-Bīrūnī, it seems, understands that the *avyakta* and *vyakta* are principles themselves. He makes no mention of *mahat* or *buddhi*, which is the principle that originates from *prakṛti*, or *avyakta*; but in his scheme, the constitutive principle emerging from *avyakta* is *vyakta*, and thus there are still twenty-five *tattva*s. This confusion, whether due to a misunderstanding on the part of al-Bīrūnī or on the part of his informants, points to an oral transmission of this account.

On the whole, however, al-Bīrūnī's description of the Sāṅkhya *tattva*s (principles) globally matches that of the *Sāṅkhyakārikā* and its commentaries. In contrast, it does not correspond to Sāṅkhya ideas which are found in the *Buddhacarita* or the *Mahābhārata*, nor those of the thirty-six *tattva*s of Kashmiri Śaivism, from which al-Bīrūnī could have hypothetically drawn his information. Therefore, the account in the *Taḥqīq mā li-l-Hind* is most likely either a summary of what al-Bīrūnī had heard from his informants about the Sāṅkhya system of thought or of a passage of the *Kitāb Sānk*. Al-Bīrūnī's confusions in this passage, however, tends to suggest that the passage was mostly orally transmitted to him, and thus a popular teaching among the people he met. Thus, these few examples tend to show that Sāṅkhya and Yoga philosophies were popular among al-Bīrūnī's informants and in the regions he visited in al-Hind.

In order to supplement the discussion of the geographical provenance of al-Bīrūnī's description, I propose to consider the question as to how al-Bīrūnī's personal inclinations may have influenced his choice of topics he covers in the *Taḥqīq mā li-l-Hind*. In the case of astronomy and mathematics, al-Bīrūnī's interests played a crucial role for the information he transmitted in the *Taḥqīq mā li-l-Hind*. Regarding Indian philosophies, it is possible that al-Bīrūnī considered the teachings found in the *Kitāb Sānk* and *Pātanǧal* to have affinities with Falsafa, or Islamic philosophy, and Sufism. He sometimes used a terminology of these systems of thought to render concepts of the metaphysics of Sāṅkhya and of the ethics of Yoga.[117] The parallel that al-Bīrūnī drew between Indian and Islamic thought may have led him to transmit Sāṅkhya and Yoga, rather than any other Indian philosophies, to his Arabic readership.

Evidence, however, shows that he was much dependent on his Indian informants and on observation of the places he visited in al-Hind for his description of Indian culture. He appears then not to have solely let his personal interests guide his work.

---

117 See for instance below pp. 143–144 and 148–149.

For instance, the *Taḥqīq mā li-l-Hind* hardly ever deals with Buddhism, despite al-Bīrūnī's knowledge of it and possible interest in it.[118] Al-Bīrūnī himself states the reasons why he did not take into account Buddhist communities:

> As I never found a Buddhist (الشمنيّة) book and none of the [Indians could] clarify [to me] their [theories] on this subject, my account of them (i.e., the Buddhists) is based on al-Īrānšahrī (الايرانشهرىّ).[119]

This passage reveals that the absence of information on Buddhism in the *Taḥqīq mā li-l-Hind* is due to al-Bīrūnī's not having access to books related to Buddhism, and not due to his lack of interest in the subject. It is however clear that al-Bīrūnī made a distinction between Buddhists and Brahmins, as the following passage illustrates:

> [Another circumstance] which intensified [the Indian] aversion and hostility [to foreigners][120] is [related to] the group known as Šamaniyya (i.e., Buddhists), [who] despite their intense hate toward the Brahmins (البراهمة) are closer to the Indians (الهند), than any other [men]. Formerly, the Khurasan (خراسان), Persia (فارس), Iraq (العراق) and Mosul (موسل)[121] up to the frontier of Syria (الشام) [belonged to] their religion until Zarathustra (زردشت) arrived from Azerbaijan (اذربيجان) and promoted Mazdeism (المجوسيّة) in Balkh (بلخ). His promotion was successful with the [king] Kuštāsb (كشتاسب), and his son Isfandiyār (إسفنديار) continued to promote it in the eastern and in the western lands, forcibly and peacefully. He erected fire temples from China (الصين) to the Greek (الروم) [empire]. The kings after him made way for their religion (i.e., Mazdeism) in Persia and Iraq. Thus, the Šamaniyyas moved from these [lands] to the east of Balkh.[122]

---

118 Sachau 1910: 1/xlv. As for Jainism, Sachau notices that a few features of al-Bīrūnī's work may go back to an encounter with Jainism (Id. 1/xl). For instance, al-Bīrūnī uses the word Jina to refer to the Buddha when he quotes the works of Varāhamihira (Id. 1/119) and Brahmagupta (Id. 1/243). See also Bhattacharyya 1964: 54.

119 *Taḥqīq* (1958), p. 206.4–5; Sachau 1910: 1/249. Īrānšahrī was a Persian scholar from Nišāpur who lived in the second half of the ninth century. He inspired al-Bīrūnī's works, but also those of the physician and philosopher Moḥammad b. Zakariyyā' Rāzi (b. 854).

120 Prior to this passage, al-Bīrūnī discusses the general antipathy of Indians toward foreigners, i.e., *mleccha* (مليج).

121 Mosul is an ancient city situated in northern Iraq.

122 *Taḥqīq* (1958), pp. 15.14–16.3; Sachau 1910: 1/21.

The term *al-šamaniyya* (الشمنية) is the Arabic for naming the Buddhists. Al-Bīrūnī did make use of this term, and in the *Taḥqīq mā li-l-Hind* he differentiates *al-šamaniyyas* from the Brahmins, namely *al-barāhimas* (البراهمة).[123] Here, he attempts to explain the decline of Buddhism in Central Asia as due to the advent of Zoroastrianism.[124] This question is a much debated one. The types of interaction between the several religious communities established in Central Asia and in the north-western parts of South Asia, such as Buddhist, Muslim, Hindu and Zoroastrian, indeed, varied regionally,[125] and it is not the place here to discuss this issue.

Nevertheless, in addition to his statement about the inaccessibility of Buddhist books for him, al-Bīrūnī did not describe any well-known Buddhist site, in the way he described Hindu temples and idols, for instance in Taneshwar, Multan and Somnāth. It is thus likely that he indeed did not directly interact with a Buddhist who could have informed him about important Buddhist sites at the time. His testimony perhaps also indicates that the significance of Buddhist sites as intellectual or cultural centres had waned in Gandhāra and Panjab and that Buddhist communities were no longer supported by the ruling dynasties at the time. In addition, as I showed above, most information that al-Bīrūnī gathered in the *Taḥqīq mā li-l-Hind* relates to a socio-religious context different from that of a Buddhist population.

Furthermore, while al-Bīrūnī generously quotes from texts linked to Sāṅkhya and Yoga philosophies in the *Taḥqīq mā li-l-Hind*, he remains silent with regard to other Indian systems of thought. For instance, he did not engage with Nyāya-Vaiśeṣika tradition, nor with the Mīmāṃsā or Vedānta, generally considered predominant in India at the time. Why the *Taḥqīq mā li-l-Hind* is silent about Advaita-Vedānta philosophy, for example, is another relevant question to consider.[126] One may argue that this absence could either be due to al-Bīrūnī's

123 See also Maclean 1989: 5.

124 Watters 1904: I/202. In *Al-āṯār al-bāqiya*, al-Bīrūnī also mentions the decline of Buddhism in Central Asia (Sachau 1879: 188–189; quoted in Elverskog 2010: 51).

125 For instance, the situation was different in Sind and Balkh (Maclean 1989: 22–77; Arezou 2017: 44–47), and Buddhist traditions also survived for a longer time in lower than upper Sind (Maclean 1989: 52–57).

126 S.J. Heras (1951: 119–123) argues that the idea of God conveyed by al-Bīrūnī in the *Taḥqīq mā li-l-Hind* is connected with Advaita-Vedānta. According to al-Bīrūnī, educated Indians believe in a unique and all-pervasive God who adopts plural manifestations in the world, whereas uneducated Indians only see the manifestations and are not able to conceive an abstract notion of God. The concept of God that al-Bīrūnī attributes to educated Indians, however, does not necessarily reflect that of Advaita-Vedānta. As seen in Chapter 4 below, much of al-Bīrūnī's interpretative work consists in reducing cultural gaps between his own culture and the culture he encountered. His interpretation of God in *the Taḥqīq*

particular preferences or to the fact that he did not access books related to these philosophical systems in Gandhāra and Panjab. In view of the above, it is likely that the people al-Bīrūnī met in Gandhāra and Panjab were no philosophers of Mīmāṃsā and Vedānta, nor of Nyāya-Vaiśeṣika.

The passage below contains the only explicit reference to philosophies other than Sāṅkhya and Yoga in the *Taḥqīq mā li-l-Hind*. In contrast with his comments on Buddhism, al-Bīrūnī does not explain his lack of information on these philosophies. The passage reads as follows:

> They (i.e., the Indians) have books on the jurisprudence of their religion, on theology, on asceticism, on deification and on the quest for liberation from the world,[127] such as the eponymous book composed by Gauḍa (گور) the ascetic; [the *Kitāb*] *Sānk* (سانگ), composed by Kapila (كپل), on divine subjects; [the *Kitāb*] *Pātanǧal* (پاتنجل), on the quest for liberation and for union between the soul and its object of apprehension; [the *Kitāb*] *Nyāyabhāṣya*,[128] composed by Kapila, on the Vedas (بيذ) with their interpretation, [on the position] that they (i.e., the Vedas) have been created, and on the distinction in these [works] between divine precepts and customs; [the *Kitāb*] *Mīmāṃsā* (ميمانس), composed by Jaimini,[129] on the same subject; [the *Kitāb*] *Lokāyata* (لوكايت), composed by Jupiter (المشترى),[130] which accepts sensory perception as the only [means to know] about an object of investigation; [the *Kitāb*] *Agastimata* (?) (اگست مت),[131] com-

mā li-l-Hind is also indebted to this tendency according to which the Indian God would resemble Allah in some of His characteristics (see below for al-Bīrūnī's interpretation of Īśvara, pp. 150–155). Moreover, al-Bīrūnī never explicitly mentions a work related to Advaita-Vedānta. It is, however, possible that Advaita ideas were spread among the Indian thinkers whom al-Bīrūnī met. These thinkers would have thus transmitted such ideas indirectly, which al-Bīrūnī conveyed then in his book, without identifying these ideas as such.

127 Al-Bīrūnī generally employs the Arabic term *al-ḵalāṣ* (الخلاص), meaning "liberation" or "deliverance," to refer to the Sanskrit terms *mokṣa* (liberation), *kaivalya* (isolation) or *apavarga* (emancipation). In the *Taḥqīq mā li-l-Hind*, he also provides a transliteration into Arabic script for the word "liberation," i.e., *mūkša* (موكش). See *Taḥqīq* (1958), p. 53.8–9; Sachau 1910: I/70.

128 In Sachau's edition, the reading is *nāyabhāša* (نايبهاش), instead of *nāyayahaša* (نايَيَهاش) as in the Hyderabad's edition.

129 Sachau reads the term as *ǧaymin* (چيمن), instead of *chiyaman* (جيمن) as in Hyderabad's edition.

130 The Arabic Jupiter here stands for the Indian Bṛhaspati who is considered the founder of the Lokāyata school of thought.

131 Sachau does not identify this work (1910: II/300). However, the Arabic transliteration from Sanskrit (*agasta mata*, اگست مت) seems to render *Agastimata*, which is the title of an

> posed by Canopus (سهيل),[132] on the use of the senses, alongside with the oral tradition, [as means to know] about the [object of investigation]; and the book *Viṣṇudharma* (بشن دهرم). The [general] interpretation of [the term] *dharma* [takes it to refer to] reward, but [here] it means "religion," as if the book were [titled] *The Religion of Allah*, related to Nārāyaṇa (ناراين).[133]

Thus, al-Bīrūnī is aware of the existence of a certain number of works connected to Indian philosophy and religion. Some problems and questions however arise with regard to his account. For instance, al-Bīrūnī enumerates a work which he entitles *Nyāyabhāṣya* and which he attributes to Kapila. If the work mentioned here by al-Bīrūnī refers to the extant Vātsyāyana's *Nyāyabhāṣya*, he was ill-informed when he associates its authorship with Kapila. In addition, al-Bīrūnī's description of the subject dealt with in this book does not reflect the actual main topic of the Sanskrit *Nyāyabhāṣya* that pertains to debate, logic, epistemology and metaphysics. This confusion contrasts with his relatively accurate description of the content of the books related to Sāṅkhya and Yoga. Also of note is that titles of works related to the Vaiśeṣika or Vedānta systems are not referred to in the above enumeration. This absence suggests that al-Bīrūnī did not access accurate information about these two traditions. As in the case of literature relating to Buddhism, al-Bīrūnī probably did not find books related to them; at the same time, this account rather reflects the scope of his informants' philosophical knowledge and training.

Two additional elements indicate that the selection of literature transmitted by al-Bīrūnī does not necessarily reflect his personal preferences: his criticism of the content of some texts, despite him quoting them in the *Taḥqīq mā li-l-Hind*, and the expression of his constant quest for knowledge and books.

Although al-Bīrūnī heavily quotes from purāṇic literature, specifically the *Viṣṇudharma*,[134] the *Viṣṇupurāṇa*, the *Matsyapurāṇa*, the *Vāyupurāṇa* and the *Ādityapurāṇa*, he also criticizes their content. For instance, referring to a passage he draws from the *Viṣṇudharma*, he writes:

Indian treatise on gemstones composed before the tenth century. Unless there was a further work with this title, al-Bīrūnī's description does not fit the content of this treatise.

132 Canopus corresponds to Agasti in Indian astronomy.

133 *Taḥqīq* (1958), p. 102.1–10; Sachau 1910: I/131–132.

134 Al-Bīrūnī's *Viṣṇudharma* is to be identified with the Sanskrit *Viṣṇudharmottarapurāṇa*; see Gonda 1951: 111.

> Further, the [*Viṣṇudharma*] says: "A [man] who reads this [statement regarding the celestial pole] and knows it accurately, him [indeed] Allah forgives [his] sins of the day and adds fourteen years to his predetermined age." How simple are [these] people! And among us, some [scholars] know between 1020 to 1030 stars. He (i.e., Allah) does not take their breaths, nor deduct from their age only because of this.[135]

Further, after having quoted the *Viṣṇupurāṇa*, the *Matsyapurāṇa*, the *Vāyupurāṇa*, the *Ādityapurāṇa* and the *Kitāb Pātanǧal* regarding the size of Mount Meru, al-Bīrūnī states:

> The excessive dimensions of Mount [Meru] only make sense because of the excessive dimensions of the earth which these [works] report. If conjecture has no limit, then there is open space for [further] lying based on what was presumed.[136]

More generally, al-Bīrūnī notes the following:

> This sum is more than thrice the one we mentioned according to the commentator Pātanǧal. This is a habit of the copyists in all languages. The authors of the Purāṇas (پرانات) are not free from it, as they are not adherents of scientific studying.[137]
>
> As for the authors of the Purāṇas, they represent heaven as a still dome above the earth and the stars wandering from east to west. How [then] would they have knowledge of the second motion? And if they would, how would [their] opponent let them [believe] that a unique thing (i.e., star) [can] move in two different directions by itself? We mention what has reached us from them, not because it is useful, as there is no usefulness in it.[138]

Despite his complaints regarding some ideas found in the purāṇic literature, al-Bīrūnī still mentions them. The transmission of such theories, which he considered worthy of critique, in his writing thus is not due to his personal inclination. Similarly, although al-Bīrūnī translated the *Kitāb Pātanǧal*, he disliked the presentation of cosmography by the author of this book:

135 *Taḥqīq* (1958), p. 200.3–6; Sachau 1910: 1/242.
136 *Taḥqīq* (1958), p. 205.12–14; Sachau 1910: 1/248.
137 *Taḥqīq* (1958), p. 196.14–15; Sachau 1910: 1/238.
138 *Taḥqīq* (1958), p. 237.11–15; Sachau 1910: 1/284.

> We have already found annoying the mention of the seven oceans, together with the [seven] earths, and even so this man [considers] making our burden lighter by adding earths below.[139]

Furthermore, two passages indicate that al-Bīrūnī actively searched for different kinds of sources in order to inform himself. The first passage where he describes his search for source materials occurs in his aforementioned historical account of the Late Shahis:

> [… The history of] such a lineage, [written] on a piece of silk, is found in the fortress of Nagarkot (نغركوت). I desired to find it but was prevented from doing so for [different] reasons.[140]

His constant search for written documents is also evident in the following extract:

> I have found it very hard to work my way into this [subject] (i.e., al-Hind), despite my desire which I alone possess in my time, and although I sacrificed myself generously and as much as possible in collecting their books from places where they were likely to be found and in gathering those who were on the right way to [find] them from places where they were hidden.[141]

Al-Bīrūnī's intellectual curiosity was thus not limited to the works he may have been sympathetic to. It is likely that had he discovered books related to Buddhism or to other traditions of Indian thought, he would have turned his attention to them and reported on them. This assumption concurs with the following statement by al-Bīrūnī in the preface to the *Taḥqīq mā li-l-Hind*: "[T]his book is a report [of the facts]. I convey the words of the Indians as they are."[142] All of the above indeed strongly suggests that al-Bīrūnī mostly described aspects of the Indian society as it was presented to him through personal collaborations taking place in Gandhāra and Panjab, including Brahmins belonging to the Late Shahi court.

139 *Taḥqīq* (1958), p. 195.1–2; Sachau 1910: I/237.
140 *Taḥqīq* (1958), p. 349.6–8; Sachau 1910: II/11. See the full passage above, pp. 17–18.
141 *Taḥqīq* (1958), p. 18.5–7; Sachau 1910: I/24.
142 *Taḥqīq* (1958), p. 5.11–12; Sachau 1910: I/7.

2.5 Concluding Remarks

This chapter first surveyed the intellectual framework in which al-Bīrūnī encountered Indian society, science and literature. It was possible to show that al-Bīrūnī gradually familiarized himself with Indian culture throughout his life. Having been originally trained as an astronomer and mathematician, he later on expanded his knowledge to the history of civilizations, Indian culture, gemmology and pharmacology. While his pre-existing knowledge of Indian science and culture was chiefly based on textual materials, the information found in the *Taḥqīq mā li-l-Hind* relies on his knowledge of Sanskrit literature, on his interactions with Indians and on his observations made during his visits to al-Hind.

Further, the context of Maḥmūd's court was favourable for al-Bīrūnī to learn Sanskrit and to act as an interpreter. In this context, al-Bīrūnī not only met traders or travellers who, for instance, informed him about the geography of different provinces of India. He also actively interacted with Brahmins, some of whom were well versed in religious matters, astronomy, purāṇic literature and philosophy, and who guided him in understanding such areas of study. The elite with which al-Bīrūnī interacted was certainly part of the royal Indian courts which were directly challenged by Maḥmūd's military campaigns, including that of the Late Shahis.

Two circles of literate persons—astronomers, priests, philosophers, interpreters, etc.—belonging to two distinct courts, namely that of Maḥmūd and of Indian kings, met thanks to the political context of the time. Further, in the case of the Ghaznavid court, the interacting group included Muslim thinkers and Brahmins who conducted various empirical studies and shared an interest in scientific problems.[143] The case of al-Bīrūnī and the assumed circumstances in which he interacted with Indian thinkers also illustrate how political power, commercial interests and religious context strongly influenced access to one's object of research at the time. I highlighted in Chapter 1, that a war context, also motivated by commercial concerns, contributed to the opening and the protection of roads network. The control over these roads encouraged knowledge to circulate among scholars. Chapter 2 of the present book confirmed these observations and highlighted the dynamism of intercultural and intellectual exchanges at Muslim royal courts in early medieval times.

In addition, when al-Bīrūnī studied Sanskrit and Indian sciences he had to collaborate with thinkers versed in Sanskrit, but also acquainted with Persian or Arabic. This group of scholars may also have worked with him through the

143 On literate circles see Touati 2000: 108.

intermediary of a vernacular language. Thanks to these collaborations and to the many texts available to him, al-Bīrūnī was well informed about Sanskrit phonology and terminology.

Places where al-Bīrūnī's long-term interactions with this elite occurred may have been in the region of Ghazna and Kābul—to where Indians might have been taken as prisoners or where they were employed as advisors and interpreters. In the territory of al-Hind, Fort Nandana, Nagarkot, and possibly Wayhind/Udabhāṇḍa, Peshawar, Lahore and Multan, would have hosted such collaborations.

Discrepancies between the Indian textual tradition and al-Bīrūnī's testimony addressed in this chapter may reflect his informants' confusion, thereby foregrounding the importance of orality in al-Bīrūnī's reception of the two Indian philosophies. Oral tradition indeed played an important role in al-Bīrūnī's acquisition of knowledge about India and in his translations.

Lastly, data drawn from al-Bīrūnī's writings and from numismatic evidence, if considered from a circumstantial perspective, strongly suggests that the sources of the *Kitāb Pātanǧal* and *Kitāb Sānk* were popular readings and textbooks among the Indian scholars whom al-Bīrūnī met, possibly in Gandhāra and Panjab. If these thinkers belonged to the Late Shahi court, one may infer which sciences, religions and philosophies these kings were supporting and promoting. In this way, the study and teaching of Sāṅkhya and Yoga philosophies may have been financially supported by the Late Shahi kings.

CHAPTER 3

Al-Bīrūnī's Translations within the Sāṅkhya-Yoga Traditions

3.1 Sāṅkhya-Yoga Literature Predating al-Bīrūnī's Time

In Chapter 2, I highlighted the role and the nature of al-Bīrūnī's collaborations with Indians in his learning of Indian culture, sciences and philosophies. There, I quoted a passage from the *Kitāb Pātanǧal* which shows that al-Bīrūnī had access to "books on wisdom", that is, philosophy, and at the same time consulted Indians to understand their content. This passage recounts that these books "were read to [him] letter by letter."[1] It is thus probably these books that constituted his Sāṅkhya and Yoga sources and which he translated under the titles *Kitāb Sānk* and *Kitāb Pātanǧal*.

However, as pointed out in Chapter 2, concepts assuming a Sāṅkhya-Yoga colour are found in various other Indic traditions, notably in Śaiva tantric doctrines,[2] while primary literary sources point to the existence of several schools known as Sāṅkhya. For instance, the *Carakasaṃhitā* and the *Buddhacarita* contain descriptions relating to ontology and metaphysics that bring to mind but do not match the relevant concepts elaborated in the *Sāṅkhyakārikā* and its commentaries. In addition, the terms *yoga* and *sāṅkhya* have been indeed widely used in ancient Sanskrit literature, for instance in some Upaniṣads, in the *Mokṣadharma* section of the *Mahābhārata* and the *Bhagavadgītā*.[3] In these works, which are considered to precede the compositions of the *Sāṅkhyakārikā* and the *Pātañjalayogaśāstra*, the two terms do not refer to systematized philosophies,[4] although they often occur alongside concepts that strongly evoke the two systems. Thus, Sāṅkhya and Yoga concepts are found throughout history in doctrines and texts which are different from the ideas elaborated as standardized systems of thought in their foundational works that are the *Sāṅkhyakārikā* and the *Pātañjalayogaśāstra*.

1 See p. 76 above.
2 See for instance Torella 1999: 555–556.
3 Chakravarti 1951: 11–64; Frauwallner 1973: 106–114; Motegi 2013: 43–45.
4 Edgerton (1924) convincingly and comprehensively discusses the exact meaning of these two terms in the Upaniṣads and the Epics. See also Renou & Filliozat 1953: 44 and below p. 122.

These observations render the history of Sāṅkhya and Yoga particularly delicate to retrace. They also make it necessary to define a chronological and philological framework in order to determine al-Bīrūnī's sources for his interpretation of Sāṅkhya and Yoga texts. Based on philological evidence, the present chapter situate al-Bīrūnī's *Kitāb Pātanǧal* and *Kitāb Sānk* within Sanskrit textual traditions.

There have been several attempts to periodize the historical development of Indian philosophy in general, and of these two philosophies in particular.[5] Although scholars have variously periodized their developments, I use, in the present book, a specific nomenclature exclusively to give a chronological frame to the philological discussion. I label "pre-classical" the period represented by some Upaniṣads, the Epic, the *Carakasaṃhitā* and the *Buddhacarita* in which are found Sāṅkhya-Yoga vocabulary and concepts, and "classical" the period which extends from the time of the compositions of the *Sāṅkhyakārikā* and the *Pātañjalayogaśāstra*, that is, between the fourth and fifth centuries, until when their latest commentaries were composed, namely the mid-eleventh century approximately.[6]

The above description in Section 2.4 of the twenty-five *tattva*s has already suggested that al-Bīrūnī's understanding of these concepts has more in common with classical Sāṅkhya-Yoga theories than with elaboration of some of these concepts in other texts and philosophies, such as the Upaniṣads, the Epic, the *Carakasaṃhitā*, the *Buddhacarita* and the Śaiva Tantra. Chapters 4, 5 and 6 confirm this preliminary observation taking on conceptual perspective.

Furthermore, the *Kitāb Sānk* and the *Kitāb Pātanǧal* must have been composed between the year 1017, namely when al-Bīrūnī accompanied Maḥmūd at his court, and the year 1030 when he wrote the *Taḥqīq mā li-l-Hind*. If al-Bīrūnī did indeed begin to study Sanskrit literature in a thorough manner at Maḥmūd's court, it is likely that the process of learning Sanskrit, as well as that of translating the two philosophical works from Sanskrit to Arabic, took a number of years. Al-Bīrūnī may have therefore become skilled—to whatever extent he was—in interpreting Sanskrit texts only some time following 1017; thus, he possibly composed the *Kitāb Pātanǧal* and the *Kitāb Sānk* between the years 1020 and 1030.

5 See Larson & Bhattacharya 1987: 3–14. On a discussion about periodization in the history of Indian philosophy, see Franco 2013.

6 Only these two early chronological frameworks are relevant to discuss in the present research. I do not intend to debate on the implications that the use of this terminology may have in another context than the chronological one. See for instance O'Brien-Kop 2017: 126–127.

Thus, the works which were read to al-Bīrūnī necessarily predate the year 1030. Therefore, a brief outline of the chronologies of the *Sāṅkhyakārikā* and the *Pātañjalayogaśāstra* and commentaries is necessary. It is possible to determine the *terminus ante quem* of the *Sāṅkhyakārikā* composed by Īśvarakṛṣṇa thanks to the work of Paramārtha (499–569 CE) who translated this text together with one of its commentaries into Chinese.[7] According to Chinese sources, Paramārtha reached China in 546 CE bringing texts from India with him. Thus, the composition of the *Sāṅkhyakārikā* must precede this year.[8] Erich Frauwallner situates it before 500 and Pulinbihari Chakravarti tentatively dates the *Sāṅkhyakārikā* to the end of the fourth century, while the authors of the *Encyclopedia of Indian Philosophy* argue for a date between 350 and 450.[9]

Several Indian authors, from the author of the commentary translated by Paramārtha in the sixth century up to Vācaspatimiśra in the mid-tenth century, commented upon the *Sāṅkhyakārikā*. The tradition of commentating upon this metrical work during this period indicates the limit before which the *Sāṅkhyakārikā* must have been compiled and at the same time demonstrates its popularity during these centuries. The title of the commentary on the *Sāṅkhyakārikā* translated into Chinese by Paramārtha probably between the years 546 and 569 has been reconstructed as **Suvarṇasaptati* in Sanskrit. Scholars have debated which Sanskrit source Paramārtha used for his translation. They have not reached so far any convincing conclusion.[10] The date generally agreed upon for the composition of another early commentary, the *Gauḍapādabhāṣya*, is between 500 and 600.[11] Whereas Frauwallner appears to situate the *Māṭharavṛtti*'s composition in the early sixth century, the authors of the *Encyclopaedia of Indian Philosophies* consider it as belonging to the mid-ninth century.[12]

The *Sāṅkhyasaptativṛtti* and the *Sāṅkhyavṛtti* are two other commentaries on the *Sāṅkhyakārikā* of which the dates of composition dates have yet to be

7 Takakusu 1904a and 1904b; Chakravarti 1951: 159.

8 Funayama 2010: 144.

9 Chakravarti 1951: 158; Frauwallner 1973: 225–226; Larson & Bhattacharya 1987: 149.

10 Takakusu 1904a: 2–4, 25 and 35; Belvalkar 1917: 172–173; Garbe 1917: 91–93; Keith 1924: 551; Larson & Bhattacharya 1987: 167–168.

11 Larson & Bhattacharya 1987: 209–210. Garbe (1917: 87) and Takakusu (1904a: 4) identified the author of the *Gauḍapādabhāṣya* as Gauḍapāda, the Advaita-Vedāntin, author of the *Māṇḍūkyakārikā*. This identification is, however, doubtful. The edition of the *Gauḍapādabhāṣya* by Har Dutt Sharma (1933) has been used in this study. Translations of this work are for instance Esnoul 1964 and Mainkar 1972.

12 Frauwallner 1973: 226; Larson & Bhattacharya 1987: 291–299. Keith (1924: 551) and Sastri (1944: xxx–xxxi) noticed that the *Māṭharavṛtti* quotes from Śaṅkara's *Hastāmalakastotra*.

firmly ascertained. Two manuscripts of these works were found in the *Catalogue of the Palm-Leaf Manuscripts* of the Jaina Grantha Bhaṇḍāra of Jaisalmer and edited by Esther A. Solomon (1973a and 1973b).[13] The manuscript of the *Sāṅkhyasaptativṛtti* bears an indication of its date, namely that it was "copied in about the first half of the twelfth cent. v.s." (Solomon 1973a: 5). The leaf on which the name of the author is written is, however, damaged.[14] As for the *Sāṅkhyavṛtti*, a note in the catalogue indicates that the manuscript was copied in *saṃvat* 1176; however, it does not contain the name of the author of the commentary.[15]

The five commentaries discussed above display striking similarities in content but also differ in many ways. Some scholars have considered them all to be originating from an *Ur*-commentary.[16] The *Gauḍapādabhāṣya* is the most concise among them. Whereas the *Sāṅkhyavṛtti* greatly resembles the **Suvarṇasaptati*, these two commentaries also differ from each other in some important points. In a similar way, the *Sāṅkhyasaptativṛtti* shares many common points with the *Māṭharavṛtti* and at the same time diverges from it.[17] The study of these five Sanskrit commentaries as compared to al-Bīrūnī's Arabic translation in Chapter 6 of this book confirms these remarks.

The *Yuktidīpikā* also counts among the earliest commentaries on the *Sāṅkhyakārikā*.[18] The lower limit of its composition can be ascertained thanks to some passages that contain criticism of ideas expressed in Vasubandhu's *Abhidharmakośa* and *Viṃśikā* (mid-4th c.),[19] references to Bhartṛhari's *Vākyapādīya* (5th c.) and quotations from the *Pramāṇasamuccaya* by Dignāga (480–540?).[20] Jayantabhaṭṭa (ca. 850–910) and Vācaspatimiśra (ca. 950–1000) both provide its *terminus ante quem*, by referring to the work under the title *Rājav-*

13 See also Larson & Bhattacharya 1987: 178–208.

14 Solomon 1973a: 5–6.

15 Solomon 1973b: 5.

16 Solomon 1974: 1; Larson & Bhattacharya 1987: 167. See also Keith 1924 and Chakravarti 1951: 159–160.

17 Solomon 1973b: 7; 1974: 100 and 106. Examples of the similarities and differences between the group of five commentaries are given in Chapter 6, below. On a comparison of metaphors and their use in these five commentaries, especially in the **Suvarṇasaptati*, the *Sāṅkhyavṛtti* and the *Gauḍapādabhāṣya*, see Verdon 2019c.

18 The *Yuktidīpikā* was edited for the first time by Pulinbehari Chakravarti in 1938 and critically edited by Albrecht Wezler and Shujun Motegi in 1998. For an English translation of the *Yuktidīpikā*, see Kumar & Bhargava 1990 and 1992. For a study on the reaction of the *Yuktidīpikā*'s author to Dignāga's criticism related to the concept of perception (*pratyakṣa*), see Harzer 2006.

19 Maas 2013: 66, on the basis of Franco & Preisendanz 2010: xvi.

20 Mejor 2004: 400 and 404–406.

ārttika. Albrecht Wezler and Shujun Motegi state that a quotation from the *Kāśikāvṛtti* (680–700) occurs in the *Yuktidīpikā*, and therefore place the composition of this commentary between the end of the seventh and the beginning of the eighth century.[21]

According to Johannes Bronkhorst, however, this quotation may belong to "any commentary of the Pāṇinian tradition"[22] rather than to the *Kāśikāvṛtti* solely. Bronkhorst's argument, if accepted, suggests the possibility of an earlier dating of the *Yuktidīpikā*'s composition. In addition, the absence of any reference to Dharmakīrti's works (6th c.) in this commentary lends support to this argument. Additionally, Marek Mejor argues that the fact that the author of the *Yuktidīpikā* refers to Vasubandhu's works may point to a chronological proximity between the two texts.[23] The date of the *Yuktidīpikā*'s composition can be thus placed between the mid-sixth and mid-seventh century.[24]

The *Yuktidīpikā*'s structure of four chapters (*prakaraṇa*) and eleven sections (*āhnika*), its detailed development of numerous philosophical topics, and the large number of references to Sāṅkhya teachers and to philosophical schools other than Sāṅkhya all contribute to a unique character of this work when compared with the other commentaries on the *Sāṅkhyakārikā*. These elements also determine its great significance for the history of Indian philosophy. Its author fully engages with arguments arising from different schools of thought in his discussions, thus reflecting the vitality of philosophical debates at the time.

The date of composition of the *Jayamaṅgalā* may be placed between that of the *Yuktidīpikā* and the *Tattvakaumudī*, respectively, namely between the seventh and the mid-tenth century.[25] However, it has so far been impossible to date it with more precision or to ascribe it to a particular author. Vācaspatimiśra, the author of both the *Tattvakaumudī* on the *Sāṅkhyakārikā* and the *Tattvavaiśāradī* on the *Pātañjalayogaśāstra*, was supposedly a Maithili Brahmin. Frauwallner posits that he was active in the mid-ninth century.[26] However, subsequent research has pushed this date to the second half of the tenth century.[27]

21 Wezler & Motegi 1998: xxvii–xxviii.
22 Bronkhorst 2003: 247.
23 Mejor 2004: 404–405 and 407.
24 Wezler & Motegi 1998: xxv–xxviii; Bronkhorst 2003: 246. I adopt Eltschinger's dates for Dignāga and Dharmakīrti (Eltschinger 2010: 398–400).
25 Larson & Bhattacharya 1987: 271. These two authors date, however, the *Tattvakaumudī* to the ninth century.
26 Frauwallner 1973: 226.
27 Srinivasan 1967: 54–65; Slaje 1986: 274; Larson & Bhattacharya 1987: 301–312 and 2008: 218–

Turning to the Yoga works, the date of composition of the *Pātañjalayogaśāstra* is relatively early. James Haughton Woods interpreted *sūtras* IV.15–16 as constituting an attack against the Vijñānavāda doctrine of Vasubandhu.[28] Philipp André Maas, considering that the Vijñānavāda doctrine may have pre-existed Vasubandhu, dates the composition of the *Pātañjalayogaśāstra* between 325 and 425.[29]

As for the *Pātañjalayogaśāstravivaraṇa* (hereafter the *Vivaraṇa*) commenting upon the *Pātañjalayogaśāstra*, several researchers have sought to demonstrate that it is a relatively late work.[30] The main argument for this, as presented by Rukmani for instance, is that its author explicitly refers to Vācaspatimiśra's *Tattvavaiśāradī* composed in ca. 950.[31] However, Kengo Harimoto and Maas questioned it, on the basis of their observation that no literal quotation from Vācaspatimiśra's works is found in the *Vivaraṇa*, nor any identifiable reference to an idea or a concept originally introduced by him. This has led Harimoto and Maas to refute Rukmani's statement.[32]

In addition, in 1983, Wilhelm Halbfass noted that Kumārila (7th c.) is the most recent author to whom the *Vivaraṇa* refers.[33] Albrecht Wezler in the same year and Maas in 2006 also pointed out that the text of the *Vivaraṇa* offers relatively ancient readings of the *Pātañjalayogaśāstra*, on which it comments. Further, Wezler and Bronkhorst consider that the *Vivaraṇa* is the oldest of the available commentaries on the *Pātañjalayogaśāstra*.[34] An earlier date, that is, between the seventh century (Kumārila) and the mid-tenth century (Vācaspatimiśra), for the *Vivaraṇa*'s composition appears thus reasonable. An author named Śaṅkara wrote the commentary.[35] Like the *Yuktidīpikā* in tex-

---

240; Acharya 2006: xxviii; Maas 2006: xii, n. 2 and 2013: 78. For a summary of the dating of the commentaries on the *Sāṅkhyakārikā* based on Larson and Bhattacharya, see Łucyszyna 2016: 304.

28 Woods 1914: xvii–xviii.

29 Maas 2006: xviii–xix and 2013: 65–66.

30 Gelblum 1992: 87; Rukmani 2001: I/xxv–xxix; Larson & Bhattacharya 2008: 240.

31 Rukmani 2001: I/xxv–xxix.

32 Harimoto 2004: 179–180 and 2014: 235–241; Maas 2013: 75.

33 Halbfass 1983: 120.

34 Wezler 1983: 27 and 33–34; Bronkhorst 1985: 203; Maas 2006: lxix and 2013: 77–78. See also Harimoto 2014: 237.

35 The question of whether this Śaṅkara is the Advaitin Śaṅkara remains a point of contention to this day. Addressing these questions, however, lies beyond the scope of the present investigation. The reader may refer to the secondary literature existing on this issue. Harimoto (1999: 36–136, 2014: 11–13 and 225–251) provides much material on this question and discusses it at length. See also Hacker 1968, Oberhammer 1977: 135, Wezler 1983: 34–36, Halbfass 1991: 204–207, 224–228, Gelblum 1992: 76–77, Rukmani 1998 and 2001: I/ix–xxxi, Larson & Bhattacharya 1987: 289 and 2008: 239–240, and Maas 2013: 73–74.

tual tradition of Sāṅkhya, the *Vivaraṇa* is a very comprehensive commentary of classical Yoga, as it often offers extensive comments upon the *Pātañjalayogaśāstra* and includes philosophical debates in its account.

The historical context of the *Rājamārtaṇḍa* is relatively well known in comparison with the other Sanskrit works under review here. King Bhoja of the Paramāra dynasty, composed, or commissioned, it. Bhoja was the ruler of the region of Mālava, located in present-day western Madhya Pradesh, with the city of Dhāra as his capital. His reign approximately dates to the first half of the eleventh century.[36] The *Rājamārtaṇḍa* was composed during the same period as al-Bīrūnī's works, but its exact date of composition is unknown. It actually could predate or postdate the composition of the *Kitāb Pātanǧal*. I, however, include the *Rājamārtaṇḍa* in the present research due to this chronological proximity.[37] Nevertheless, the *Rājamārtaṇḍa* is extremely concise and, in contradistinction to the other Yoga texts dealt with in the present book, this commentary only glosses the *sūtra*-part of the *Pātañjalayogaśāstra*. Its commentary does not equate the *bhāṣya*-part of the *Pātañjalayogaśāstra* but at times evokes its content.

Thus, Sanskrit works which the philological and conceptual studies of the present books include, are, for Sāṅkhya, the **Suvarṇasaptati*, the *Gauḍapādabhāṣya*, the *Māṭharavṛtti*, the *Sāṅkhyasaptativṛtti*¸ the *Sāṅkhyavṛtti*¸ the *Yuktidīpikā*, the *Jayamaṅgalā* and the *Tattvakaumudī*, and, for Yoga, the *Pātañjalayogaśāstra*, the *Vivaraṇa*, the *Tattvavaiśāradī* and the *Rājamārtaṇḍa*.

3.2 Tenets of Sāṅkhya and Yoga

Before turning to the analyses of the above-mentioned Sanskrit commentaries in relation to al-Bīrūnī's translations, I provide an outline of the teachings of classical Sāṅkhya and Yoga as they are elaborated in the *Sāṅkhyakārikā* and the *Pātañjalayogaśāstra*, respectively. The following outline is not exhaustive, as it discusses the tenets of these systems in the context of al-Bīrūnī's Arabic translations. It, however, constitutes a crucial starting point in contextualizing the scholar's *Kitāb Sānk* and *Pātanǧal*, and an essential introduction to the subsequent analyses in Chapters 4, 5 and 6.[38]

---

36 Pingree 1981: 336–337.

37 Larson & Bhattacharya 1987: 4, 313 and 2008: 266; Maas 2006: xvii and 2013: 73.

38 For a more developed exposition of these two philosophical systems, the reader can refer to the extensive secondary literature on the topic (Chakravarti 1951: 171–325; Frauwallner 1973: 274–315; Larson 1979 and Torella 2011: 76–77). A special edition of the periodical *Asi-*

Broadly speaking, the authoritative text *Sāṅkhyakārikā* and its commentaries elaborate metaphysical and ontological aspects of the Sāṅkhya system, while the *Pātañjalayogaśāstra* provides a thorough description of mind-states, levels of concentration and practical means to master the mental.

The first assumption of classical Sāṅkhya is the existence of a threefold suffering (*duḥkhatraya*) in life, and its aim is to provide theoretical teachings on how to eliminate it.[39] I describe below twenty-five fundamental principles (*tattva*) that play a constitutive part in the creation of the world. Every living being is essentially connected to a conscious self, which is defined in the *Sāṅkhyakārikā* as inactive, being pure consciousness and not subject to change, as it only passively observes the emanated world.[40] The world originates from the basic primordial original cause, that is, *mūlaprakṛti*, also referred to as *pradhāna*, the primary matter,[41] or *avyakta* (unmanifest), which is conversely active, unconscious and liable to change.

According to the *Sāṅkhyakārikā*, the original cause is unique and undetectable by the common organs of senses. It constitutes the only creative source of the world and gives birth to, or emanates, twenty-three principles that will shape the material and phenomenal world. In the same way as all principles emanate from the original cause, they also merge back into it at the time of final liberation.

The first principle emanating from the cause is the intellect (*buddhi*, also called *mahat*, meaning "great"),[42] which produces the "I" maker or "I" consciousness (*ahaṃkāra*).[43] The latter in turn causes the emanation of eleven instruments, that is, the five senses of perception, related to the intellect (*bud-*

atische Studien / Études Asiatiques, published in 1999, is devoted to Sāṅkhya. On Yoga, see for instance Frauwallner 1973: 321–348, Feuerstein 1979, Weiss 1986, Larson & Bhattacharya 2008 and Mallinson & Singleton 2017. Mass' publications on the subject have also significantly added to our knowledge of Yoga. See further Brill's *Encyclopaedia of Hinduism* on the two philosophies.

39 Kā. 1.

40 The Sanskrit term *puruṣa* literally signifies a man or person, or the soul. For references to the concept of *puruṣa*, see kā. 2, 3, 11, 17, 18, 19, 20, 21, 31, 55, 57, 61, 62, 65 and 66, and its commentaries.

41 For the descriptions of *prakṛti* and its derivatives see kā. 3, 8, 10, 11, 22, 37, 42, 58–59, 60–64 and 66, as well as the commentaries. See also PYŚ I.3, PYŚ (1904) II.6, II.21 and IV.23.

42 Some commentaries on kā. 46 provide further synonyms and characterizations of *buddhi* (YD, p. 238.8: *pratyaya, niścaya, adhyavasāya*. GPBh, p. 43.7: *pratyaya, adhyavasāya, dharma, jñāna*). See also PYŚ I.11, PYŚ (1904) II. 6 and II.21.

43 See kā. 22, 24, 25 and 35, and PYŚ I.45 and PYŚ (1904) III.48 on this concept. In the *Pātañjalayogaśāstra*, the concept of *asmitā*, or individuality, overlaps with that of *ahaṃkāra*. On these specific concepts in classical Sāṅkhya and Yoga, see Hulin 1978: 72–90.

dhīndriya), namely sight (*cakṣus*), hearing (*śrotra*), smelling (*ghrāṇa*), tasting (*rasana*), touching (*tvac*), the five organs of action (*karmendriya*), namely the voice (*vāc*), the hands (*pāṇi*), the feet (*pāda*), the anus and the organs of procreation (*pāyūpastha*), and finally the mind (*manas*).[44] The "I" consciousness also produces the five subtle elements (*tanmātra*).[45] From these subtle elements originate the five gross elements (*mahābhūta*).

Among the Sāṅkhya principles, the original cause is only a producer and not produced by anything else. Seven principles originating from it are at the same time producers (*prakṛti*) and products (*vikṛti*). Sixteen of these principles (*tattva*) are described as being only produced (*vikṛti*), not producers (*prakṛti*). They are the five senses of perception, the five organs of action, the mind and the five gross elements. These sixteen principles taken altogether are also qualified as transformations (*vikāra*).[46] As mentioned above, the *puruṣa* stands outside of this evolutionary scheme as a mere witness of it.

The *puruṣa* and the *prakṛti* share among other things the quality of not being produced, of being permanent and omnipresent. These two, however, also differ from each other. Whereas the conscious self is inactive, the original cause produces other elements.[47] Sāṅkhya philosophy thus offers a worldview that is fundamentally dualist: the world is constituted of twenty-four active principles, while the conscious self, as the twenty-fifth, is inactive. The notion that the conscious self is actively involved in the world and is connected to the products of the original cause is erroneous and results in the triple suffering.

At the end of the *Sāṅkhyakārikā*, the relationship between the two of them is likened to that between an audience and a female dancer (also possibly an actress) performing in front of it.[48] The original cause reveals itself to the conscious self, in just the same way as a female dancer or actress does to her audience. Once she has been seen by the spectators, she stops to produce anything and does not return to the audience, which then becomes separated from her. In the same way, when the original cause disappears from the sight of the conscious self, the latter becomes aware of its distinctness from the cause. This state is called *kaivalya*, translated as "isolation" in this book, and the conscious self is called "isolated" (*kevala*).[49]

---

44 On the senses of perception and the organs of action see kā. 26 to 28.

45 On the *tanmātras* see, for instance, kā. 22, 24, 25 and 38.

46 See the schema in Larson & Bhattacharya 1987: 52.

47 Some characteristics of the original cause in comparison to those of the conscious self are, for instance, outlined in the *Gauḍapādabhāṣya* on kā. 11.

48 This analogy is referred to in kā. 42, 59, 61, 65 and 66.

49 On *kaivalya* in the *Sāṅkhyakārikā* see specifically kā. 55 to 69.

In the metaphysics of the *Sāṅkhyakārikā*, three additional elements play an important role. They are the three constituents (*guṇa*): *sattva* characterized by the properties of whatever is good and enlightenment, *rajas* defined by the properties of passion and movement, and *tamas* associated with apathy and immobility. These elements exist as constituents in every principle (*tattva*)—except for the *puruṣa*—from the non-manifest subtle original cause (*prakṛti*) to the manifest gross elements (*mahābhūta*). Each principle contains a unique proportion and combination of these three constituents. The original cause, for instance, is constituted by their perfect balance. When their proportion changes, other principles are produced from it. The multiplicity of the phenomenal world thus exists by virtue of the respective combination of the three constituents in each principle.[50]

The knowledge of the twenty-five principles (*tattva*), and the distinction between the original cause (*prakṛti*) and the conscious self (*puruṣa*) consists in correct discriminative knowledge, or discernment (*vivekakhyāti*). It leads to the elimination of the threefold suffering and to isolation (*kaivalya*), or emancipation (*apavarga*), which implies the escape from karmic retribution and from the cycle of rebirths.

As the original cause is unmanifest (*avyakta*), that is to say imperceptible by the common organs of senses, one needs other means to know its existence. The *Sāṅkhyakārikā* accepts the existence of three means of knowledge (*pramāṇa*): direct perception (*pratyakṣa*), authoritative tradition (*āgama*) and inference (*anumāna*). It is through inference that one may grasp the entirety of the metaphysical concepts developed in this *Sāṅkhyakārikā*.[51]

A causal link connects twenty-four of these principles, as each principle that is produced is the effect (*kārya*) of what produces it, namely its cause (*kāraṇa*). Thanks to this link, it is possible to infer the existence of the imperceptible principles, even if it is not comprehensible through direct perception. The *Sāṅkhyakārikā* and its commentaries elaborate on the theory that an effect pre-exists in its cause. The well-known example of the pot and the clay in Indian philosophy is used to explain this causal link. According to the Sāṅkhya tradition, the pot exists in its cause, the clay, before its production. The existence of clay can therefore be inferred by the observation of its effect, the pot. The quality of the cause has changed or evolved due to the specific combination of the constituents, while its substance remains. This theory is called *satkāryavāda*, which signifies "the doctrine of the effect [pre-]existing [in the cause]."[52]

50 On the constituents with regard to *puruṣa* and *prakṛti*, see kā. 11, 16, 23 and 27.

51 Kā. 2.

52 Bronkhorst 2011b: 50.

The *Pātañjalayogaśāstra* accepts the metaphysics elaborated in the *Sāṅkhyakārikā* and its commentaries. It acknowledges the existence of three types of suffering,[53] the same twenty-five principles, the three constituents and the three means of knowledge. At times, however, its author uses a terminology different from that of Sāṅkhya to convey these concepts. For instance, whereas the *Sāṅkhyakārikā* tradition refers to the mind as *manas*, the *Pātañjalayogaśāstra* designates it as *citta*. The Yoga text equally accepts the theory of causality called *satkāryavāda*, advocates the distinction between the conscious self and the original cause, and uses the same terminology for designating the state of deliverance as *kaivalya*.

The *Pātañjalayogaśāstra* diverges from Sāṅkhya in the psychological domain, as it develops a sophisticated and complex theory on mental and meditative states. Its author considers that the mind (*citta*) has a flow of several dispositions (*vṛtti*). Different mental practices and types of meditation are then described and prescribed with the aim to hinder or suppress these dispositions (*cittavṛttinirodha*), thus enabling one to approach isolation (*kaivalya*). In the *Pātañjalayogaśāstra*, the term *yoga* is defined as *samādhi*, a type of meditative state, which can be rendered in English as "absorption."[54] The work distinguishes two types of absorption. The first one is a cognitive absorption, centred on an object (*saṃprajñāta samādhi*), while the second type is a non-cognitive absorption, namely not centred on any object (*asaṃprajñāta samādhi*). The latter is a meditative state in which the mind has not only reduced or ceased its different dispositions but also lacks some anchor for meditation. It is this second type of absorption that leads to *kaivalya* or isolation.

The *Pātañjalayogaśāstra* also develops the theory of the eight components (*aṣṭāṅga*). This refers to eight successive practices that include a set of specific modes of ethical behaviour, the control of one's breath and three meditative techniques. These have to be followed in order to reach absorption (*samādhi*), the eighth and last component of the Yoga path. The eight components are: keeping to ethical rules (*yama*), observances (*niyama*), postures (*āsana*), breath control (*prāṇāyāma*), withdrawal [from one's own senses of perception (*buddhīndriya*)][55] (*pratyāhāra*), fixation (*dhāraṇā*), meditation (*dhyāna*), and eventually the aforementioned twofold absorption (*samādhi*).[56]

53 PYŚ I.31, pp. 49.3.

54 PYŚ I.1–2. The term *samādhi* is derived from the verbal form *sam-ā-√dhā* which literally means "to put together" or "to hold together."

55 See PYŚ (1904), pp. 115.5–117.2 and *Vivaraṇa* (1952), pp. 231.2–232.22 on sū. II.54–55.

56 The eight components are described in PYŚ (1904) II.29–55 and III.1–8; Woods 1914: 177–208.

According to the *Pātañjalayogaśāstra*, the consequences ("ripening") of karma, meaning "action," that lead to a cycle of rebirths are rooted in five afflictions (*kleśa*). These afflictions exist in the mind (*citta*), but are actually attributed to the conscious self, as the latter experiences their consequences.[57] Therefore, in order to free the conscious self from afflictions and thus from the cycle of rebirths, one needs to weaken these afflictions. The last component, absorption, reduces them.[58]

Finally, whereas, in Sāṅkhya, knowledge (*jñāna*) leads to isolation, in Yoga, it is reached through repeated practice (*abhyāsa*) and dispassion (*vairāgya*).[59] The Sāṅkhya-Yoga concepts explained in this outline play important role in these philosophies, as they do in al-Bīrūnī's interpretations of them.

3.3 Authorships and Titles

3.3.1 *The Sāṅkhyakārikā*

The Sāṅkhya tradition acknowledges several teachers that preceded the composition of the *Sāṅkhyakārikā*.[60] At the end of this work, several of these Sāṅkhya teachers are cited:

> This secret treatise, in which the existence, the production and the dissolution of beings are considered, was formulated by the supreme sage [Kapila] for the sake of [explaining] the goal of the conscious self (*puruṣa*).[61] [Moved] by compassion, the sage bestowed upon Āsuri [this]

57 See also Maas' (2009: 266) description of the interconnectedness between the mind and the conscious self.

58 On the concept of afflictions, see PYŚ (1904) II.2–12.

59 On several discrepancies between the Sāṅkhya and Yoga systems, see for instance Larson 1999: 728–731, Larson & Bhattacharya 2008: 45–52 and Rukmani 1999.

60 For literature on Sāṅkhya teachers, see Chakravarti 1951: 111–155, Frauwallner 1973: 222–225 and Larson & Bhattacharya 1987: 107–146. On passages in the *Mahābhārata* dealing with Sāṅkhya and Yoga teachers, see also Brockington 1999. For the possible identification of Īśvarakṛṣṇa with Vārṣagaṇya and Vindhyavāsin, see Takakusu 1904a: 37–60, Bronkhorst 1985: 205–210 and Larson & Bhattacharya 1987: 131–146 and 149. The *Yuktidīpikā* mentions several Sāṅkhya-Yoga teachers, such as Paurika, Pañcādhikaraṇa, Patañjali (a Sāṅkhya teacher) and Vindhyavāsin. Vindhyavāsin's ideas are to be gathered from references to him in different works, as no work by him is extant. Frauwallner (1973: 315–320) treats some of his views. On Patañjali, the Sāṅkhya teacher, see Larson & Bhattacharya 1987: 129–130.

61 An alternative reading for *puruṣārthārtham* is *puruṣārthajñānam*, i.e., "the knowledge of the *puruṣa*'s goal" or "the knowledge for the sake of *puruṣa*." See for instance the reading of kā. 69 in GPBh, p. 61, V1, p. 78 and V2, p. 66.

> excellent means of purification. Āsuri also [bestowed it] upon Pañcaśikha, [who] propagated this system. And this [system], having been transmitted by a succession of disciples, was summarized in the form of *āryā* [verses] by Īśvarakṛṣṇa, whose thoughts are noble, after he had correctly understood the doctrine. The topics [developed] in the 70 [verses] (i.e., the *Sāṅkhyakārikā*), are the [same] topics as [that of] the entire *Ṣaṣṭitantra* (i.e., the treatise of 60 topics), [but] deprived of short narratives and free from [discussions of] opponents' views.[62]

Kapila, who is referred to as "the supreme sage" in the above quotation, is considered the founder of the Sāṅkhya system.[63] Kapila's name does not appear in the *Sāṅkhyakārikā* itself, only in its commentaries. Kapila and Āsuri are legendary figures related to the early transmission of Sāṅkhya teachings and no specific philosophical concepts can be attributed to them with certainty. The *Mokṣadharma* section of the *Mahābhārata* and Sāṅkhya-Yoga literature, attributes several points of view to Pañcaśikha. These views, however, do not present any uniformity or coherence and are rather chaotic.[64] In addition, some of these attributions are clearly erroneous. It is therefore possible that the name Pañcaśikha was used by the Sāṅkhya philosophers as an instrument to provide authoritativeness to certain views, but without reference to an actual historical figure.

According to the *Sāṅkhyakārikā*, Īśvarakṛṣṇa summarized the doctrine transmitted via Āsuri and Pañcaśikha. The last *kārikā* explains that the topics described in 70 strophes are the same as those of the *Ṣaṣṭitantra,* a work considered as having served as the foundation of the *Sāṅkhyakārikā*. This latter work is known on the basis of reference to it in other sources, which are, however, inconsistent and which attribute the work variously to Kapila, Pañcaśikha or Vārṣagaṇya.[65] Vārṣagaṇya, another early teacher associated with Sāṅkhya who is not mentioned in the above *kārikās*, appears to be the best candidate for the authorship of the *Ṣaṣṭitantra*.[66]

---

62 SK 69–72: *puruṣārthārtham idaṃ śāstraṃ guhyaṃ paramarṣiṇā samākhyātam | sthityutpattipralayaś ca cintyante yatra bhūtānām* ||69|| *etat pavitram agryaṃ munir āsuraye 'nukampayā pradadau | āsurir api pañcaśikhāya tena ca bahudhā kṛtaṃ tantram* ||70|| *śiṣyaparamparayāgatam īśvarakṛṣṇena caitad āryābhiḥ | saṃkṣiptam āryamatinā samyag vijñāya siddhāntam* ||71|| *saptatyāṃ kila ye 'rthās te 'rthāḥ kṛtsnasya ṣaṣṭitantrasya | ākhyāyikāvirahitāḥ paravādavivarjitāś cāpi* ||72||.

63 On Kapila, see for instance Jacobsen 2008.

64 See Larson & Bhattacharya 1987: 118–123 and Motegi 1999.

65 Larson & Bhattacharya 1987: 117–118 and 127.

66 Oberhammer 1960; Bronkhorst 2008: 79.

The title *Sāṅkhyakārikā* perhaps postdates the composition of the work as such. Among the editions of its commentaries available to me, only that of the *Gauḍapādabhāṣya* by Har Dutt Sharma (1933) provides this title in its colophon. According to Junjiro Takakusu, the work commonly referred to as the **Suvarṇasaptati* also bears the title *Sāṅkhyaśāstra*, which is the result of a transliteration from Sanskrit to Chinese.[67] The *Yuktidīpikā*, the *Sāṅkhyavṛtti*, the *Sāṅkhyasaptativṛtti*, the *Māṭharavṛtti* and the *Jayamaṅgalā* all have the designation *sāṅkhyasaptati* in their respective colophons, which can be translated as "the seventy [verses] of Sāṅkhya."[68] As for the *Tattvakaumudī*, no specific title is provided for the text it glosses. Given this fact, *Sāṅkhyakārikā* may not be the original title of the work attributed to Īśvarakṛṣṇa, and further investigation into the available manuscripts of the commentaries would perhaps result in a reconsideration of the original title of this work.[69] Even so, the title *Sāṅkhyakārikā* appears to have been adopted as the designation of this fundamental and hegemonic text of Sāṅkhya in the secondary literature.

The etymological meaning of the word *sāṅkhya* is "related to numbers." It is, however, reasonable to follow Edgerton's translation and understand it as meaning "(the method of salvation) based on reckoning or calculation"[70] when it is used to refer to the philosophical system. The synonyms of this term which are offered in the *Amarakośa*[71] are *carcā* ("repeating over in thought," "considering") and *vicāraṇa* ("consideration"), a definition that rather concurs with the interpretation of Edgerton. In addition, the term *sāṅkhya*, especially in the plural, can refer to an adherent of the philosophical system, rather than solely to the doctrinal system itself.

The above outline on questions related to the authorship and title of the *Sāṅkhyakārikā* aims at contextualizing the *Kitāb Sānk* within this tradition.

3.3.2 *The Kitāb Sānk*

In the *Taḥqīq mā li-l-Hind*, al-Bīrūnī mentions the "[*Kitāb*] *Sānk* which Kapila composed" (سانگ عمله کپل).[72] He thus conceives Kapila as the author of the original cause of the *Kitāb Sānk*, not as the founder of a philosophical sys-

67 Takakusu 1904a: 4.

68 Except for sections seven to nine, the section-colophons of the *Yuktidīpikā* read (*sāṅkhya*) *saptati*.

69 This was suggested to me by Maas in a personal communication. However, such an investigation lies beyond the scope of the present study.

70 Edgerton 1924: 36–37.

71 Quoted in Chakravarti 1951: 2, n. 2.

72 *Taḥqīq* (1958), p. 102.2–3; Sachau 1910: I/132. The exact Latin transliteration of the Arabic word here is *sānga*.

tem.[73] Apart from this statement, al-Bīrūnī does not provide any additional information on Kapila and never refers to Īśvarakṛṣṇa or to the other teachers mentioned in the *kārikās*, such as Āsuri or Pañcaśikha. As already mentioned, Kapila's name only appear in the commentaries on the *Sāṅkhyakārikā* and is absent from the *kārikās* themselves. These observations lead to two possible hypotheses: either al-Bīrūnī's interlocutors supplemented his knowledge and attributed the source of the *Kitāb Sānk* to Kapila, or al-Bīrūnī worked with a commentary which explicitly mentioned Kapila.[74] The two possibilities are not mutually exclusive. At any rate, as I show in Chapter 6, al-Bīrūnī made use of a written commentary on the *kārikās*.

Moreover, al-Bīrūnī does not mention any other name in relation to the composition of the *Kitāb Sānk* that reminds any of the known commentators on the *Sāṅkhyakārikā*, such as Māṭhara or Vācaspatimiśra. He, however, enumerates another book in the *Taḥqīq mā li-l-Hind*, which is most probably connected to Sāṅkhya, namely "the eponymous book composed by Gauḍa the ascetic" (کتاب عمله گور الزاهد و عرف باسمه).[75] Al-Bīrūnī does not provide any description of the topic of Gauḍa's book, which would have helped in the identification of this text. Whether this Gauḍa is Gauḍapāda, the author of the Sāṅkhya commentary or not is uncertain.[76] However, as al-Bīrūnī enumerates this book beside the *Kitāb Sānk*, he certainly considered them as two distinct works. Thus, the mention of Gauḍa here does not point to the authorship of the *Kitāb Sānk*, but to that of another Sāṅkhya, or any other Sanskrit, text.

Furthermore, al-Bīrūnī entitles his translation *Kitāb Sānk*, literally *The Book Sānk*. As already mentioned, the manuscript of its text is not extant. However, in the *Taḥqīq mā li-l-Hind*, al-Bīrūnī explicitly refers to it by name eleven times, less than to the *Kitāb Pātanǧal*. The following table illustrates all places where the *Kitāb Sānk* is mentioned in the *Taḥqīq mā li-l-Hind*:

73 The context in which this account occurs makes it clear that al-Bīrūnī refers to a book and not to a philosophical system. This statement by al-Bīrūnī is found in his account where he enumerates several Sanskrit philosophical works and system. See the full passage above, pp. 91–92.

74 A work identified as the *Māṭharavṛtti* by Shripad Krishna Belvalkar (1917: 171) is for instance attributed to Kapila in the catalogue (*Sanskrit manuscripts from Gujarat, Cutch, Sindh and Khandesh*, compiled under the supervision of G. Bühler, Bombay 1873) where it was recorded.

75 See above pp. 91–92. *Taḥqīq* [1958], p. 102.2.

76 Sachau posed the question of whether this Gauḍa was the author of the *Gauḍapādabhāṣya*, without, however, finding an answer (Sachau 1910: II/267). As there is no information about the content of Gauḍa's book, it is difficult to provide a definitive answer as to whether this figure can be identified with the Advaita-Vedāntin Gauḍapāda, or with the

TABLE 6 List of references to the *Kitāb Sānk* in the *Taḥqīq mā li-l-Hind*

No.	Arabic expression	English translation and references
1	اسمه سانك	its name is Sānk (*Taḥqīq* [1958], p. 6.2; Sachau 1910: I/8)
2	فى كتاب سانك	in the book Sānk (*Taḥqīq* [1958], p. 22.12; Sachau 1910: I/30)
3	فى كتاب سانك	in the book Sānk (*Taḥqīq* [1958], p. 36.15; Sachau 1910: I/48)
4	صاحب كتاب سانگ	the author of the book Sāng (*Taḥqīq* [1958], p. 47.13; Sachau 1910: I/62)
5	فى كتاب سانگ	in the book Sāng (*Taḥqīq* [1958], p. 48.16; Sachau 1910: I/64)
6	فى كتاب سانك	in the book Sānk (*Taḥqīq* [1958], p. 57.5; Sachau 1910: I/75)
7	فى كتاب سانگ	in the book Sāng (*Taḥqīq* [1958], p. 62.1; Sachau 1910: I/81)
8	قيل فى سانگ	a statement in [the book] Sānk (*Taḥqīq* [1958], p. 63.7; Sachau 1910: I/83)
9	فى كتاب سانگ	in the book Sāng (*Taḥqīq* [1958], p. 67.11; Sachau 1910: I/89)
10	عن سانگ	according to [the book] Sāng (*Taḥqīq* [1958], p. 69.15–16; Sachau 1910: I/92)
11	سانگ	[the book] Sāng (*Taḥqīq* [1958], p. 102.1; Sachau 1910: I/132)

A clear phonetic correspondence can be established between al-Bīrūnī's *sānk* and the Sanskrit designation *sāṅkhya*, which is found in titles of works related to the Sāṅkhya philosophy, such as *Sāṅkhyakārikā, Sāṅkhyasaptativṛtti* and *Sāṅkhyavṛtti*. Al-Bīrūnī employs the Arabic *sānk* when referring to both the title of the work he has translated and to the theories elaborated upon in it. The long *ā* is always respected, whereas the aspirated consonant *kh* in the original Sanskrit is either transcribed as *k* or as *g* in the main Arabic printed edition used in this research, losing its original aspiration. The difference between the characters *k* and *g* is minor, as only diacritic marks distinguish the two Semitic letters.[77] In al-Bīrūnī's works, the aspiration of consonants was not always rendered in the Arabic transliterations, at least according to the available editions. The final *ya*, on the other hand, often disappears in the Arabic transliterations of original Sanskrit terms in al-Bīrūnī's writings.[78] Al-Bīrūnī does not provide a meaning for the term *sāṅkhya* transliterated by him as *sānk*.

Lastly, as will become apparent in Chapter 6, he was in the habit of translating the aphoristic text together with the commentary on them, so that the title *Kitāb Sānk*represents the translation of a work whose Sanskrit title included the word *sāṅkhya* (*sānk* in the Arabic transliteration) as the first member of a Sanskrit compound and a term such as *-vṛtti* or *-śāstra* (*kitāb*) as its second member. Thus, the above discussion showed how the information regarding

author of the Sāṅkhya commentary. Al-Bīrūnī does not display direct acquaintance with Advaita-Vedānta in the *Taḥqīq mā li-l-Hind*. See also below on Gauḍa, p. 197.

77 Sachau 1888: 10–11.

78 Sachau 1888: 33–34.

the authorship and title of the *Sāṅkhyakārikā* and of the *Kitāb Sānk* overlaps. The following two sections deal with the Yoga works and contextualize al-Bīrūnī's *Kitāb Pātanǧal* within the Indian textual tradition in a similar manner.

3.3.3 *The Pātañjalayogaśāstra*

Contrasting with the *Sāṅkhyakārikā*, the *Pātañjalayogaśāstra* itself does not offer literary evidence for the history of its textual transmission. However, the name Hiraṇyagarbha is connected to the transmission of Yoga. This figure is, for instance, honoured in the initial laudatory verses of the *Tattvavaiśāradī* and of the later Yoga text *Maṇiprabhā* relating to PYŚ I.1.[79] The author of the *Vivaraṇa*, when glossing PYŚ (1904) III.39, also refers to Hiraṇyagarbha, stating that his work, or the method described in it, explained the means of controlling one's breath in detail.[80] Thus, the role of Hiraṇyagarbha in the transmission of Yoga is not as clear as that of Kapila for Sāṅkhya.

Furthermore, two different points of view co-exist among ancient and modern scholars regarding the authorship of the *Pātañjalayogaśāstra*. These two opinions arise from the fact that this work is composed of two conflated layers of text. This is evidenced by the last member of the Sanskrit compound making up the title of the work: *śāstra*, or treatise, encompassing both a series of aphorisms (*sūtras*), the first layer of text, and a relatively concise commentary (*bhāṣya*) that constitutes the second layer.

The first opinion supports the idea that two different authors composed the two layers of text, respectively, so that the *sūtra*-part, referred to as the *Yogasūtra*, is believed to have been compiled by Patañjali, while the *bhāṣya*-part, the so-called *Yogabhāṣya*, was supposedly penned by [Veda]vyāsa, the legendary compiler of the *Mahābhārata*.[81] According to the second opinion, the *Pātañjalayogaśāstra* is conceived as a whole that a single author composed, generally identified with Patañjali. The following section summarizes the current state of research, in order to situate the evidence provided by al-Bīrūnī within this debate.

Hermann Jacobi, followed by Bronkhorst, was the first to question the attribution of the alleged *Yogabhāṣya* to Vyāsa. Jacobi points out that Vyāsa is not mentioned in the chapter-colophons of the work as available to him. He further

---

79 TVŚ, pp. 2.7 and 31.20, MP, p. 2.20 See also Woods 1914: 5 and 26.

80 *Vivaraṇa* (1952), p. 294.14. The commentary in fact uses an adjective derived from this personal name, i.e., *hairaṇyagarbha*.

81 Garbe 1896: 40–41; Dasgupta 1920: i, 1922: 212, 1924: vii and 1941: 181; Strauss 1925: 178 and 191; Renou & Filliozat 1953: 46; Tucci 1957: 99; Angot 2008.

notes that, in the chapter-colophon appearing in several editions of the work, the derivative Sanskrit adjective *pātañjala*, meaning "of Patañjali" or "related to Patañjali," qualifies the expression *sāṅkhyapravacana yogaśāstra*, that is, "the Yoga treatise expressive of Sāṅkhya."[82] This remark implies that several editors of the work considered Patañjali the author of the whole *śāstra*. In his attempt to establish the oldest reading of these chapter-colophons, Maas supports Jacobi's and Bronkhorst's observations, when he remarks that there is no mention in these chapter-colophons of Vyāsa as having been involved in the composition of the *Pātañjalayogaśāstra*.[83]

According to the chapter-colophons, thus, it is not only the *sūtra*-part that is attributed to Patañjali, but the work as a whole. The adjective *pātañjala* ("of Patañjali") indeed qualifies the compound *yogaśāstra* ("Yoga treatise"). Thus, these chapter-colophons indicate that the scribes of the various copies of the *Pātañjalayogaśāstra* did not dissociate the *sūtra*-layer of the text from its commentary. Another clue to the text's being an integrated whole is the fact that the *sūtra*s do not boast their own chapter-colophons and were thus not considered independent from the *bhāṣya*.[84] Furthermore, since "in the early classical period of Indian philosophy the terms *sūtra* and *bhāṣya* did not designate different literary genres but compositional elements of scholarly works (*śāstra*),"[85] it is likely that the *Pātañjalayogaśāstra* was also conceived as a single whole when it was originally compiled.

In addition to the evidence drawn from the colophons of the manuscripts of the *Pātañjalayogaśāstra*, several classical Indian thinkers appears to have considered that the entire treatise (*śāstra*) was composed by a single author whose name was not Vyāsa, such as Śaṅkara, the author of the *Vivaraṇa*. The first edition of this commentary, in 1952, followed by Rukmani's edition in 2021, refers to the commentary either as the *Pātañjalayoga**sūtrabhāṣya**vivaraṇa* or as the *Pātañjalayoga**śāstra**vivarṇa* (my emphases).[86] This indicates a lack of clarity on this question among the scribes and editors of the text. Nevertheless, Bronkhorst and Wezler have drawn the attention of Indologists to the fact that the former reading may not have been an accurate rendition of the original title.[87]

82 Jacobi 1970: 683 and 685; Bronkhorst 1985: 203. Maas thoroughly discusses the question in several of his publications; see Maas 2006: xii–xix, 2009: 264, and 2013: 57–59 and 62–65.

83 Maas 2006: xx–xxi. Chapter-colophons of the *Pātañjalayogaśāstra* are provided in Table 9, below, in comparison to the corresponding titles of the chapters in the *Kitāb Pātanǧal*.

84 Maas 2013: 58.

85 Maas 2013: 65.

86 Sastri & Sastri 1952: 1, 119, 232 and 370; Rukmani 2001: I/204, 377, 2001: II/211.

87 Wezler 1983: 17 and 37, nn. 1 and 2; Bronkhorst 1985: 203, n. 12.

Harimoto's critical edition on the *Vivaraṇa* confirms these preliminary observations and offers another reading of the commentary's title based on the colophons, that is the *Pātañjalayoga**śāstrabhāṣya**vivaraṇa* (my emphasis).[88] This reading may indicate that the author of this commentary considered Patañjali's work an integral treatise (*śāstra*) and did not necessarily dissociate the *sūtras* from their *bhāṣya*. Accordingly, the *Vivaraṇa* comments on the *Pātañjalayogaśāstra* in its entirety. As it is one of the earliest extant commentaries on the *Pātañjalayogaśāstra*, it may be regarded as a faithful witness of the classical understanding of the *Pātañjalayogaśāstra*'s structure and authorship.

Further, as several scholars have already noted, other authors, such as Śrīdhara in the *Nyāyakandalī* (dated to 991), Abhinavagupta in some of his works (Kashmir, second half of the 11th c.) and Malliṣeṇa in the *Syādvādamañjarī* (end of the 13th c.) appears to have considered that Patañjali was the author of both layers of texts.[89]

The conception of the *Pātañjalayogaśāstra* as being a work composed by two distinct authors perhaps found its origin in the *Tattvavaiśāradī*, written by Vācaspatimiśra in the mid-tenth century. This commentary on the *Pātañjalayogaśāstra*, at least according to the printed editions, calls the author of the *bhāṣya* Vyāsa, in both the laudatory verses and the chapter-colophons. However, as Bronkhorst and Maas note, Vācaspatimiśra's attitude on this question is ambiguous, and his different works offer contradictory evidence: at least one passage of his *Nyāyavārttikatātparyaṭīkā* indicates that Vācaspatimiśra attributed a portion of the *bhāṣya* found in the *Pātañjalayogaśāstra* to Patañjali.[90]

The *Rājamārtaṇḍa*, in contrast with the other two discussed commentaries on the *Pātañjalayogaśāstra*, only comments upon the *sūtras* as aforementioned. By doing so, the *Rājamārtaṇḍa* may also have influenced the textual tradition on the composition of the *Yogasūtra* and its *bhāṣya*. The authors of later commentaries on the *sūtras*, such as Vijñānabhikṣu who wrote the *Sāṅkhyapravacana* (mid-16th c.),[91] Rāmānanda Sarasvatī who composed a *Maṇiprabhā* (late 16th c.),[92] or Nāgeśa (or Nāgojī) Bhaṭṭa, the author of the *Vṛtti* (early 18th c.),[93] also seem to have considered the *sūtras* and the *bhāṣya* as two separate entities.[94]

---

88 Harimoto 1999: 36, 350, n. 6 and 2014: 9, n. 3.

89 For further references and detailed studies on these works, see Jacobi 1970: 685, Raghavan 1980: 78–87, Bronkhorst 1985: 203–207, and Maas 2006: xii–xv and 2013: 57.

90 Bronkhorst 1985: 204–207, and Maas 2006: xiii–xiv and 2013: 68.

91 Maas 2006: xiii.

92 Larson & Bhattacharya 2008: 54 and 282–283.

93 Id.: 355–356.

94 Jacobi 1970: 685.

Lastly, the word *vyāsa*, supposedly referring to the author of the *bhāṣya* called [Veda]vyāsa, is only found in some chapter-colophons of late manuscripts of the *Pātañjalayogaśāstra*. The only mention of Vyāsa in the *Vivaraṇa* actually occurs in connection with a quotation drawn from the *Mahābhārata* and does not refer to the author of the *bhāṣya* at all.[95] Maas offers an alternative interpretation of the occurrence of the word *vyāsa* in the more recent manuscripts of the *Pātañjalayogaśāstra* and in the commentary by Vācaspatimiśra. The word may be understood as a derivative of the verbal root *vi-as* ("to dispose of something, to arrange something") formed with the *uṇ-ādi* suffix.[96] In this sense, it would thus simply mean "compiler." This interpretation implies that the word *vyāsa* may have originally been used as a generic designation, and not as a proper name. If this is correct, it is possible that Vācaspatimiśra interpreted the term differently from its original meaning and ascribed the work to the author [Veda]vyāsa.[97]

Two main opinions therefore emerge concerning the authorship of the *Pātañjalayogaśāstra* in the Indian textual tradition. Some scribes and authors considered Patañjali as having penned both the *sūtra*s and the *bhāṣya*, while others regarded him as the author of the *Yogasūtra* and [Veda]vyāsa as the author of the *Yogabhāṣya*. It further appears from this survey that there was a certain confusion as to who had written what. The above disparity of opinions among ancient and medieval Indian thinkers is probably at the root of the division of opinions that exists in modern scholarship. However, a number of sources, notably early works on Yoga, supports the position that the *Pātañjalayogaśāstra* was written as a single entity by a single author. This suggests that the confusion arose later on in the textual transmission.

3.3.4 *The Kitāb Pātanǧal*

According to the *Kitāb Pātanǧal*, Hiraṇyagarbha played a role in the transmission of the philosophical system elaborated in its Sanskrit source. The laudatory introduction to al-Bīrūnī's translation indeed explains that his source follows the "method of Hiraṇyagarbha."[98] Barring this figure, al-Bīrūnī does not specify any personal name for the authorship of the *Kitāb Pātanǧal*. However, several observations lead to the conclusions that the Arabic term *pātanǧal* refers to the author, to the title of the book and to a protagonist of the narrative created by al-Bīrūnī.

95 Maas 2006: xv and 2013: 58–59.
96 Tubb & Boose 2006: 49.
97 Maas 2013: 68.
98 Pines & Gelblum 1966: 310. See below Section 5.3.

To begin with, al-Bīrūnī is unaware of the tradition that considers Vyāsa the author of a work related to Yoga; the name Vyāsa never appears in the *Kitāb Pātanǧal*. The text of the manuscript bears the letters لارناص (*lā-r-nā-ṣ*)—which have no meaning—in a passage of the *Kitāb Pātanǧal* dealing with cosmography and corresponding to the group of question/answer 46 (KP 46). Hellmut Ritter emends them to read لوياص (*li-wyāṣa*), literally meaning "for/of Vyāsa," thus making the Arabic word render a transliteration of the Sanskrit *vyāsa* and artificially associating the name Vyāsa with the *Kitāb Pātanǧal*. Shlomo Pines and Tuvia Gelblum point out, however, that in the *Taḥqīq mā li-l-Hind* al-Bīrūnī does refer to Vyāsa with وياس or بياس (*wyāsa* or *byāsa*), that is, with a final voiceless plain sibilant س (*s*) instead of the voiceless emphatic sibilant ص (*ṣ*).[99] Further, Pines and Gelblum propose the reading الاراضى (*al-ʾārāḍī*; "earths"). This solution suits the context of this section of KP 46, considering the general topic of the passage dealing with cosmography.[100]

When al-Bīrūnī mentions Vyāsa in the *Taḥqīq mā li-l-Hind*, he refers to him as the son of Parāśara and the author of the *Kitāb Bhāraṯa*. Occasionally, he attributes to Vyāsa a role in the transmission of the Vedas.[101] Thus, al-Bīrūnī never associates the name Vyāsa with the *Kitāb Pātanǧal*, just as the name Vyāsa is never explicitly connected to the composition of the *bhāṣya* in the *Pātañjalayogaśāstra* or in the *Vivaraṇa*.

In addition, the full title of the *Kitāb Pātanǧal* is: *The Book by Pātanǧal the Indian, on the Liberation from the Burdens,* [*being*] *a Translation into Arabic by Abū l-Rayḥān Muḥammad bin Aḥmad al-Bīrūnī*. Given this title, it is clear that al-Bīrūnī regarded Pātanǧal as the author of the book.[102] In order to delve further into the question of who al-Bīrūnī believed the author was, the numerous references to Pātanǧal in the *Taḥqīq mā li-l-Hind* are provided in the following table:

---

99 KP 46, p. 185, n. 6; Pines & Gelblum 1966: 304 and 1983: 275, n. 88; Maas 2013: 59. On KP 46 of the *Kitāb Pātanǧal*, see below Table 10.

100 For the full passage, see below p. 173.

101 For the mention of Vyāsa (or *vyāsa*) in the *Taḥqīq mā li-l-Hind*, see *Taḥqīq* (1958), pp. 34.2, 78.14, 82.10, 97.8, 101, 102.10, 102.15, 104.4, 134.5, 196.7, 286.15, 296.16, 310.9, 331, 334.4 and 334.10; Sachau 1910: I/44, 104, 107, 126, 131, 132, 134, 171, 238, 340, 341, 352, 369, 394, 397 and 398.

102 The edition by Ritter (KP, p. 167.1–2) reads the plural word meaning "metaphors" or "images" (الامثال) instead of "burdens" (الاثقال), which is the reading proposed by Pines and Gelblum (1966: 308, n. 51). Massignon (1954 [1922]: 97) and Hauer (1930: 276) concur with Ritter on his reading. However, Pines' and Gelblum's reading, i.e., "burdens," appears appropriate, as al-Bīrūnī uses this word to translate the concept of afflictions (*kleśa*). According to the *Pātañjalayogaśāstra*, mental absorption not only weakens these afflictions but also brings about isolation (*kaivalya*) of the conscious self (*puruṣa*). The title of al-Bīrūnī's translation would refer to this specific idea.

TABLE 7 List of references to Pātanğal in the *Taḥqīq mā li-l-Hind*

No.	Arabic expression	English translation and references
1	يعرف بپاتنجل	[the book] is known as Pātanğal (*Taḥqīq* [1958], p. 6.3; Sachau, 1910: I/8)
2	في كتاب پاتنجل	in the book Pātanğal (*Taḥqīq* [1958], p. 20.9–10; Sachau 1910: I/27)
3	فى كتاب پاتنجل	in the book Pātanğal (*Taḥqīq* [1958], p. 42.7–8; Sachau 1910: I/55)
4	قال صاحب كتاب پاتنجل	the author of the book Pātanğal said (*Taḥqīq* [1958], p. 52.5; Sachau 1910: I/68)
5	فهذا ما قال پاتنجل	and this is what [the book] Pātanğal said (*Taḥqīq* [1958], p. 53.8; Sachau, 1910: I/70)
6	فى كتاب پاتنجل	in the book Pātanğal (*Taḥqīq* [1958], p. 58.5; Sachau 1910: I/76)
7	في خاتمة كتاب پاتنجل	at the end of the book Pātanğal (*Taḥqīq* [1958], p. 61.16–17; Sachau 1910: I/81)
8	فى كتاب پاتنجل	in the book Pātanğal (*Taḥqīq* [1958], p. 62.10; Sachau 1910: I/82)
9	إلى طريق پاتنجل	in the [same] manner as [the book] Pātanğal (*Taḥqīq* [1958], p. 66.12; Sachau 1910: I/87)
10	فى كتاب پاتنجل	in the book Pātanğal (*Taḥqīq* [1958], p. 70.13; Sachau 1910: I/93)
11	ومثل پاتنجل	and like [the book] Pātanğal (*Taḥqīq* [1958], p. 102.3; Sachau 1910: I/132)
12	عن پاتنجل	according to [the book] Pātanğal (*Taḥqīq* [1958], p. 150.9; Sachau 1910: I/189)
13	لمفسّرِ كتاب پاتنجل	for the commentator in the book Pātanğal (*Taḥqīq* [1958], p. 191.1; Sachau 1910: I/232)
14	لكن مفسّر كتاب پاتنجل	but the commentator in the book Pātanğal (*Taḥqīq* [1958], p. 192.6–7; Sachau 1910: I/234)
15	مفسّر پاتنجل	the commentator in [the book] Pātanğal (*Taḥqīq* [1958], p. 193; Sachau 1910: I/235)
16	مفسّر كتاب پاتنجل	the commentator in the book Pātanğal (*Taḥqīq* [1958], p. 194.6; Sachau 1910: I/236)
17	عن مفسّر كتاب پاتنجل	according to the commentator in [the book] Pātanğal (*Taḥqīq* [1958], p. 196.15; Sachau 1910: I/238)
18	كمفسّر كتاب پاتنجل	like the commentator in the book Pātanğal (*Taḥqīq* [1958], p. 205.14; Sachau 1910: I/248)
19	عن تفسير پاتنجل	according to the commentary in [the book] Pātanğal (*Taḥqīq* [1958], p. 393.5; Sachau 1910: II/62)

The word *pātanğal* is invariably written with a long *ā* in the initial syllable, in both the *Taḥqīq mā li-l-Hind* and the *Kitāb Pātanğal*. Pines and Gelblum suggest that al-Bīrūnī consistently uses the long *ā* in order to make sure that his readership would read the correct vowel.[103] This transliteration probably renders the Sanskrit adjective *pātañjala*, that is, the *vṛddhi* ablaut of the first vowel in the proper name Patañjali, which is found in the title of the Sanskrit text. In general, al-Bīrūnī's Arabic transliterations of Sanskrit words preserve the correct length of vowels, as shown Table 5 in Chapter 2 of this book. Nevertheless, in Sanskrit, the word *pātañjala* is an adjective compounded in the title of the work and refers to its author; *patañjali* is the actual name of the author. In Arabic, adjectives are not formed by lengthening vowels in the original word.

According to Carl Edward Sachau, the Arabic word *pātanğal* may refer to both the author and the title of the book, while for Surendranath Dasgupta, it is nothing more than the title of al-Bīrūnī's translation.[104] Three occurrences of this word in context support Sachau's contention. The first example of the above listed items reads, "[the book] is known as Pātanğal" (no. 1). Item number 4 has the expression "the author of the book Pātanğal"; and finally, the instance found in number 11 occurs within an enumeration of different titles of Indian works, thus suggesting that there the word *pātanğal* is understood as the title of the text.

In the other cases, the Arabic expression can be freely interpreted as meaning either "the book [entitled] Pātanğal" or "the book by Pātanğal." Al-Bīrūnī may not have felt the need to specify the author's name, for the simple reason that it was already provided in the title of his translation. In contrast, he needed to provide the name of the author of the *Kitāb Sānk* as it was not evident from the title of his translation. It is then likely that al-Bīrūnī did not distinguish between the adjectival form of the name (*pātañjala*) and the proper name itself (*patañjali*); he seems to have used the same form, namely *pātanğal*, to transliterate both Sanskrit words. If this hypothesis is correct, it lends evidence to the assertion that al-Bīrūnī's knowledge of Sanskrit grammar was relatively superficial.

In Table 7, there are six mentions of a specific commentator (*mufassir*) of, or in, the *Kitāb Pātanğal* (nos. 13, 14, 15, 16, 17 and 18) and one instance refers to a commentary (*tafsīr*) of, or in, the *Kitāb Pātanğal* (no. 19). Al-Bīrūnī merged a text and a commentary on it in his *Kitāb Pātanğal*.[105] Furthermore, as Maas points out, the Arabic term *kitāb* (book)—in the same way as in the title *Kitāb*

103 Pines & Gelblum 1966: 308, n. 50.
104 Sachau 1910: II/257; Dasgupta 1930: 60.
105 See below Section 5.2.

Sānk—may well have been used as a translation of the Sanskrit *śāstra* (treatise), thus referring to two layers of text, and not only to the *sūtras*.[106] Consequently, the commentary mentioned by al-Bīrūnī may already have formed part of the source of the *Kitāb Pātanğal* instead of being a commentary on it. Indeed, neither the grammatical study of the expressions in Table 7 nor the analysis of specific passages of al-Bīrūnī's translation in Chapters 4 and 5 below excludes this possibility.

On the contrary, the discussion in Section 5.2 of the present book suggests that it is appropriate to understand the commentary as being part of the source of the *Kitāb Pātanğal*. Therefore, the translation "the commentator in the *Kitāb Pātanğal*" is preferred, rather than "the commentator on the *Kitāb Pātanğal*" in Table 7 above.

In addition, it is evident that al-Bīrūnī conceived Pātanğal as a protagonist of the narrative about the origin of the *Kitāb Pātanğal*, as the beginning of this work indicates:

> [Question] 1. The ascetic roaming in deserts and jungles questioned Pātanğal and said to him: [...]
> [Answer]. Pātanğal said: [...].[107]

The explicit naming of this figure appears to be an innovation on al-Bīrūnī's part. He himself explains that he had reshaped the text into a dialogue in his translation, which led him to introduce a new character, the wise Pātanğal, who answers to questions.[108]

Furthermore, in the case of the *Kitāb Pātanğal*, if al-Bīrūnī based his translation on the *Pātañjalayogaśāstra*, he elided the crucial word *yoga* in his interpretation of the title of his source. As a matter of fact, he never transliterates the word *yoga* in the context of the Indian philosophy, in both the *Kitāb Pātanğal* and the *Taḥqīq mā li-l-Hind*. Instead of including this word in the title of his translation, he appears to have preferred to define the topic of the work he translated. He explains it as dealing with "the liberation from the burdens" (في خلاص من الاثقال) in the title of the work and, elsewhere, as being "the means leading to the perfection of the soul by the liberation from these bonds and to

106 Maas 2013: 59–60.

107 KP 1, p. 169.10 and 169.15; Pines & Gelblum 1966: 313. The reading of this passage is uncertain. Ritter (KP, p. 169, n. 4) proposes two possibilities: "jungle" (الغياض) or "wasteland, desert" (الفيافى). The first reading has been chosen here. See also Pines & Gelblum 1966: 313, n. 92.

108 On the dialogical form of the *Kitāb Pātanğal*, see below Section 4.3.1.

the attainment of eternal happiness" (الاسباب المؤدّية الى كمال النفس بالخلاص عن هذا الوثاق و الوصول الى السعادة الابدية).[109] As seen in Section 4.4.3 of the present book, al-Bīrūnī often supplemented his sources with definitions, in order to help his readership understand the text. In this particular case, he would have glossed the topic of the translated work instead of transliterating the word *yoga* in its title.

Thus, the above discussion showed different functions that al-Bīrūnī attributed to the Arabic term *Pātanǧal*, namely author, title and a protagonist. It also highlighted how al-Bīrūnī conceived the Sanskrit source of his Arabic translation as made up of two layers of text composed by one author, in the same manner as several ancient Indian thinkers understood the nature of the *Pātañjalayogaśāstra*. In addition, thanks to an examination of the instances of the expression *Kitāb Pātanǧal* in the *Taḥqīq mā li-l-Hind*, I suggested that the commentary which al-Bīrūnī associates with the title *Kitāb Pātanǧal* may be integrated into the Sanskrit source of this work, rather than a gloss that comments upon it.

Lastly, al-Bīrūnī used the Arabic word *kitāb* as a generic term to render a Sanskrit word similar as *śāstra*, thus including under this label a fundamental text and its commentary. Furthermore, the two titles of al-Bīrūnī's translations indicate that he drew upon works whose title included the words *pātañjala* and *sāṅkhya*. Whereas the Arabic words *pātanǧal* and *sānk* are relatively accurate renderings of the corresponding Sanskrit words, the Arabic *kitāb* (كتاب), that is, "book," is a generic term.

In general, the pieces of philological information presented so far tend to show that the Sanskrit sources of the *Kitāb Sānk* and *Kitāb Pātanǧal* respectively belonged to the *Sāṅkhyakārikā* and *Pātañjalayogaśāstra* traditions. As a foretaste of the discussions in the subsequent chapters, I also highlighted al-Bīrūnī's proclivity to transform his source-texts and creativity, with the examples of the protagonist that he added into his translation and the elision and definition of the term *yoga*.

3.4 Intersections and Disagreements of the Two Philosophical Systems

3.4.1 *The Indian Sāṅkhya and Yoga*

In order to further contextualize the *Kitāb Sānk* and *Kitāb Pātanǧal* within the Sanskrit tradition of Sāṅkhya and Yoga, the next two sections discuss the nature

109 KP, p. 168.11–12; Pines & Gelblum 1966: 311.

of the relationship of the Sāṅkhya and Yoga philosophies, on the one hand, and of al-Bīrūnī's interpretations on the other.

Henry Thomas Colebrooke, who provides the "first academic publication on Yoga philosophy based on primary sources,"[110] conceived Patañjali's *Yogaśāstra* and Kapila's Sāṅkhya as belonging to the same doctrine, while conceding that they also display distinct features.[111] Along similar lines, Erich Frauwallner interpreted Yoga as "a second direction of the School" of Sāṅkhya.[112]

The terms *sāṅkhya* and *yoga* as they appear in epic literature, such as the *Bhagavadgītā*, refer respectively to "the way of salvation by pure knowledge, the intellectual method" and to a "disciplined, unselfish activity" producing "none of the evil results which action otherwise produces;"[113] both practices share a common aim, that is, salvation, but employ two different methodologies.[114]

The terms *nirīśvara-sāṅkhya* meaning "Sāṅkhya without [a creator] God," and *seśvara-sāṅkhya* meaning "Sāṅkhya with [a creator] God," have been used in Sanskrit literature since at least the eighth century to distinguish between two different systems of thought. The common view holds that the adjective *nirīśvara* was used to refer to the Sāṅkhya system, while *seśvara* qualified the Yoga system. Refuting this view, Bronkhorst argues that at an early date the expression *nirīśvara-sāṅkhya* actually stood for both Sāṅkhya and Patañjali's Yoga. However, Mādhava's *Sarvadarśanasaṃgraha* (15th c.) certainly makes use of *seśvara-sāṅkhya* to refer to the Yoga philosophy of the *Pātañjalayogaśāstra* and of *nirīśvara-sāṅkhya* to designate the system elaborated upon in the *Sāṅkhyakārikā* by Īśvarakṛṣṇa.[115] It is unknown whether Mādhava created this specific terminological distinction or whether he followed an earlier tradition. Nevertheless, as Bronkhorst shows, no evidence has been found that would suggest that Indian thinkers explicitly dissociated the two systems by using these terms prior to Mādhava's distinction.[116]

Further, the phrasing of the chapter-colophons of the *Pātañjalayogaśāstra* puts the two compounds *sāṅkhya-pravacana* ("expressive of Sāṅkhya") and *yoga-śāstra* ("Yoga treatise") in apposition in such a way that *sāṅkhyapravacana* qualifies *yogaśāstra*. The chapter-colophons thus indicate that some scribes

110 Maas 2013: 55.
111 Colebrooke 1824: 38, quoted in Maas 2013: 55. See also Renou & Filliozat 1953: 2.
112 Frauwallner 1973: 224.
113 Edgerton 1924: 4.
114 Edgerton 1924: 19–20.
115 Bronkhorst 1981: 316; Hattori 1999: 616.
116 Bronkhorst 1981.

of copies of the *Pātañjalayogaśāstra* considered his work as belonging to the teachings of Sāṅkhya, or at least as being strongly related to this philosophical system.[117]

Another example of the interconnection between Sāṅkhya and Yoga is found in the *Nyāyabhūṣaṇa* by Bhāsarvajña (early or mid-10th c.),[118] who quotes *sūtras* from the *Pātañjalayogaśāstra* referring to them as belonging to the "doctrine of the followers of Sāṅkhya" (*sāṃkhyānāṃ matam*).[119]

Furthermore, there are references to the Pātañjala Yoga tradition in *Sāṅkhyakārikā*'s commentaries. For instance, Sāṅkhya describes eight states (*bhāva*) of the intellect (*buddhi*).[120] The first four are righteousness (*dharma*), knowledge (*jñāna*), dispassion (*virāga* or *vairāgya*) and mastery (*aiśvarya*), whereas the last four consist in their opposites.[121] The *Gauḍapādabhāṣya* on kā. 23, in its explanation of righteousness, supplements its description with a quotation from the *Pātañjalayogaśāstra* and reads:

> This intellect (*buddhi*) is eightfold, due to the variety of forms related to *sattva* and *tamas*. Among these, the form of the intellect pertaining to *sattva* is fourfold: righteousness, knowledge, dispassion and mastery. [Among these states of intellect, the one] named righteousness is characterized by compassion, generosity, [fulfilment of] ethical rules (*yama*) and observances (*niyama*). Among these, the [fulfilment of] ethical rules and the observances have been defined in [the work] of Patañjali: "The [fulfilment of] ethical rules are non-violence, truth, abstaining from thievery, chastity and abstaining from possession; the observances are purity, contentment, religious austerity, practice of recitation and profound contemplation on Īśvara."[122]

---

117 Bronkhorst 1981: 309, 1985: 203 and 209; Larson 1999: 727 and 731; Maas 2006: xvi, xx–xxi and 2013: 58. The Sanskrit compound *sāṅkhyapravacana* can be interpreted as a *bahuvrīhi*-compound ("whose teaching is Sāṅkhya" or "expressive of Sāṅkhya"), which serves as an adjective, or as a *tatpuruṣa*-compound ("the teaching of Sāṅkhya"), as a noun apposition to the following compound *pātañjala-yogaśāstra* ("the Yoga treatise related to Patañjali"). In this study, the first interpretation has been chosen.

118 Torella (2011: 36) dates him to the second half of the ninth century. See however the discussion in Slaje 1986 and the concluding notes in Muroya 2011: 358–359.

119 Yogīndrānanda's edition of 1968, p. 442, quoted in Torella 2011: 91.

120 On the theory of *bhāva*s in classical Sāṅkhya, see Frauwallner 1973: 267–271.

121 On al-Bīrūnī's treatment of these concepts, see below pp. 219–221.

122 GPBh, p. 26.1–5 on kā. 23: *sā ca buddhir aṣṭāṅgikā, sāttvikatāmasarūpabhedāt. tatra buddheḥ sāttvikaṃ rūpaṃ caturvidhaṃ bhavati—dharmaḥ, jñānam, vairāgyam, aiśvaryaṃ ceti. tatra dharmo nāma dayādānayamaniyamalakṣaṇaḥ. tatra yamā niyamāś ca pātañjale 'bhihitāḥ. "āhiṃsāsatyāsteyabrahmacaryāparigrahā yamāḥ"* (sū. II.30). *"śaucasantoṣatapaḥsvādhyāyeśvarapraṇidhānāni niyamāḥ"* (sū. II.32).

The author of the *Gauḍapādabhāṣya* explains righteousness as including the fulfilment of ethical rules (*yama*) and observances (*niyama*), which are to be counted among the eight components (*aṣṭāṅga*) of the path that leads to liberation in the *Pātañjalayogaśāstra*. The authors of the *Māṭharavṛtti* and the *Jayamaṅgalā* also provide the two quotations from the *Pātañjalayogaśāstra* in the context of this *kārikā*. Vācaspatimiśra, in the *Tattvakaumudī*, on the other hand, only refers to the Yoga of the eight components (*aṣṭāṅgayoga*) in his commentary on kā. 23.[123]

In the *Sāṅkhyasaptativṛtti* and the *Māṭharavṛtti* on kā. 19, the conscious self (*puruṣa*) is compared to a religious mendicant (*bhikṣu*).[124] These two commentaries also qualify such a person as being "devoted to [the fulfilment] ethical rules and observances" (*yamaniyamarata*), as well as a "master of Sāṅkhya and Yoga" (*sāṅkhyayogācārya*). Their authors thus associated the practice of ethical rules and observances with both systems of thought.

Lastly, as outlined above in Section 3.2, Sāṅkhya and Yoga of the classical period have similar metaphysical, ontological and epistemological views, while they offer different means of reaching isolation (*kaivalya*), or final liberation. The *Sāṅkhyakārikā* and its commentaries are concerned with the acquisition of the theoretical knowledge of a specific metaphysics and ontology. The *Pātañjalayogaśāstra*, on the other hand, chiefly describes the psychological and mental conditions of the human being, as well as different meditative states, the last of which brings about isolation (*kaivalya*).[125]

Thus, although the exact relationship between Sāṅkhya and Yoga is difficult to establish, the evidence outlined above shows that their respective doctrines share essential features, to the extent that some scribes of copies of the *Pātañjalayogaśāstra* have considered the work as belonging to Sāṅkhya.

---

123 The authors of the *Sāṅkhyasaptativṛtti*, the *Sāṅkhyavṛtti* and the *Yuktidīpikā* also refer to the yogic ethical rules and observances when glossing kā. 23. However, the listed items in these commentaries do not correspond to those enumerated in the *Pātañjalayogaśāstra*, the *Gauḍapādabhāṣya*, the *Māṭharavṛtti* or the *Jayamaṅgalā* ad loc. Indeed, the three former commentaries do not include profound contemplation on Īśvara (*īśvarapraṇidhāna*) among the observances. Instead, they list: abstaining from anger, obedience to one's master(s), purity, moderation with food and abstaining from negligence. As a corresponding excerpt from the *Kitāb Sānk* is not extant, it is not possible to draw conclusions about al-Bīrūnī's possible source on the basis of this passage.

124 In Section 6.3.2, below, this analogy will be discussed in contrast to a passage drawn from the *Kitāb Sānk*.

125 Renou & Filliozat 1953: 45; Rukmani 1999: 733 and 735; Whicher 1999: 779–780. See above Section 3.2.

3.4.2 *The Kitāb Sānk and the Kitāb Pātanǧal*

As for the *Kitāb Sānk* and *Kitāb Pātanǧal*, an examination of how al-Bīrūnī—and thus his informants—regarded them in their formal aspects lends short insight into the question of their relationship. Nevertheless, in the same way as the two Indian philosophies are interconnected in the Sanskrit tradition, the *Kitāb Sānk* and *Kitāb Pātanǧal* appear to have shared common features. First, al-Bīrūnī often mentions and quotes from these two works alongside each other.

The first reference to these two books appears in his preface to the *Taḥqīq mā li-l-Hind* and reads:

> I have translated two books into Arabic: the first of them on fundamental elements (المبادئ)[126] and a description of what exists (صفة الموجودات),[127] named Sānk (سانك); the second on the liberation of the soul from the fetters of the body (فى تخليص النفس من رباط البدن), known as Pātanǧal (پاتنجل). These two [books] contain most principles (الاصل)[128] around which their (i.e., the Indians) faith revolves, without the subdivisions of their religious laws (دون فروع شرائعهم).[129]

This passage suggests a connection between the topics of the two books, because al-Bīrūnī mentions them together, but chiefly because he describes both of them as containing "most principles around which their faith revolves, without the subdivisions of their religious laws."

Further, Chapter 7 of the *Taḥqīq mā li-l-Hind*, entitled "On the nature of liberation from the world and of the path leading to it" (فى كيفيّة الخلاص من الدنيا و صفة الطريق المؤدّى إليه), includes interwoven quotations from the *Kitāb Sānk* and *Kitāb Pātanǧal*, combined with quotes drawn from the *Kitāb Gītā* and some Purāṇas.[130] In this way, al-Bīrūnī stresses the correlations between the *Kitāb Sānk* and *Kitāb Pātanǧal*, especially in regard to Hindu soteriology.

---

126 Sachau translates "origines" (*sic*). The plural Arabic word مبادئ can also signify "principles" or "fundamental elements." Al-Bīrūnī refers here to the *tattvas*, i.e., principles.

127 Sachau translates "created beings." In its descriptions, classical Sāṅkhya aims to encompass the imperceptible and phenomenal worlds. Therefore, "what exists" suits well here.

128 The reading of this word follows Sachau's edition (*Taḥqīq* [1887], p. 4.19) of the Arabic text, as the spelling is more accurate than in the Hyderabad edition (Taḥqīq [1958]).

129 Sachau translates: "most of the elements of the belief of the Hindu, but not all the single rules derived therefrom" (1910: 1/8). Al-Bīrūnī maybe contrasts theoretical principles (الاصل) with practical applications or derivatives (فروع); see *Taḥqīq* (1958), p. 6.1–4. See also Mario Kozah (2016: 176–179) on this passage.

130 *Taḥqīq* (1958), p. 51.15–67.8; Sachau 1910: 1/68–88.

Al-Bīrūnī's account also reflects the position of his Indian informants, who obviously regarded these two works as fundamental treatises on the subject of religion. Moreover, these informants and/or al-Bīrūnī himself assigned a common definition to the *Kitāb Sānk* and the *Kitāb Pātanğal*, and thus recognized an inherent connection between them.

In the above extract, al-Bīrūnī also provides distinct descriptions for the *Kitāb Sānk* and the *Kitāb Pātanğal*, differentiating them in this way. On the one hand, the *Kitāb Sānk* is "on fundamental elements and a description of what exists,"[131] which is a reference to the metaphysics developed in classical Sāṅkhya. Al-Bīrūnī's definition thus fits the emphasis this system puts on the enumeration, description and explanation of the twenty-five principles that constitute the world. On the other hand, the *Kitāb Pātanğal* deals with "the liberation of the soul from the fetters of the body," which refers to the Sanskrit *kaivalya*. In classical Sāṅkhya and Yoga, as mentioned above, the *puruṣa* needs to be liberated, not from the "fetters of the body," but from the false idea that it plays an active part in the evolutionary scheme of the world. In this particular case, the Arabic word for "soul" (نفس) translates the Sanskrit word *puruṣa*.

Both al-Bīrūnī and his informants conceived the two works as describing different aspects of fundamental Indian religious beliefs, namely metaphysical and psychological principles. This understanding contrasts with the meaning of the terms *yoga* and *sāṅkhya* in the Epics and Upaniṣads. The philosophical descriptions in the *Kitāb Pātanğal* and the *Kitāb Sānk* provided by al-Bīrūnī therefore rather correspond to the subjects dealt with in the *Pātañjalayogaśāstra* and in the *Sāṅkhyakārikā* together with its commentaries.

3.5 Concluding Remarks

This chapter showed that al-Bīrūnī was in some way aware of the tradition that considers Kapila as playing a role in the transmission of Sāṅkhya, without however understanding him as the founder of the system of thought or crediting Īśvarakṛṣṇa as the author of the source of the *Kitāb Sānk*. Al-Bīrūnī associated the method taught in the Sanskrit source of the *Kitāb Pātanğal* with that

131 Kozah (2016: 82) interprets this description as indicating that al-Bīrūnī solely used classical Sāṅkhya metaphysics. Several passages drawn from the *Kitāb Sānk* in the *Taḥqīq mā li-l-Hind* are, however, dealing with other topics, such as the rewards of attaining heaven (Appendix, passage XI), the eight powers (Appendix, passage XIII) and the nine rules how to conduct one's life (Appendix, passage XV).

of Hiraṇyagarbha. This latter view can be connected with the portrayal of Hiraṇyagarbha, in some Sanskrit commentaries on the *Pātañjalayogaśāstra*, as a figure who transmitted the philosophy of Yoga.

Arabic *sānk*, in the title of al-Bīrūnī's *Kitāb Sānk*, is a relatively faithful transliteration of the Sanskrit word *sāṅkhya*, which probably appeared in the title of the work he translated. In Indian tradition, the term *sāṅkhya* refers to the school of thought to which this work belongs. It is impossible to know whether al-Bīrūnī considered *sānk* as the designation of a philosophical system. In the case of the title *Kitāb Pātanǧal*, the Arabic *pātanǧal* seems to express both the adjective *pātañjala* in the title of the translated work and the proper name Patañjali.

Al-Bīrūnī considered the source of the *Kitāb Pātanǧal* as one text composed by a single author, whom he did not conflate with Vyāsa. His Indian interlocutors most probably influenced this understanding. Thus, this view on the authorship of the *Kitāb Pātanǧal* suggests that its source was also considered to have been written by one single author, as the *Pātañjalayogaśāstra* most probably was.

More generally, I also shed light on al-Bīrūnī's understanding of the authorship and title of the *Kitāb Sānk* and *Kitāb Pātanǧal*. His testimony on this subject reflects how ideas on these works circulated and were transmitted during the early eleventh century in Gandhāra and Panjab. Al-Bīrūnī's description of the *Kitāb Sānk* and *Kitāb Pātanǧal* indicates that he and his informants regarded the two systems of thought as sharing common features; this equally mirrors the Indian textual tradition on the relationship between the philosophies of Sāṅkhya and Yoga.

Thus, in this chapter, I proposed a philological perspective in situating al-Bīrūnī's translations, demonstrating that his references to the *Kitāb Sānk* and *Kitāb Pātanǧal* are connected to the *Sāṅkhyakārikā* and *Pātañjalayogaśāstra* traditions, respectively, rather than to any other Indian schools that had included Sāṅkhya-Yoga concepts in their teachings. Therefore, I selected works belonging to these two traditions and predating the composition of the *Taḥqīq mā li-l-Hind* in 1030, in order to examine the content of al-Bīrūnī's translations and their possible Sanskrit sources. I put then forward in Chapters 4, 5 and 6 of the present book that, when al-Bīrūnī translated works related to *Sāṅkhyakārikā* and *Pātañjalayogaśāstra* literature into Arabic, he made extensive changes to his sources, by way of translation strategies.

CHAPTER 4

Al-Bīrūnī's Interpretative Strategies

4.1 Al-Bīrūnī's Methods through the Lens of Translation Studies

Several observations made during this research led to the decision of examining al-Bīrūnī's *Kitāb Sānk* and *Kitāb Pātanğal* by way of translation strategies. First, several scholars have explored the relationship between these two translations and Sanskrit literature, with the aim of determining their original Sanskrit sources. Mostly looking for literal correspondences between Arabic and Sanskrit works, they generally observed important parallels between the sources and the translations along with substantial differences.[1] Among them, Garbe, Pines and Gelblum observe that al-Bīrūnī was creative in his translations, yet they continued to analyse his work as if his translations were more or less literal.[2] N.A. Baloch writes that al-Bīrūnī "was more concerned with ideas than with words,"[3] while in 2016 Mario Kozah, in his discussion of the Indian concept of liberation in al-Bīrūnī's writings, concludes that the latter "was actively engaged in a process of creatively reading and interpreting these Sanskrit texts rather than merely translating and citing from them."[4]

Thus, scholars who studied al-Bīrūnī's works have already foreseen the significance of his choices of interpretation for analysing his translations. They however have not deepened or developed their work in that direction. In this context, the present chapter supports and expands Baloch's and Kozah's observations.

Second, several passages of the *Taḥqīq mā li-l-Hind* and of the *Kitāb Pātanğal* already quoted in the present book suggest transformations (or confusion) on the part of al-Bīrūnī in his rendering of Indian philosophy into Arabic. Third, much of the discussion in Chapter 2 showed the essential role of the socio-intellectual context and of Indian informants for al-Bīrūnī's learning of Indian culture and philosophies. I thus argue that one has to comprehend the question of al-Bīrūnī's sources from different angles.

1 Reviews on relevant scholarship are available in Sections 5.1 and 6.1 below.

2 Garbe 1896: 41–42; Pines & Gelblum 1966: 305 and 307.

3 Baloch 1973: 12.

4 Kozah 2016: 170. See also Kozah (2016: 178) who argues that the role of al-Bīrūnī's interpretative work must be taken into account in future studies about his translations. In the introduction to his edition and translation of the *Kitāb Pātanğal*, Kozah (2020: XIX-XXII) briefly discusses questions related to its composition and al-Bīrūnī's challenges for interpretating the Yoga work.

In Chapter 4, for methodological reasons, I chiefly base the analyses of al-Bīrūnī's interpretative choices on the *Kitāb Pātanǧal*, as the entire work is extant in contrast with the *Kitāb Sānk*, with the aim of exploring the concept of translational strategies and the processes through which al-Bīrūnī prepared his translations. This conceptual framework bears fruits, as it not only helps have better and refined insight into the relationship between the Sanskrit and Arabic works, as well as exclude some commentaries from being the sources of these Arabic works, but also enables to explore different reasons for al-Bīrūnī's interpretative choices.

In order to do so, I adopt the working hypotheses that 1) the *Kitāb Sānk* was based on the *Sāṅkhyakārikā* and a commentary and 2) the *Kitāb Pātanǧal* was based on the *Pātañjalayogaśāstra*. Throughout this chapter, I make use of these working hypotheses, which find support in the course of the discussion, and in the subsequent chapters of this research. Chapters 5 and 6 indeed confirm the two above working hypotheses, by discussing the possible sources of the *Kitāb Sānk* and *Kitāb Pātanǧal*, by providing additional examples taking into account all commentaries that belong to these two Sanskrit literary traditions.

Before delving into textual analysis, I want to stress the importance of taking into consideration challenges al-Bīrūnī faced as a translator. In order to render the two Sāṅkhya and Yoga texts into Arabic, he needed to translate not only from one language to another, but also from one cultural and historical context to another. He had to accomplish the dual task of understanding Brahminical concepts that had been systematized hundreds of years before his time, that is, during the fourth-fifth century CE, and of conveying these Indian ideas to the Muslim audience of the early eleventh century. Further, the nature itself of these philosophical concepts certainly made the process of al-Bīrūnī's translation difficult. Moreover, Sanskrit and Arabic belong to two distinct language groups. The lack of common linguistic roots of the two languages rendered the translation of these concepts even more arduous. Thus, al-Bīrūnī had to bridge considerable temporal, cultural, conceptual and linguistic gaps in his effort to transfer the two Indian philosophies to the Islamic world of his time. In this context, thus, the field of Translation Studies provides an essential framework within which al-Bīrūnī's translations can best be understood.

Reflecting on the parallels and discrepancies between a translation and an interpretation, Hans-Georg Gadamer theorizes that substantial discrepancies are in accordance with the degree of difference between the translated—or interpreted—text and its original source.[5] Considering this idea as funda-

5 Gadamer 1976: 406–409.

mental for discussing processes of translation, Umberto Eco conceptualizes it as being the "difference in degree of intensity" (Fr. différence en *degré d'intensité*)[6] between the target-text, that is, the translation or interpretation, and the source-text, that is, the original work. For example, a literal translation would differ from its source in a low degree of intensity, while an interpretative work would depart from it in a high degree of intensity.

A high degree of intensity in difference between an original and its translation can partly be due to major gaps, temporal, cultural, conceptual, linguistic, etc., which differentiate the source-text and culture from the target-text and culture. When one compares the *Kitāb Sānk* and the *Kitāb Pātanǧal* to the *Sāṅkhyakārikā* and *Pātañjalayogaśāstra* textual tradition, respectively, it becomes clear that al-Bīrūnī's Arabic translations display discrepancies in a high degree of intensity in relation to their possible original sources. The various gaps he had to face necessitated an adaptation of his sources. These observations raise the question of whether it would even have been an option for al-Bīrūnī to provide a word for word translation. His aim was to promote intellectual exchange across the Indian and Islamic cultures[7] and he intended to provide an effective and meaningful translation for his readership. In the preface to the *Kitāb Pātanǧal*, he also explicitly specifies his wish to avoid a literal translation that might affect the meaning of his translation and announces three transformations that he made to his source text, as I show below.[8]

An article by Maas and Verdon for the first time considered in detail al-Bīrūnī's interpretative choices. In this article, the authors describe these transformations, and elaborate on the concept of translational strategies.[9] This concept refers to the various possible choices of interpretation for translating a text and relates to the way in which a translator negotiates between the source-text and the target-text. Al-Bīrūnī, far from providing literal translations, interprets his sources and in doing so resorts to translational strategies.

---

6 Eco 2010: 293.

7 *Taḥqīq* (1958), pp. 6.4–5 and 547.17–18; Sachau 1910: I/8 and II/246. Whereas al-Bīrūnī expresses this desire several times in the *Taḥqīq mā li-l-Hind* and indeed transfers much information about India in his book, he did not deem that an exchange was possible with regard to the medical science, as noted by Fabrizio Speziale (2014: 785, n. 4).

8 See al-Bīrūnī's preface to the *Kitāb Pātanǧal* below pp. 132–133 (KP, pp. 167.21–168.5; Pines & Gelblum 1966: 310). The question of the reception of al-Bīrūnī's works among his peers is broached in Maas & Verdon (2018: 290–291), as well as in the conclusion to the present book, see below p. 237.

9 Maas & Verdon 2018: 315–328.

Maas and Verdon analyse al-Bīrūnī's translations utilizing a model established by the linguist Vladimir Ivir, which emphasizes the process of translation between cultures rather than between languages. Ivir proposes seven procedures that a translator may deploy: borrowing, definition, literal translation, substitution, lexical creation, omission and addition.[10] According to Ivir, these translational strategies are utilized to reduce cultural gaps and render a translation as effective as possible in terms of its communicative goal. In al-Bīrūnī's case, the model enables an analysis of his translations from a cultural perspective. It also allows to evaluate their efficiency in communicating a message rather than words. Lastly, it provides analytical tools to consider al-Bīrūnī's translations from a different perspective than a direct comparison between the source-texts and the target-texts.

Nearly all translation procedures defined by Ivir are evident in passages of the *Kitāb Pātanğal* and *Kitāb Sānk* when being confronted to their possible Sanskrit source-texts. While Maas and Verdon provide a detailed discussion of these strategies, together with their drawbacks and benefits and expound specific examples drawn from al-Bīrūnī's *Kitāb Pātanğal*,[11] the present book focuses on the strategies that al-Bīrūnī used the most. The preliminary analysis by Maas and Verdon demonstrates that a clearly discernible difference between the *Kitāb Pātanğal* and the *Pātañjalayogaśāstra* does not necessarily prove that al-Bīrūnī used another Sanskrit work besides the *Pātañjalayogaśāstra* as the main source of his translation. Providing several examples of translational strategies used by al-Bīrūnī, Maas and Verdon argue that "[u]nderstanding al-Bīrūnī's motives for deviating from his source as well as determining other reasons for differences between the *Kitāb Pātanğal* and its sources then [leads] to a fuller picture of al-Bīrūnī's literary activity and creativity."[12]

The following exposition builds upon their study and takes a further step through the identification and classification of the possible reasons for al-Bīrūnī's choices of translational strategies with the help of additional examples. It also draws attention to the fact that his transformations are often intertwined throughout his translations. It also includes, when possible, analyses of passage drawn from the *Kitāb Sānk*.

Lastly, Eco explains that the target-language, in combination with the worldly knowledge of the translator, determines the process of interpretation.[13] Al-Bīrūnī extensively uses his intellectual background in his interpretat-

10 Ivir 1987: 37–45.
11 Maas & Verdon 2018: 321–328.
12 Maas & Verdon 2018: 329.
13 Eco 2010: 38.

ive work. He therefore exemplifies Eco's statement. Al-Bīrūnī was, for instance, knowledgeable about Greek literature and science, via Arabic translations. In the *Taḥqīq mā li-l-Hind*, he quotes Ptolemy, Plato, Galen, Proclus and Aristotle.[14] In his treatment of different theological and philosophical themes, he uses his knowledge of Islamic religion and philosophy (Falsafa) in order to transfer Sāṅkhya-Yoga concepts into Arabic. Thanks to his research on Indian sciences and literature and due to his interaction with Indian thinkers, al-Bīrūnī also acquired a significant intellectual background on these subjects. His encyclopaedic knowledge plays a significant role in his interpretation of Sāṅkhya-Yoga concepts, as I show in the present and following chapters.

The thorough understanding of al-Bīrūnī's worldly knowledge, alongside that of his translational strategies, leads to group the reasons for discrepancies between the source-text and the target-text into several categories: 1) al-Bīrūnī's didactic intentions, 2) his own logic, 3) his intellectual and religious education, and 4) the influence of Indian sources, oral and written. The determination of the underlying reasons for al-Bīrūnī's adaptations of the originals allows for a further understanding of the multi-faceted relationship between the translations and their original sources. It also makes it possible to establish a similar pattern of his hermeneutics in both the *Kitāb Sānk* and the *Kitāb Pātanǧal*. Examining these in detail thus enables us to understand the motives behind such changes, as well as different sources of information from which al-Bīrūnī drew in order to prepare his translations. Lastly, as some of these modifications affected the form of his source, while some others had an impact on its content, I also classify al-Bīrūnī's modifications into formal and substantial transformations.

4.2 Three Explicit Transformations

To begin with, this section discusses the statement of al-Bīrūnī that he made three modifications to his sources. Al-Bīrūnī announces these in the preface to the *Kitāb Pātanǧal*, which precedes the translation of his Sanskrit source per se:

> Their books (i.e., the books of the Indians) are composed according to metres, and their texts are commented in such a way that a complete and accurate translation is difficult, because the commentators are concerned

14 Sachau 1910: I/xli–xliii.

> with grammar (نحو) and etymology (اشتقاق) and other [matters] which are of use only to [someone] who is versed in their literary languages (لغاتهم الفصيحة), but not in the vernacular (المبتذلة) [ones]. Therefore, in the translation I was forced to merge the text (نصّ) with that lengthy commentary (ذلك التفسير المزيد), to arrange the work in a way which resembles [a dialogue consisting of] questions and answers, and to omit [the parts] which are concerned with grammar (نحو) and language (لغة). This is an apology which I offer for the difference in size of the book in the two languages, if such a comparison is made between them. [I do this] in order that no one should think that this [difference] is due to negligence in [the rendering of] the meaning. Indeed, [readers should be assured] that it is due to a revision of what [otherwise] would impair [the translation]. May God bestow His favour upon the good. This is the beginning of the *Kitāb Pātanǧal*, its text [being] interwoven with its commentary.[15]

According to this passage, al-Bīrūnī deliberately transformed his source-text in three ways: 1) combining a text (نصّ) and a commentary (تفسير; شرح), 2) recasting these two layers of text into a dialogue, and 3) omitting elaborate literary and etymological expositions. Thus, the explicit formulation of these three types of transformation indicates the importance of taking into account al-Bīrūnī's input in the process of translating the Sanskrit works into Arabic. As for the *Kitāb Sānk*, there is no information regarding adaptations al-Bīrūnī may have made to its Sanskrit source. However, when one considers certain extracts of the *Kitāb Sānk*, it is possible to outline how he adapted his original Sanskrit source, in the same way as he did in the case of the source of the *Kitāb Pātanǧal*.

The preface to the *Kitāb Pātanǧal* became accessible to academic circles when Ritter edited the text in 1956. Al-Bīrūnī's personal remarks with regard to his work were thus unavailable to Sachau, Garbe, Dasgupta and Filliozat. Pines and Gelblum did note the combination of two layers of text and the dialogue form in the *Kitāb Sānk*. They suggest that al-Bīrūnī may have "systematized this form (i.e., the dialogue form) into a series of questions and answers"[16] and that this specific form as found in the Arabic translation may reflect that of the Sanskrit source.[17] They also state that both the combination of two layers of text and the systematic introduction of a dialogue form may be "an adaptation

15 KP, pp. 167.21–168.5; Pines & Gelblum 1966: 310.
16 Pines & Gelblum 1966: 303.
17 Pines & Gelblum 1966: 303 and 1989: 265.

based on an Arabic usage."[18] Maas and Verdon describe these three transformations. They conclude that both the dialogue form and the combination of two layers of text constitute features that exist in the hypothesised main source of the *Kitāb Pātanǧal*, namely the *Pātañjalayogaśāstra*, and that al-Bīrūnī may have enhanced and systematized these pre-existing characteristics.[19]

4.3 Al-Bīrūnī's Reshaping of the Original Texts

4.3.1 *Pedagogical Motives*

Two of the three transformations announced by al-Bīrūnī particularly affect the form of his source: the combination of two layers of text and the systematic introduction of a dialogue form. With regard to the former, it is worth noting that the *Pātañjalayogaśāstra* itself, the extant Sanskrit work which is the closest to the *Kitāb Pātanǧal*, does not always clearly distinguish between the aphorisms (*sūtra*) and the commentary (*bhāṣya*), as the case of PYŚ 1.5 shows.[20] Moreover, as I set forth in Chapter 3 above, several Indian thinkers and Sanskrit sources regarded the *Pātañjalayogaśāstra* as constituting a single entity made up of two layers of text. In general, it is indeed sometime difficult to dissociate a foundational text from its first commentary in Sanskrit manuscripts tradition. This feature may have led al-Bīrūnī to enhance a characteristic present in his Sanskrit source, when he fully intermingled a text and its commentary.

The case of the *Kitāb Sānk* show similar features. First, even though the name Kapila is only found in the commentaries on the *Sāṅkhyakārikā*, al-Bīrūnī is aware of the tradition that holds Kapila to be the founder of the Sāṅkhya system of thought (Section 3.3.2). Furthermore, several passages drawn from this book indicate that al-Bīrūnī made use of a commentary. For instance, the *Kitāb Sānk* contains a passage that deals with different views on action and agent (Section 6.3.3) and that relates to the discussion on the source of the phenomenal world in the *Sāṅkhyakārikā* and commentaries (kā. 61). However, only some commentaries on the *Sāṅkhyakārikā* enumerate these positions, not the *kārikās*. Lastly, several passages of the *Kitāb Sānk* found in the *Taḥqīq mā li-l-Hind* deal with analogies that are used in Sāṅkhya in order to illustrate abstract concepts of its philosophy. Some of these analogies rendered in the *Kitāb Sānk* are only found in commentaries on the *Sāṅkhyakārikā*, while others are hinted at in the *kārikās* and contextualized in the commentaries (Sections 6.3.2,

18 Pines & Gelblum 1966: 303.
19 Maas & Verdon 2018: 315–321.
20 Maas & Verdon 2018: 317.

6.3.4 and 6.3.5). Without using a Sāṅkhya commentary, al-Bīrūnī could not have provided explanations of these analogies. The Sanskrit source of the *Kitāb Sānk* was thus also composed of both a basic text and a commentary, which al-Bīrūnī appeared to have merged in his translation.

In this context, it is worth mentioning al-Bīrūnī's treatment of the *Brāhmasphuṭasiddhānta* composed by Brahmagupta and its now-lost commentary by Balabhadra. About this text, David Pingree observes that the distinction between the root-text (*mūla*) and the commentary (*ṭīkā*) is not clearly marked in the numerous quotations found in the *Taḥqīq mā li-l-Hind*. Pingree concludes that 1) al-Bīrūnī could not consult the manuscript of the original astronomical work or 2) had insufficient knowledge of Sanskrit, and that 3) the Indian thinkers he encountered influenced him in the combination of the root-text and the commentary.[21]

Al-Bīrūnī exhibits a relatively high proficiency in the relevant Sanskrit terminology (as discussed in Chapters 2, 5 and 6 of the present book). His interpretation of the Yoga source-text is relatively faithful to the message of the *Pātañjalayogaśāstra*. In my opinion, thus, al-Bīrūnī attained a significant knowledge of the Sanskrit lexicon in general by the time he composed the *Taḥqīq mā li-l-Hind*, whereas his proficiency in Sanskrit grammar might have been low.

Pingree's two other points, that is, the inaccessibility of the original astronomical work and the influence of Indian thinkers on a-Bīrūnī, may be valid. Nevertheless, al-Bīrūnī may only have enhanced an existing feature of the original Sanskrit astronomical work or have adopted the very conception of the works on the part of the Indian thinkers he had met with, in the same way as he did for the *Kitāb Pātanğal* and *Kitāb Sānk*. The *Brāhmasphuṭasiddhānta* by Brahmagupta was, however, definitely commentated upon by a distinct person, Balabhadra, and belongs to a scientific field quite different from that of the Sanskrit sources of the *Kitāb Pātanğal* and *Kitāb Sānk*. Due to this discrepancy, it is problematic to draw a precise parallel between al-Bīrūnī's ways of dealing with the *Brāhmasphuṭasiddhānta*, on the one hand, and the Sanskrit sources of his *Kitāb Pātanğal* and *Kitāb Sānk*, on the other hand.[22] Nonetheless, in the case of the two philosophical works the combination of two layers of text appears to have been a common procedure employed by al-Bīrūnī.

21 Pingree 1983: 356, 356, n. 29 and 360. See also Pingree 1969.

22 For a comparison between the table of contents of the Sanskrit *Brāhmasphuṭasiddhānta* as edited by S. Dvivedin (1901) and that of al-Bīrūnī's Arabic *Brāhmasiddhānta*, see Verdon & Yano 2020: 68–71.

The second modification that al-Bīrūnī formulated is the systematic organization of the discourse in the form of questions and answers. The first protagonist of the narrative is an ascetic (زاهد) who asks (سايل) questions, while the second protagonist, Pātanğal himself, is the one who answers (مجيب), as the initial question of the *Kitāb Pātanğal* shows.[23]

Pines and Gelblum point out an apparent contradiction in al-Bīrūnī's statements on the dialogue form of the *Kitāb Pātanğal*. The last sentence of al-Bīrūnī's *Kitāb Pātanğal* states that his Sanskrit source, originally consisted of "one thousand and a hundred questions in the form of verse" (و هو كله الف و مائة سؤال من الشعر).[24] In their view, this statement contradicts al-Bīrūnī's initial remark in the preface to the *Kitāb Pātanğal*, namely that he reshaped the text of his source into a dialogue.[25] They also note that the Arabic word "question" (سؤال) perhaps goes back to a wrong transcription of an original reading *šlūka* (شلوك), which would be a transliteration of the Sanskrit word *śloka*, meaning "stanza."[26]

Despite the resemblance between the two Arabic spellings, the word سؤال is, in my view, not a mistaken transcription of the Sanskrit *śloka*. Al-Bīrūnī only occasionally transliterates Sanskrit words into Arabic in the *Kitāb Pātanğal*. Moreover, this hypothesis does not solve the aforementioned apparent contradiction. Two explanations for the use of "question" (سؤال) in this final sentence of the *Kitāb Pātanğal* are possible. If the wrong transcription is accepted, it may actually relate, as Maas suggests, to the number of units called *śloka* or *grantha*, which are sometimes noted on Sanskrit manuscripts in order to indicate the units on which the price of copying the manuscript is based.[27] However, al-Bīrūnī may also simply have used the word "question" in a more figurative sense of "topic," referring to the many topics that are dealt with in the original work.

Furthermore, some questions found in the Arabic translation reflect introductory questions to certain *sūtra*-parts of the *Pātañjalayogaśāstra*, whereas some others do not parallel each other. On the one hand, Maas notes that, for instance, KP 12 is an almost literal translation of the introductory question to PYŚ I.24. On the other, Maas and Verdon analyse in detail KP 2 and 3, which correspond to PYŚ I.3, and observe that KP 2 can be paralleled to the introductory question to *sūtra* I.3, while KP 3 is a new question formulated by al-Bīrūnī,

23 See above p. 120.
24 KP, p. 199.1–2; Pines & Gelblum 1966: 303.
25 Pines & Gelblum 1989: 271.
26 Pines & Gelblum 1989: 304, n. 155.
27 Maas 2013: 60.

TABLE 8 Correspondences between question-and-answer sets in the *Kitāb Pātanğal* and the *Pātañjalayogaśāstra*, based on Pines' and Gelblum's annotations

Ch. 1, Q. 1–23		Ch. 2, Q. 24–41		Ch. 3, Q. 41–56		Ch. 4, Q. 57–78	
Q	PYŚ	Q	PYŚ (1904)	Q	PYŚ (1904)	Q	PYŚ (1904)
1	~ I.1–2	24–25	II.1–2	41	III.1–8	57	IV.1
2	I.3	26	II.3–4	42	~ III.9; III.11	58	~ IV.2
3	I.3	27	II.5–10	43	III. 16	59	~ IV.3
4	I.3–4	28	II.11–12	44	III. 13–15	60	IV.4–5
5	I.5–11	29	II.13	45	III.17–20	61	IV.6
6	I.12–16	30	II.14	46	III.21–32; 34–35	62	~ IV.7–8
7	I.17–18	31	II.15	47	~ III.36–38	63–64	~ IV.9–10
8–10	I.19–22	32	II.16	48	~ III.39–42	65	IV.10–11
11	I.23	33	II.17, II.24	49	~ III. 43–48	66	IV.12–13
12–14	~ I.24–I.25	34	II.18(19)	50	~ III.49–50	67	IV.14
15	~ I.27	35	II.18	51–52	~ III.51	68	~ IV.15–16
16–17	~ I.25–26	36–38	~ II.20–26	53	~ III.52	69	IV.19
18	~ I.28–29	39	II.27	54	~ III.53	70	~ IV.19–24
19	I.30	40	II.28	55	III.54	71	~ IV.25
20	~ I.31	41	II.29–55	56	III.55	72	IV.25–26
21	I.32					73	IV.27
22	I.33–34					74	IV.29–30
23	~ I.40–51					75	IV.31–32
						76	IV.33
						77	IV.33
						78	IV.34

whose answer is provided by the quasi-literal translation of *sūtra* I.3.[28] This type of formal modification of the potential source makes it difficult at times to find exact correspondence between questions and answers in the *Kitāb Pātanğal* and *sūtra* and *bhāṣya* parts of the *Pātañjalayogaśāstra*.

However, thanks to the edition and English translation of the complete text of the *Kitāb Pātanğal*, it is possible to match some questions and answers of

28 Pines & Gelblum 1966: 314, n. 104; Maas & Verdon 2018: 317–320. For other such instances, see below Table 11.

the *Kitāb Pātanǧal* with the *sūtra*- and *bhāṣya*-parts of the *Pātañjalayogaśāstra*. Table 8 outlines these correspondences, despite the different form of the two works. Interestingly, in some cases, one set of question and answer includes the topics of several *sūtra*s, with attached *bhāṣya*-parts, as in the case of KP 46 that covers PYŚ (1904) III.21–35.[29] In other cases, the topics of a single *sūtra* with its *bhāṣya*-part are distributed across several sets of question and answer, as is the case with the two questions and answers in KP 2 and 3, which relate to PYŚ I.3.

Pines and Gelblum ascertained that some *sūtra*s are not represented in al-Bīrūnī's Arabic translation.[30] Given the high degree of modifications to the text, in form and substance, made by al-Bīrūnī in his translation, the apparent absence of topics addressed by particular *sūtra*s in the *Kitāb Pātanǧal* does not entail their actual absence from al-Bīrūnī's original Sanskrit source.

As for the *Kitāb Sānk*, three passages take the form of a dialogue, involving a hermit (ناسك) and a sage (حكيم), whose names are not given (passages I, XVII and XX of the Appendix and of Table 12). The corresponding Sanskrit passages, kā. 61, 67 and 53 with commentary, respectively, are not presented in the form of a dialogue. However, the available commentaries at times introduce *kārikā*s by a question (see for instance the **Suvarṇasaptati*, the *Gauḍapādabhāṣya*, the *Sāṅkhyavṛtti*, the *Sāṅkhyasaptativṛtti* and the *Māṭharavṛtti* on kā. 3).[31]

Thus, if al-Bīrūnī translated the whole of the *Sāṅkhyakārikā* and a related commentary into Arabic, he may have enhanced a characteristic that already existed in the original Sanskrit source of the *Kitāb Sānk*, in the same way as he did for the *Kitāb Pātanǧal*. In the absence of the complete text of the *Kitāb Sānk*, it is, however, not possible to determine whether his translation was systematically characterized by this form or not.

The dialogical form is also a common characteristic of Sanskrit scholastic works, in which the authors present opposing opinions, the *siddhānta*, that is the established position of the tradition to which a work belongs, and the *pūrvapakṣa*, that is the position of some opponent to its tradition, from a polemical perspective. This form of dialogue is meant to eventually refute all arguments that are presented to invalidate the position of the author of the work, or of the followers of the school of thought formulated in the work. It

29 See also Kozah 2020: 108–110 for a similar comparison of questions-and-answers set of the *Kitāb Pātanǧal* and the *Yogasūtras*. On KP 46 see below Table 10.

30 According to Pines & Gelblum, the missing *sūtras* are: I.35–39, I.46–47, I.50–51, II.32, II.41, II.50–51, III.8–9, III.12–13, III.15, III.33, IV.17 and IV.20–22 (1966: 323, n. 217, and 325, n. 241, 1977: 522, 1983: 258 and 1989: 265).

31 The systematic structure of the *Yuktidīpikā* in the form of the presentation of the position of an opponent followed by that of the position of the proponent sets this commentary apart from the other ones.

might be argued that the form of both the *Kitāb Pātanǧal* and the *Kitāb Sānk* reflects such a polemical dialogue. However, the dialogue in these two works is not a polemical one. Al-Bīrūnī rather organizes his translations in a didactic, or epistemic, dialogue, in which the questioner yearns to learn about the concepts exposed by the respondent (see for instance Table 11).

More generally, dialogue constitutes a common literary *genre*. For instance, the *Dharma Pātañjala*, an Old Javanese work related to Yoga, was composed in a similarly didactic fashion as the *Kitāb Pātanǧal* and the *Kitāb Sānk*. In the Old Javanese version, however, the two protagonists are Kumāra and the Lord (Śiva), and the text contains 39 questions, as against the 78 questions of the *Kitāb Pātanǧal*.[32] Dialogues in Arabic works also occur in different literary *genres*, the *Quran*, *Ḥadiṯs*, *Adab*-literature and poetry. Medical treatises, in particular, made use of the *genre* of dialogue in a didactic way.[33] The dialogue *genre* was thus common at the time in both the source- and the target-culture.

Further, the first-person speaker involved in the narrative of the *Kitāb Pātanǧal* is an ascetic "roaming in deserts and jungles" (الزاهد السايح فى الصحارى والغياض).[34] This type of figure is commonly found in medieval Arabic literature concerned with spiritual quests. Roaming in deserts came to symbolize the austerity that accompanies the spiritual journey of saints and mystics.[35] Thus, by creating a systematic dialogue form and by including this special type of figure into his narrative, al-Bīrūnī may have adjusted his source-text to the cultural sphere of his readership. This approach may also have provided his translations with a sense of authority on account of their formal parallel to a literary *genre* that his readers acknowledged as valid.

Lastly, at least three of al-Bīrūnī's own works were written in the form of a dialogue: *Answers to the Questions of the Indian Astronomers*, *Answers to the Ten Kashmiri Questions*,[36] and the epistolary exchange with Ibn Sīnā are also presented in the form of questions and answers (*Questions Asked to Ibn Sīnā*; مسائل سأل عنها إبن سينا).[37]

Al-Bīrūnī appears not only to have drawn his inspiration from the Sanskrit sources he consulted, but also from existing Arabic literature. An additional significant advantage of the dialogue form over the form of basic text and com-

32 Acri 2011: 193–339 and 2012: 260.
33 Touati 2000: 21; Daiber 2012.
34 KP 1, p. 169.10. See above p. 120.
35 Touati 2000: 187–192.
36 Boilot 1955: 199 and 200, nos. 71–72; see above p. 44.
37 Boilot 1955: 227, no. 147; Nasr & Mohaghegh 2005. See also no. 28 in Boilot 1955: 186.

mentary is that it easily captures the attention of the reader. It thus constitutes an effective didactic means to transmit knowledge and encourages the reader to step into the questioner's shoes. Al-Bīrūnī's choice of this form was perhaps not an arbitrary decision but was rather led by his goal to facilitate the transmission of the Indian works in his own cultural sphere.

4.3.2 *Changes due to Logic*

In addition to the transformations indicated by al-Bīrūnī in the preface to the *Kitāb Pātanǧal*, he made changes to his source-text in other ways without explicitly stating it. As a matter of fact, many discrepancies between the two texts can be explained through al-Bīrūnī's hermeneutics. First, a relatively common, but silent, formal transformation is the rearrangement in the description of certain concepts. Al-Bīrūnī appears to have made these modifications according to his own logic.

For instance, KP 5 lists and describes five different kinds of faculties of the soul (قوى النفس) that correspond to the five mental dispositions (*cittavṛtti*) discussed in PYŚ I.5 to I.11. The first one is apprehension (ادراك), the Arabic translation of "means of knowledge" (*pramāṇa*). As for the second and third mental dispositions, the *Kitāb Pātanǧal* appears to present an inverted sequence of the two items when compared with the sequence documented in the extant Sanskrit texts. Indeed, the second faculty of the soul listed in the *Kitāb Pātanǧal*, that is imagination (تخيّل) in the sense of forming an image in the mind, shall correspond to the mental disposition of conceptual thinking (*vikalpa*), which is however listed as the third item in the *sūtra*s themselves. The third faculty of the soul in the *Kitāb Pātanǧal*, [false] assumption (ظنّ), is related to the mental disposition of error (*viparyaya*), however enumerated as the second item. This inversion is perhaps an example of rearrangement of the source-text by al-Bīrūnī. Numbers four and five are then listed in parallel and correspond well to each other. The Arabic "dream" or "vision" (رؤيا), as number four, parallels the Sanskrit "deep sleep" (*nidrā*) and memory (ذكر) as number five is the rendering of "memory" in Sanskrit (*smṛti*).[38]

In addition to the inversion in the sequence of two of the mental dispositions, al-Bīrūnī appears to have gathered several *sūtra*- and *bhāṣya*-parts, namely from I.5 to I.11, in one set of question and answer, KP 5. He has thus given a slightly different structure to the description of these items. The author of the *Pātañjalayogaśāstra* first introduces the mental dispositions in *sūtra* I.5,

38 Woods 1914: 17–32; Maas 2006: 10–21; KP 5, p. 171.1–13; Pines & Gelblum 1966: 315–316. This passage also constitutes an example of the possible integration of the *sūtra*- and *bhāṣya*-parts of the *Pātañjalayogaśāstra* in al-Bīrūnī's translation (Maas & Verdon 2018: 317).

second enumerates all of them in *sūtra* I.6, and then dedicates five *sūtras*, I.7 to I.11, to the individual explanation of each of the five dispositions. In the *Kitāb Pātanǧal*, al-Bīrūnī does not provide the initial enumeration, but only the explanations.

In KP 41, al-Bīrūnī also rearranges the order in which eight qualities (خصلة), corresponding to the eight components (*aṅga*) of Yoga, are discussed in PYŚ (1904) II.29–55 and III.1–8. He adds definitions of these concepts immediately after their respective mention, whereas in the *Pātañjalayogaśāstra* lists the components in II.29 and discusses them one by one in the subsequent *sūtra-* and *bhāṣya*-parts.[39] Al-Bīrūnī appears to have reorganized KP 5 and KP 41 in a similar systematic way.

In the *Kitāb Sānk*, observations of similar rearrangements do not emerge from the analysis of the various excerpts found in the *Taḥqīq mā li-l-Hind*, but one cannot rule out this possibility as most of this work is unavailable to us. However, in one of its excerpts al-Bīrūnī appears to have reorganized the content of his source-text. The passage entitled "births depending on virtue and vice" (Appendix, passage XII), which corresponds to kā. 39, describes two conditions of a future life resulting from one's actions. Living a virtuous life leads to the divine sphere, whereas a present existence characterized by vice leads to a future reincarnation in the animal or vegetable kingdom. The **Suvarṇasaptati*, the *Gauḍapādabhāṣya*, the *Sāṅkhyasaptativṛtti* and the *Māṭharavṛtti* have an analogous passage on kā. 39, in which the commentators explain that the subtle body (*sūkṣmaśarīra*) is reborn in an animal or plant, or in a divine being, depending upon one's conduct. Al-Bīrūnī may have inverted the order in which the two conditions of existence are described in the mentioned commentaries,[40] which first address the consequences of vicious conduct, and then those of a virtuous life. These discrepancies are relatively minor, and they mostly affect the outer form of the text, but al-Bīrūnī did not explicitly mention this kind of transformation. The above formal reorganizations likely constituted, in al-Bīrūnī's view, a more coherent way to express the themes developed in his original Sanskrit sources.

39 On the components (*aṅga*), see also below pp. 175–177.

40 See below p. 204–205.

4.4 Al-Bīrūnī's Transformations of Content

4.4.1 *Omission of Technical Notions and of Redundancies*

In addition, al-Bīrūnī modified the source-text in substance and content. These substantial modifications could be linked to four translational strategies which Vladimir Ivir describes. They are omission, substitution, addition and definition.[41] Omission is the third transformation mentioned by al-Bīrūnī in his preface to the *Kitāb Pātanǧal*. He decided to "omit [the parts] which are concerned with grammar and language" (اسقاط ما يتعلق بالنحو واللغة) which are, as he believes, of no use for those who are not versed in Indian literary languages.[42] This statement indicates that his aim was to simplify the narrative of his translation and that the original Sanskrit source contained grammatical and linguistic explanations, as is commonly the case in scholarly Sanskrit literature.

Maas and Verdon note that the *Pātañjalayogaśāstra*, for instance, sporadically provides linguistic explanations and give several examples found in PYŚ I.1. The *bhāṣya* explains the function of the adverb "now" (*atha*), the meaning of the word "authoritative exposition" (*anuśāsana*) and the etymology of the term *yoga*, elements that are all absent from the *Kitāb Pātanǧal*. Moreover, PYŚ I.13, starts with the word *tatra*. This word ending with the adverbial locative suffix is often translated in the locative or partitive sense, as "there" or "among them," but in this particular case it should be understood in a causal sense referring to the motive. The author of the *Vivaraṇa* states that "this is the seventh (i.e., locative) [case ending in the sense of] the cause" (*sā ca nimittasaptamī*),[43] which in his view is required in order to properly interpret the *sūtra*-part of PYŚ I.13. When al-Bīrūnī wrote his comment on omitting some parts of his source, he most probably had such explanations in mind, which, as a matter of fact, are unintelligible without knowledge of Sanskrit grammar.

Furthermore, Maas and Verdon also observe that the *Pātañjalayogaśāstra* does not contain many such explanations and that their omission would not account for a substantial difference in size between the *Kitāb Pātanǧal* and its probable source, the *Pātañjalayogaśāstra*, despite al-Bīrūnī's statement in his

---

41 Borrowing was not a translational strategy that al-Bīrūnī frequently used in his two translations, in contrast to the *Taḥqīq mā li-l-Hind*. In the *Kitāb Pātanǧal*, he appears to have transliterated only some proper names (Maas & Verdon 2018: 309–311 and 322–323). He also transliterated the technical term *mahāvideha* (used as an adjective in the nominative singular feminine) into the Arabic script (مهابده: *mahābidaha*; PYŚ (1904) III.43, p. 160.15; KP 49).

42 See above p. 131–132.

43 *Vivaraṇa* (1999), p. 215.3.

preface. Nevertheless, the above suggests that if the *Kitāb Pātanǧal* does not include certain passages found in the *Pātañjalayogaśāstra*, al-Bīrūnī may have simply omitted them, and not necessarily that he had consulted an entirely different Sanskrit work as his source-text.[44]

Maas and Verdon further point out that al-Bīrūnī may also have omitted parts of the source-text other than pure linguistic explanations. For instance, PYŚ I.2 defines cognitive absorption (*saṃprajñāta samādhi*) as a characteristic of all mental states.[45] Al-Bīrūnī does not provide such an explanation in the corresponding passage of the *Kitāb Pātanǧal*, namely KP 1. Neither does he broach the topic of absorption in this particular passage. Maas and Verdon suggest that al-Bīrūnī remains silent on this psychological definition because he regarded it as being "of no interest to his readership."[46]

Furthermore, PYŚ I.1 contains an enumeration of five mental states, namely scattered, confused, distracted, one-pointed and ceased (*kṣiptaṃ mūḍhaṃ vikṣiptam ekāgraṃ niruddham iti cittabhūmayaḥ*),[47] which al-Bīrūnī does not address at all. In my view, he considered this specific categorization too technical for translation, and thus to include it in the *Kitāb Pātanǧal*. As I argue in the following paragraphs, if he indeed used the *Pātañjalayogaśāstra* as his main source, al-Bīrūnī frequently omitted the treatment of technical Sāṅkhya-Yoga—or Indian—notions in his translation, as well as statements he may have regarded as redundant.

For instance, in KP 1–2, which appears to be a rough summary of PYŚ I.1–2, al-Bīrūnī does not include technical terms relating to absorption (*samādhi*), despite their mention in PYŚ I.1. Neither does he include the four aspects of cognitive absorption (*saṃprajñāta samādhi*), namely, reasoning (*vitarka*), pondering (*vicāra*), joy (*ānanda*) and individuality (*asmitā*).

However, another passage of the *Kitāb Pātanǧal* deals with the Yoga concept of absorption. In order to avoid redundancy and complex discussions on this meditative state, al-Bīrūnī probably decided to introduce the topic only in KP 7, which is his interpretation of PYŚ I.17–18. In this passage, he interprets "absorption" with the Arabic term *taṣawwur* (تصوّر), which literally means "imagining" or "conceiving."[48] This Arabic term refers to a technical concept that is used in at least two different contexts. Islamic philosophy, for instance employs the

44 Maas & Verdon 2018: 320–321.

45 Maas 2009: 267–268.

46 Maas & Verdon 2018: 321.

47 The final kind of mental state refers to the cessation of mental dispositions (*cittavṛttinirodha*).

48 KP 7, p. 172.11–13; Pines & Gelblum 1966: 318.

term in the sense of "conception" or "conceptualisation" with reference to the mental apprehension or comprehension of an object.[49] In a Sufi context, the term can refer to the visualization of one's master (*šeik̲*; شيخ) or to a contemplative technique and is thus translated either as "visualization" or "contemplation." It is not clear from which of the two ideological contexts al-Bīrūnī drew this term used by him for translating "absorption" (*samādhi*). However, the Yoga concept does not relate to the mental representation of an object, as the notion of "conception" or "conceptualisation" drawn from Islamic philosophy suggests, but to a mental state. The meaning carried by the Sufi concept translated into English as "contemplation," by contrast, would be rather close to the original idea in the supposed source-text.

In KP 7, al-Bīrūnī would not provide a literal translation of his Sanskrit source. His interpretation rather looks like an attempt to transfer the message of the possible source-text by using technical terms known to his readership when he describes two types of contemplation (تصوّر). The *Pātañjalayogaśāstra* defines the two types of absorption. The first type of absorption is cognitive (*saṃprajñāta*) and is described as having a support (*sālambana*), that is, as having an objective support, in PYŚ I.17. The second type, the non-cognitive (*asaṃprajñāta*) one, is said to be void of an object (*sa cārthaśūnyaḥ*) and without support (*nirālambana*), that is, without objective support, in PYŚ I.18. PYŚ I.2, for instance, describes it in the following terms: "[t]his (i.e., the just described state) is absorption without seed. 'Non-cognitive' means that in this [state], nothing is being cognized" (*sa nirbījaḥ samādhiḥ. na tatra kiṃcit saṃprajñāyata ity asaṃprajñātaḥ*).

In KP 7, al-Bīrūnī conveys a general distinction between two kinds of contemplation (تصوّر): the contemplation of the perceptible with matter (تصوّر المحسوس ذى المادة), and the contemplation of the intelligible free from matter (تصور المعقول المجرد عن المادة). Al-Bīrūnī thus not only simplifies the two Yoga concepts of cognitive and non-cognitive absorption, but also interprets them within his own intellectual and philosophical framework. He distinguishes two types of contemplation, just as the Sanskrit work does describe two types of absorption. In both cases, the difference lies in the object of the contemplation or absorption. The *Pātañjalayogaśāstra*, however, considers the second type of absorption to be independent from any object, whereas al-Bīrūnī conveys the idea that both types of contemplation focus on an object, which is of two kinds, that is, perceptible or intelligible.

49 See Goichon 1933: 63 and 1938: 191–193, Finianos 1975: 12 and 210 on Ibn Sīnā, and Forcada 2014 on al-Fārābī and Ibn Bājja.

The interpretation of the two types of absorption described in the *Pātañjalayogaśāstra* has been the subject of much discussion in contemporary scholarship.[50] Al-Bīrūnī's account only deals with these concepts in KP 7, which provides a very concise definition of the two kinds of contemplation and omits the four aforementioned aspects related to this concept. The *Kitāb Pātanǧal* thus reflects the difficulty of transmitting such sophisticated concepts. Lastly, al-Bīrūnī's rather simple rendering of these puzzling concepts into Arabic is not only a result of his desire to avoid complex explanations, but also reveals his own idiosyncratic understanding of these ideas.

Al-Bīrūnī adopts a similar attitude in regard to another meditative state described in the *Pātañjalayogaśāstra*, namely the contemplative state (*samāpatti*) (KP 23; PYŚ I.42–46). He extensively summarizes and rephrases the text of his presumed Sanskrit source in his interpretation. In this passage, the author of the *Pātañjalayogaśāstra* describes the contemplative state as well as its four subdivisions: with reasoning (*savitarka*), without reasoning (*nirvitarka*), with pondering (*savicāra*) and without pondering (*nirvicāra*). The subject matter of this passage is similar in the Arabic and Sanskrit versions, as they both deal with different types of mental apprehension of objects. However, al-Bīrūnī does not use any specific technical terminology that could be linked to Sanskrit terms. He rather describes four different stages corresponding to the aforementioned subdivisions which a person, in all likelihood an ascetic, can gradually reach.[51]

Thus, when dealing with the meditative states of Yoga, al-Bīrūnī probably decided to omit some technical notions, to paraphrase the content of his source and to use a terminology known to him and his readership, in this way altering the meaning of the source-text. Embedded in the Indian tradition of developing ideas about meditation, the author of the *Pātañjalayogaśāstra* elaborates upon complex theories regarding states of concentration or meditation and describes their characteristics and interrelations. Since al-Bīrūnī was not acquainted with such cultural framework, these specifically Indian concepts would have presented a true challenge to him in terms of interpretation.

50 Reference to Buddhist terminology is particularly helpful to interpret these two categories of meditative states. See, for instance, the discussion of the translation of PYŚ I.17–18 and of the Yoga notion of absorption in Maas 2009: 271–280. See also Bronkhorst 1993: 46–49. See also O'Brien-Kop 2021.

51 KP 23, p. 177.1–9; Pines & Gelblum 1966: 324–325. The manuscript is damaged at the place where al-Bīrūnī discusses the third stage of contemplative state (*samāpatti*). However, al-Bīrūnī quotes this passage in the *Taḥqīq mā li-l-Hind* (no. 5 in Table 7 above), where he lists four types of knowledge, the last of which leads to emancipation (*Taḥqīq* [1958], p. 53.1–9; Sachau, 1910: I/70).

An additional example of such omission is that of the five mental dispositions (*cittavṛtti*). PYŚ I.5 states that some mental dispositions are afflicted (*kliṣṭa*), while others are non-afflicted (*akliṣṭa*). The two notions are technical and specific to Yoga philosophy. Al-Bīrūnī may not have attempted to translate them, although in KP 5 he describes the five mental dispositions relatively faithfully. The *Kitāb Pātanǧal* does not deal with the complicated notion of the twofold "action, with impetus and without impetus" (*sopakramaṃ nirupakramaṃ ca karma*) expounded in PYŚ (1904) III.22, despite al-Bīrūnī's detailed treatment of the topics of PYŚ (1904) III.21–34 in KP 46.[52]

A different case of omission may have occurred in KP 46, in which al-Bīrūnī provides numerous Sanskrit terms transliterated into the Arabic alphabet. However, in the specific portion of the *Kitāb Pātanǧal* that deals with Mount Meru, the scholar does not provide the names of the mountains, kingdoms, rivers and seas located on its four sides. He explains that "there is no use in enumerating them because they are unknown, nor in naming them because the [names] are in the Indian [language]" (لا فايدة فى تعديدها لانها ليست بمعروفة و لا فى تسميتها لانها بالهندية).[53]

Omissions of technical terms or ideas also seem to occur in the *Kitāb Sānk*. For instance, in passage XVIII (See the Appendix) drawn from the *Kitāb Sānk* the type of knowledge leading to liberation, namely that of the twenty-five principles (*tattva*), is not specified, whereas it is described in most passages that correspond to this quotation in the commentaries on the *Sāṅkhyakārikā*.[54] Further, classical Sāṅkhya considers three categories of entities that constitute the world: the manifest (*vyakta*), the unmanifest (*avyakta*) and the knower (*jña*), that is, the conscious self (*puruṣa*). Every *tattva* belongs to either one of these categories. In the passage entitled "Six views on action and agent" (Appendix, passage I; see also Section 6.3.3), according to al-Bīrūnī's account, one opinion is that time is the agent, or the cause according to the Sanskrit texts.

When refuting this view, most of the commentators explain that time is included in one of these categories, namely the manifest, and thus cannot produce the world.[55] In this case, the corresponding passage of the *Kitāb Sānk* appears to be a rather literal translation of the Sanskrit original. The Arabic

52 See below no. 3 in Table 10.

53 KP 46, p. 187.6–7; Pines & Gelblum 1983: 261. See below p. 189.

54 See chapter 6, section 3.4.

55 GPBh, p. 55.3: *vyaktāvyaktapuruṣās trayaḥ padārthāḥ, tena kālo 'ntarbhūto 'sti. sa hi vyaktaḥ*. See also V1, p. 73.11–12; V2, p. 60.12–13, and MV, p. 56.16–17, which expose the same idea in a slightly different wording on kā. 61. The *Suvarṇasaptati departs from this idea and states that there is nothing beyond the three categories (the manifest, the unmanifest and the knower), and since time is not part of them, it does not exist (Takakusu 1904b: 1051).

text does not, however, contain the argument according to which time is not a separate category but is included in the three accepted categories of the unmanifest, the manifest and the knower. Al-Bīrūnī probably omitted this argument in his translation intentionally, as it is specifically connected to Sāṅkhya ideas and particularly challenging to transfer into Arabic.

The excerpt from the *Kitāb Sānk* which discusses different births according to one's behaviour (Appendix, passage XII) diverges from the corresponding passage in the source-text, namely kā. 39 and its commentary. Here, al-Bīrūnī may have deliberately omitted several specific notions. The Sanskrit commentaries on kā. 39 deal with the particulars (*viśeṣa*). These particulars are said to be threefold: the subtle (*sūkṣma*) body, the one born from mother and father (*mātāpitṛja*) and the gross one (*prabhūta*). Among these three, only the subtle body (*sūkṣmaśarīra*) is constant (*niyata*), and, according to some commentaries, transmigrates in the three worlds, divine, human and animal. In the corresponding passage of the *Kitāb Sānk*, al-Bīrūnī avoids a discussion of the notions of the particulars and transmigration. He rather explains that a man can either become a spiritual being or an animal, based on his behaviour. He probably foregoes the involved technical terms that are not only difficult to understand and explain but would also be foreign to his Muslim readership.

In addition to omitting technical concepts, al-Bīrūnī appears to have excluded from his translations what he considered redundant and regarded as unnecessary explanations. For instance, PYŚ I.7 in its *bhāṣya*-section quite extensively discusses direct perception (*pratyakṣa*), one of the three means of knowledge (*pramāṇa*) accepted by Sāṅkhya and Yoga. In contrast with the Sanskrit text, the *Kitāb Pātanǧal* does not contain any definition or explanation of direct perception but simply refers to apprehension by means of the five senses (بالحواسّ الخمس); it seems as if al-Bīrūnī considered this notion clear enough.[56]

Omission due to al-Bīrūnī's wish to avoid redundancy perhaps also occurred in some quotations drawn by him from the *Kitāb Sānk*. When al-Bīrūnī deals with the eight states (*bhāva*) of the intellect (*buddhi*), he only defines three of them, namely, righteousness (*dharma*), dispassion (*vairāgya*) and mastery (*aiśvarya*). However, he refers to the fourth, the state of knowledge (*jñāna*), in his discussion on dispassion and perhaps did not deem it necessary to deal with this concept separately, as is done in the corresponding Sanskrit passage. The remaining four states, which are the opposite notions of these

56 The question of different means of knowledge was also an object of discussion among Muslim thinkers (Touati 2000: 16–18, 25–35 and 123–128).

four states, namely, unrighteousness (*adharma*), ignorance (*ajñāna*), passion (*rāga*) and lack of mastery (*anaiśvarya*), are not taken into consideration by al-Bīrūnī, although the Sanskrit source he consulted most probably described them.[57]

As already underlined, however, since only some quotations of the *Kitāb Sānk* scattered in the *Taḥqīq mā li-l-Hind* are extant, it is difficult to ascertain whether al-Bīrūnī omitted the treatment of the remaining states of the intellect in the quotations only or whether this was also the case in his translation of the whole Sāṅkhya work. Nevertheless, in general, and in the case of the *Kitāb Pātanǧal* at least, omissions constituted a common translation strategy on the part of al-Bīrūnī.

4.4.2 *Substitution due to al-Bīrūnī's Cultural Background*

The next substantial transformation that Ivir expounds is the translational strategy of substitution. Ivir explains it as the process of substituting a source-concept with another concept belonging to the target-culture. This approach can be adopted when the source- and target-concepts involve "a partial overlap rather than a clear-cut presence *vs.* absence of a particular element of culture."[58] Substitution enables the translator to transmit a concept in words that are not completely unknown to his audience and that reduce the foreignness of the source-concept for the target-culture. The primary drawback of substitution, however, is that significant discrepancies between the two concepts may be overlooked. This strategy finds its way throughout al-Bīrūnī's works.

As seen in Section 2.4, al-Bīrūnī defines the unmanifest (*avyakta*) original cause (*prakṛti*) with the term *hayūlā* (هيولى), an Arabic word derived from the Greek *hule* which refers to primary matter but does not equate *prakṛti* of Sāṅkhya. In this way, however, al-Bīrūnī relates the Indian concept of *avyakta* to a concept originally described in Aristotle's *Metaphysics* and further elaborated by Falsafa.[59] Further, Sāṅkhya connects the principles of unmanifest (*avyakta*) and manifest (*vyakta*) with its specific notion of causality so as to explain the creation of, and change in, the phenomenal world. In order to transfer the notion of cause and effect—the *satkārya* theory of Sāṅkhya—al-Bīrūnī makes use of the Aristotelian terminology of potentiality and actuality.[60] Des-

57 See below Section 6.3.4.
58 Ivir 1987: 41.
59 Book Z (VII), parts 10–11 (Duminil & Jaulin 2008: 263). On a definition of *hayula* in the Falsafa context, see Goichon 1938: 413–414.
60 Al-Bīrūnī makes use of the same terminology when he explains that the conscious self

pite the conceptual discrepancies between the two theories, metaphysics—to be understood here as the description of what is beyond the perceptible world and how it changes—covers both Indian and Greek concepts.

Another interesting example of substitution is al-Bīrūnī's use of the Arabic term "faculty" or "force" (sg. قوة; pl. قوى).[61] Here the scholar adopts the same term for two different key-concepts of Yoga-Sāṅkhya, one related to psychology and the other to metaphysics. As aforementioned, he translates the concept of mental disposition (*cittavṛtti*) with the Arabic expression "faculties of the soul" (قوى النفس). The *Pātañjalayogaśāstra* prescribes the cessation or suppression of these dispositions with the goal of reaching a state close to liberation from karmic retribution and cycle of rebirths. Al-Bīrūnī explains that the ascetic must compress these faculties within himself and prevent them from spreading out (KP 2: قبص الانسإن اليه قوى نفسه و منها عن النتشار); the ascetic then reaches an intermediary state between attachment to the material world and complete liberation.

The faculties of the soul are treated in Greek and Islamic philosophies. Aristotle elaborated his own theory on the hierarchy of the soul's faculties, according to which plants would have only one or two faculties, animals may have more, and human beings would possess all of them. Ibn Sīnā further developed his position on the question.[62] Al-Bīrūnī uses the same terms, "faculty" or "force" (قوى), to translate the technical Sāṅkhya concept of the three constituents (*guṇa*). He refers to them by using the Arabic expression "three (primary) forces" (القوى الثلاث الاول or قوى ثلاث). According to Sāṅkhya-Yoga, each of the three constituents not only possesses its own qualities, but also dominates its own world, *sattva* the divine sphere, *rajas* the human sphere and *tamas* the animal and plant sphere.

Perhaps here again, al-Bīrūnī used the specific Arabic term "faculties" in reference to the different faculties of the soul that were conceived by the Greeks and developed in Islamic philosophy (i.e., vegetal, animal and human). However, in contrast with the definition of the faculties in Greek and Arabic sources, the Sāṅkhya constituents are not found only in one element, such as the soul in Aristotle's philosophy, but are present in twenty-four of the Sāṅkhya principles (*tattva*). The conscious self (*puruṣa*), which al-Bīrūnī renders as soul, is described as being devoid of the constituents (*nirguṇa*).[63]

(*puruṣa*) is ignorant in actuality and intelligent in potentiality. See above p. 85 and Verdon 2019b: 6–8.

61 For the polysemous Arabic expression "forces" or "faculties," see Boer & Arnaldez 2012.

62 Hasse 2010: 305–310.

63 This is mentioned in kā. 60.

Al-Bīrūnī thus uses terminology that originates from Greek philosophy to transmit two different Indian concepts. In each case, it is possible to observe shared attributes between the two original concepts. The overlaps, however, remain partial, and the fact that al-Bīrūnī used the same term for these two Sāṅkhya-Yoga ideas indicates that he utilized this term as a heuristic tool, rather than as a comparative tool.

In the domain of theology, al-Bīrūnī translates the Sanskrit word *īśvara* by the Arabic *allāh* (الله), both referring to God. However, the notion of God does not play the same role, or has the same significance, in the two respective cultural contexts. Modern scholarship has not yet thoroughly examined al-Bīrūnī's interpretation of the concept of God in the *Kitāb Pātanǧal* and *Kitāb Sānk*. Carl Edward Sachau, Junjiro Takakusu, Richard Garbe and Jean Filliozat remain silent on this subject. Surendra Nath Dasgupta describes al-Bīrūnī's conception of God in the *Kitāb Pātanǧal* and observes that God has become "the only object of meditation and absorption in him is the goal."[64] This observation led Dasgupta to assume that al-Bīrūnī's Sanskrit source was influenced by later theistic developments of Yoga philosophy.[65]

Pines and Gelblum, in addition to Dasgupta's assumption, consider that al-Bīrūnī may have been conditioned by his own socio-cultural background when translating Sanskrit texts into Arabic. They do not, however, settle between the two hypotheses, that is, whether al-Bīrūnī's idiosyncratic interpretation of the notion of God is due to later theistic developments of Yoga philosophy known to him or to his socio-cultural background.[66] Maas mentions the influence of al-Bīrūnī's background on his translation, but does not offer any further analysis.[67] Kozah, who examined extracts of the *Kitāb Pātanǧal* found in the *Taḥqīq mā li-l-Hind*, notes that God, as al-Bīrūnī describes Him there, bears a strong resemblance to the Islamic God and that the terminology is typically Islamic.[68] Without dealing at length with the concept of God, Maas and Verdon claim that the word *allāh* (الله) used for translating *īśvara* operates as a substitution according to Ivir's model. They point out some of the striking common points and discrepancies between the two associated concepts:

> Both concepts refer to the idea of a supreme being. In the case of Pātañjala Yoga, this supreme being is a special kind of Subject (*puruṣa*) that

64 Dasgupta 1930: 62.
65 Dasgupta 1930: 60–62.
66 Pines & Gelblum 1966: 305.
67 Maas 2013: 59.
68 Kozah 2016: 41–55.

> mainly serves as an object of meditation and whose role in the world is rather limited. In contrast, on an ontological level, Allah is unique. Allah is the God of judgment and retribution who determines the post-mortem fate of all human beings. In contradistinction to this, Yoga philosophy and religion takes the quasi mechanism of karmic processes to determine the welfare or otherwise of human beings in their next existences.[69]

The concepts of Allah and Īśvara certainly share some common features, while at the same time they have their own specific characteristics, as they originate from two distinct socio-cultural contexts. Two passages refer to God in the *Kitāb Sānk* and the *Kitāb Pātanǧal*. The one found in the *Kitāb Sānk* (Appendix, passage I) relates to kā. 61 and a commentary.[70] It faithfully transfers the viewpoint of classical Sāṅkhya that Īśvara (Allah) is not the cause of the world, but that the original cause (*prakṛti*) fulfils this function. In this passage, however, al-Bīrūnī offers no further definition of God.[71]

Therefore, the present analysis focuses on the second passage, which corresponds to KP 11–18 and PYŚ I.23–I.29.[72] A comparison between the *Kitāb Pātanǧal* and its possible source, the *Pātañjalayogaśāstra*, shows that al-Bīrūnī would have profoundly reworked it in terms of structure and content. The Arabic passage begins with KP 11 with the question whether there is another way than habituation (تعويد)—that is training oneself on preventing the faculties of the soul from spreading out—[73] and asceticism (زهد) in order to reach liberation (خلاص). Similarly, PYŚ I.23 is introduced by the question whether or not there is another means than repeated practice (*abhyāsa*)—that is of seeing the true nature of the conscious self as pure and different from that of the original cause—[74] and dispassion (*vairāgya*), for attaining the state of absorption. The third means is devotion (عبادة) to Allah in the *Kitāb Pātanǧal*, which is said

---

69 Maas & Verdon 2018: 325–326.

70 *Taḥqīq* (1958), pp. 22.12–23.5; Sachau 1910: I/30–31. See below Section 6.3.3.

71 The observations made in the present chapter show that a thorough and well-grounded study of al-Bīrūnī's interpretation of several Indian deities (*īśvara, mahādeva, nārāyaṇa,* etc.) in the *Taḥqīq mā li-l-Hind* would be an important further step toward understanding the Indian society he encountered.

72 *Taḥqīq* (1958), pp. 20.9–21.16; Sachau 1910: I/27–29. KP 12 to 18, with the exception of KP 13, are rephrased in the *Taḥqīq mā li-l-Hind*. See Table 8 of the present book for understanding the correspondences between these specific sets of question and answer of the *Kitāb Pātanǧal* and the *Pātañjalayogaśāstra*.

73 See KP 6 (Ritter 1956: 171.15–20; Pines & Gelblum 1966: 316–317).

74 PYŚ I.12, p. 22.6: *vivekadarśanābhyāsena*; and I.16, p. 25.2: *puruṣadarśanābhyāsāt*.

to lead to liberation, and profound contemplation on God (*īśvarapraṇidhāna*) in the *Pātañjalayogaśāstra* which prescribes it in order to obtain absorption (*samādhi*).

KP 6 and PYŚ I.12–16 had already discussed the two first means habituation, or repeated practice, and asceticism, or dispassion.[75] Al-Bīrūnī, however, deems it necessary to specify them again in KP 11, whereas the author of the *Pātañjalayogaśāstra* only implicitly refers to the two means in this place. Further, the *Pātañjalayogaśāstra* subsequently explains that "the attainment of absorption and the fruit of absorption become extremely near to the yogi."[76] Thus, the state that one attains by this third means would not be exactly that of final liberation. Al-Bīrūnī does not, however, transmit this nuance.

KP 12 provides a general description of Allah as possessing eternity (ازلية) and oneness (وحدانية), two concepts inherent in the Islamic conception of Allah. The *Pātañjalayogaśāstra* also expresses Īśvara's transcendence of time. The *bhāṣya*-part of PYŚ I.24 states that "Īśvara's connection to the [three kinds of bondage][77] is not past, [and] not future"[78] and concludes with "[h]e, however, is certainly always liberated, certainly always Īśvara."[79] For the *Pātañjalayogaśāstra*, Īśvara is beyond the limits of time. Thus, as the transcendence of time is a characteristic shared by Īśvara and Allah, al-Bīrūnī equally ascribes eternity to the God described by him in the *Kitāb Pātanǧal*.

Furthermore, al-Bīrūnī perhaps interprets the notion of Īśvara "being a special kind of *puruṣa*" (*puruṣaviśeṣa īśvaraḥ*) in *sūtra* I.24 as referring to God's oneness (وحدانية). The Yoga idea that Īśvara is a special kind of conscious self is a technical one and alien to Islamic culture. Al-Bīrūnī, thus, may have interpreted it in the sense of God's being unique and, therefore, he attributed this property to God (Allah) in KP 12, without, however, reflecting the exact description of Īśvara found here in the *Pātañjalayogaśāstra*. Further, Allah is conceived as being unique to such an extent that one cannot compare Him in any way to the human soul (i.e., *puruṣa*). Therefore, the technical nature of this aspect of Īśvara's description and the absence of such a concept in Islamic thought, together with the contradiction with Islamic belief which it implies, could explain al-Bīrūnī's free interpretation here.

---

75 On the three paths leading to emancipation, see also Kozah 2016: 104–106.

76 PYŚ I.23, p. 35.5: *yogina āsannataraḥ samādhilābhaḥ samādhiphalaṃ ceti.*

77 The three kinds of bondage are not explained in the *Pātañjalayogaśāstra* itself. The *Vivaraṇa* briefly defines them as related to *prakṛti*, to *vikṛti* and to sacrificial fee (Harimoto 2014 p. 49.3: *prākṛtavaikṛtadākṣiṇāni*; see id. 91, n. 102).

78 PYŚ, p. 36.7: *īśvarasya tatsaṃbandho na bhūtaḥ, na bhāvī.*

79 PYŚ, p. 37.9: *sa tu sadaiva muktaḥ sadaiveśvara iti.*

The second part of KP 12 is devoted to God's knowledge. It describes God in the following terms: "knowing eternally by nature" (العالم بذاته سرمداً), and "to whom ignorance does not belong in any way, at any time or in any state" (ليس الجهل بمتّجه عليه في وقت ما او حال). In terms of content, it corresponds to the first part of PYŚ I.25. The *sūtra* of this passage states that "in [Him] the seed of the omniscient is unsurpassed" (*tatra niratiśayaṃ sarvajñabījam*). The *bhāṣya*-part ad loc. qualifies Him as omniscient (*sarvajña*). It appears that al-Bīrūnī left out the obscure and technical idea of the "seed of the omniscient" (*sarvajñabīja*) but transferred the idea of God's absolute and eternal knowledge to this passage.

PYŚ I.24 is introduced by the question of how Īśvara differs from the conscious self and from the primary matter. Because Īśvara is a special kind of *puruṣa* untouched by deposits of afflictions (*kleśa*), of karma (*karman*) and its ripening (*vipāka*), He has always been liberated and will always be. In contrast, KP 13 asks how God differs from the one who is liberated.[80] The response discusses the difference between God and the one who is liberated (متخلّص), which is principally grounded in the fact that God is eternally liberated and does not depend upon time.

Thus, even if the object from which Īśvara/Allah differs in the Arabic and Sanskrit works, the implication, that is, eternal liberation, is same in both works. Furthermore, in KP 13, al-Bīrūnī does not refer to the notions of affliction, of karma and its ripening that are however dealt with in PYŚ I.24, but he evokes them in KP 14. He may have rephrased the idea that "Īśvara is untouched by deposits of afflictions, karma and [its] ripening"[81] in the *Kitāb Pātanǧal* in the following terms: "He is the knowledge free from pollution by heedlessness and ignorance" (هو العلم الخالص عن دنس السهو الجهل). Al-Bīrūnī would have focused on the absence of afflictions in God in this part of his translation. Ignorance reflects here the first affliction, as PYŚ I.24 states that "afflictions start with ignorance."[82]

PYŚ I.27 explains that the syllable *aum* (*praṇava*) is the signifier of Īśvara (*vācaka*) and Īśvara that which is signified (*vācya*) by it.[83] The *Kitāb Pātanǧal*, on the other hand, is very concise in this place and never mentions the syllable

80 Pines & Gelblum 1966: 320, n. 170.

81 PYŚ, sū. I.24, p. 35.2: *kleśakarmavipākāśayair aparāmṛṣṭaḥ* [...] *īśvaraḥ*.

82 PYŚ, p. 35.3: *avidyādayaḥ kleśāḥ*. See further the description of the afflictions in PYŚ (1904) II.3–12 and KP 26–28. Maas and Verdon (2018: 296–297) point out the fact that ignorance appears as the first and primary affliction in both the *Pātañjalayogaśāstra* and the *Kitāb Pātanǧal*.

83 On the syllable *aum* see Feuerstein 1987: 392–393 and Maas 2009: 277–278, as well as Gerety 2015: 367–368 and 377 on its specific use in the *Pātañjalayogaśāstra*.

aum nor implicitly refers to it.[84] Neither does it contain any mention or discussion whatsoever on the signifier and the signified of *aum* seen in *Pātañjalayogaśāstra* I.27. KP 15, which corresponds to PYŚ I.27, however, ascribes speech (كلام) to God, which the corresponding Sanskrit passage does not ascribe to Īśvara.

Thus, Pines and Gelblum note that al-Bīrūnī perhaps understood the Sanskrit word *vācaka* "as referring to speech as an attribute of God [...] and not to the sacred syllable 'Om' (*praṇava* in the *sūtra*) as expressive of God."[85] Such misunderstanding is possible. It is also possible that al-Bīrūnī had recourse to a readjustment of ideas, in the same way as he did in his interpretation of Īśvara's being a special kind of *puruṣa* as referring to God's oneness. The connection between the syllable *aum* and God being very alien to al-Bīrūnī's intellectual and religious framework, he may have simply ascribed to God another attribute, namely speech (كلام), and thus avoided an explanation of this specific technical concept.

As for the last three sets of question and answer in the *Kitāb Pātanǧal* with regard to God, al-Bīrūnī appears to have simplified, transformed and reorganized their content in comparison with the corresponding passage of his possible source-text the *Pātañjalayogaśāstra*. KP 16–17 focus on God's knowledge and its transmission and can be likened to some sections of PYŚ I.25–26. KP 18 describes the way by which God, who is imperceptible by the senses, can be worshipped. This roughly corresponds to the content of PYŚ I.28–29. Lastly, KP 15 reflects to some extent the content of PYŚ I.27.

In view of the above it is pertinent to reconsider the question raised by Dasgupta, Pines and Gelblum whether al-Bīrūnī's rendering is an Islamization of the concept of God, or whether he based his translation on a later theistic Yoga text. Moreover, the fact that al-Bīrūnī frequently refers to an elite class of Indians who interacted with him and who were able to conceptualize a unique and abstract God may also account for his (mono)theistic rendering of Indian religion.

However, the foregoing analysis shows two different ways in which al-Bīrūnī transferred Indian thought to his readership. In some cases, such as that of God and His relation to time, al-Bīrūnī's interpretation mirrors the original Yoga conceptions, because he saw in this characteristic a common point with his conception of Allah. In other cases, however, such as that of the speech attributed to the God—and of Īśvara's relation to the sacred syllable *aum*—he modi-

84 Al-Bīrūnī deals with the syllable *aum* in the *Taḥqīq mā li-l-Hind* (*Taḥqīq* [1958], pp. 56 and 135; Sachau 1910: I/75 and 173).

85 Pines & Gelblum 1966: 320, n. 178.

fied the content in such a way that two unconnected ideas were actually adjusted to each other. By doing so, al-Bīrūnī in a sense avoided that Indian unorthodox religious concepts would enter their way in his translation and contradict Islamic religion.[86] Thus, I argue that the discrepancies highlighted above are due to a great part to him adjusting the concept of Īśvara to that of Allah.

Moreover, Īśvara does not have much impact on the world, in contrast with Allah who is considered to be the creator of the world and the final judge regarding the destiny of men in the afterlife. Sāṅkhya does not consider Īśvara the cause of the world, nor does present *Pātañjalayogaśāstra* anything that would suggest such an idea.[87] As both Sāṅkhya and Yoga regard the original cause (*prakṛti*) the creative force, there is not much room for a creator deity. The *Kitāb Pātanǧal*, like the *Pātañjalayogaśāstra*, remains silent on a creative role of Allah or Īśvara. Given the above observation that al-Bīrūnī attempted to adjust the concept of Īśvara to that of Allah, it is likely that he would have seized this additional opportunity to parallel Īśvara to Allah if his source had attributed the creation of the world to the former.

Thus, the portion of the Arabic text under discussion, that is, KP 11–18, relates to PYŚ 1.23–28, even though al-Bīrūnī significantly reorganized the content of his assumed source-text in this passage on God. Accepting this connection between the two portions of texts made it possible to select some main characteristics attributed to Īśvara in the *Pātañjalayogaśāstra* and compare them with al-Bīrūnī's translation. In this way, I highlighted that some of the involved issues overlap in both Islamic and Brahminical traditions, such as the notions of a divine sphere, eternity and knowledge. In case of other not overlapping concepts, al-Bīrūnī misunderstood, reinterpreted or simplified them,[88] leading to interpretations which mostly reflect his understanding of God grounded in his religious background.

4.4.3 *Addition and Definition from Pre-existing Knowledge*

Al-Bīrūnī also appears to have employed addition and definition as translational strategies.[89] Although they belong to two different categories in the

86 In a slightly different context Kozah (2016: 78) makes a similar observation.

87 Bronkhorst 1981: 316.

88 An example of misunderstanding appears in KP 28, corresponding to PYŚ (1904) II.12, in al-Bīrūnī's translation of the Sanskrit expression *devānām indraḥ*, which in this context means "the chief of the gods," and not "Indra among the gods" (Pines & Gelblum 1977: 537, n. 63; Maas & Verdon 2018: 300).

89 Pines and Gelblum already noticed that al-Bīrūnī had to define and specify notions that were clear for Indian thinkers but obscure to a foreign audience (Pines & Gelblum 1966: 308).

model elaborated by Ivir, I deal with them together, considering definition a a type of addition. Furthermore, as I show below, both also originates from the same source of information, that is the encyclopaedic knowledge of al-Bīrūnī which was in most cases based on textual or oral sources of Indian origin.

For instance, KP 46, most probably a rendering PYŚ III.29, has a passage on the transformation of food in the human body that is absent from the *Pātañjalayogaśāstra* and from other commentaries. Maas and Verdon discuss this example in detail.[90] As the two authors argued, al-Bīrūnī in all likelihood drew this addition from some Indic sources—written and/or oral—other than a Yoga work that he had at his disposal. In general, it is natural to assume that al-Bīrūnī in his additions and definitions made use of elements drawn from the Sanskrit works available to him and from the information received from the Indian thinkers he had met.

In KP 5, which translates PYŚ I.7, al-Bīrūnī appears to have added an analogy in order to illustrate apprehension (ادراك) by oral tradition (بالسماع), which corresponds to authoritative tradition (*āgama*) as a means of knowledge (*pramāṇa*). In contrast with the *Pātañjalayogaśāstra* and with all its known Sanskrit commentaries, the *Kitāb Pātanǧal* is the only work that provides an example for this notion. It reads: "Just as our knowledge that the city of Kanauj is on the bank of the Gaṅgā River, this [knowledge] results from [oral] report but stands for its direct apprehension by eyesight" (كمعرفتنا ان بلد كنوج على شط نهر كنك فانها حاصلة بالخبر و قايمة مقام ادرا كه لذلك بالبصر).[91] Interestingly, the transmission of knowledge by way of oral tradition (بالسماع) and by way of verbal communication in the form of authoritative tradition (*āgama*) is fundamental in both the Islamic and the Indian culture, respectively. The provenance of the above example is, however, uncertain; it may have been orally communicated to al-Bīrūnī or actually created by himself.[92]

The analysis of the excerpts from the *Kitāb Sānk* also indicates that al-Bīrūnī may have resorted to addition when dealing with his source-text. When he enumerates the different views on action and agent, the view according to which "action is nothing but the recompense for a preceding deed" (ليس الفعل سوى المكافاة على العمل المتقدّم)[93] is absent from all available Sanskrit sources under consideration in the present book.[94] If al-Bīrūnī added this opinion to the other

90 Maas & Verdon 2018: 303–306; see below pp. 171–172.
91 KP 5, p. 171.4–5; Pines & Gelblum 1966: 315.
92 See above, p. 81.
93 *Taḥqīq* (1958), p. 23.1–2; Sachau 1910: I/31.
94 See below Section 6.3.3.

views—a hypothesis that cannot be ascertained—, he certainly drew the relevant information from his knowledge of Indian culture.

Additions can also be detected when al-Bīrūnī narrates an analogy to illustrate the four hierarchical states of the intellect (*buddhi*): error (*viparyaya*), inability (*aśakti*), satisfaction (*tuṣṭi*) and accomplishment (*siddhi*). The analogy stages four disciples who ascertain one after each other the identity of a far-away indiscernible object upon their master's request. When the fourth disciple, who illustrates the accomplished state of knowledge, is about to ascertain the identity of the object, al-Bīrūnī's account of the disciple's reflections is very detailed when compared with that of the Sanskrit commentaries ad loc.[95] The reasons why al-Bīrūnī felt the need to further explain this portion of text are unknown, as is the provenance of this complementary information. However, these few examples show that al-Bīrūnī chiefly formulated supplementary definitions and other explanatory additions based on his intellectual background which includes his pre-existing knowledge of Indian thought.

4.5 Concluding Remarks

This chapter revealed that although significant discrepancies exist between al-Bīrūnī's translations and his probable sources, these differences do not mean that he used Sanskrit works unknown to us when composing the *Kitāb Sānk* and the *Kitāb Pātanǧal*. It equally showed that the *Kitāb Sānk* must have been subject to similar modifications as the *Kitāb Pātanǧal*.

Furthermore, this chapter directed our attention to the necessity of providing an analysis of al-Bīrūnī's translations from a perspective different from a mere literal comparison between the Arabic and Sanskrit texts of his potential sources. It also pointed to the need to look at al-Bīrūnī's work through the lens of the possible influence of his intellectual background, be it that of his own culture or of his knowledge of Indian culture. I thus argue that his interpretations are mostly indebted to this intellectual framework and that his adaptations strongly reflect this debt.

In this respect, Ivir's concept of translational strategies has been particularly helpful. I posit that al-Bīrūnī made abundant use of omission, substitution and addition (including definition), especially when great cultural gaps needed to be bridged—in other words, when the notions to transmit are very technical and specific to Sāṅkhya-Yoga or Indian culture. Al-Bīrūnī's treatment of his

95 See below Section 6.3.4.

sources thus constitutes a rather clever manner of interpreting and transferring these Indian ideas. Furthermore, considering his desire to transfer Indian ideas to a Muslim readership and to facilitate the exchange between these two cultural spheres, such transformations in fact constitute a relatively natural process.

When formally transforming his Sanskrit source-texts on account of his didactic intentions and own logic, al-Bīrūnī obviously omitted parts of them in his translation, as a result of his desire to avoid technical content or repetition. He also took recourse to substitution. For this latter translational strategy, al-Bīrūnī employed his knowledge of the world, specifically that of his own socio-cultural and intellectual background, be it in the domain of religion, philosophy or mysticism.

His idiosyncratic interpretations also played a role in some of his substitutions. In any case, it appears unlikely that his Indian informants may have suggested substitutions. On the other hand, additions were either due to his own creativity, to specific Sanskrit sources or to the Indian tradition as such. In addition, these observations and the complexity both of the source language and the concepts used in the various domains—astronomical, philosophical, mythological, etc.—confirm the previous remarks that al-Bīrūnī must have received assistance from Indian thinkers.

Interestingly, al-Bīrūnī's interpretation of some of Īśvara's characteristics indicate that he transformed the source-text in his translation to the extent of a distortion of the original concepts. When no common points of reference existed between notions of his own intellectual culture and specific Indian notions, such as in the case of the nature and function of the syllable *aum* and the description of Īśvara as a special kind of *puruṣa*, I suggest that he has filled the gap and created his own descriptions of the God of the *Kitāb Pātanǧal*. Lastly, al-Bīrūnī's modifications meant to make possible, or at least to ease, the understanding of the Sanskrit originals for himself and his readership.

Thus, I have put forward in the present chapter preliminary analysis of al-Bīrūnī's interpretative choices with the aim of laying the groundwork for the detailed studies of Chapters 5 and 6. The subsequent discussion confirm these observations. I argue that keeping in mind such approach enables us to narrow down the possibilities regarding the Sanskrit originals of the *Kitāb Pātanǧal* and the *Kitāb Sānk*.

CHAPTER 5

On the Kitāb Pātanǧal and Its Sources

5.1 Scholarship Review

In Chapter 4 of this book, I have investigated the reasons why former attempts to identify al-Bīrūnī's Sanskrit sources were unsuccessful despite the existence of Ritter's edition of the complete manuscript of the *Kitāb Pātanǧal*. Based on a new methodological framework, in the present chapter, I further examine the relationship between passages of the *Kitāb Pātanǧal* and possible corresponding passages in Sanskrit sources. Maas and Verdon have thoroughly assessed previous scholarly arguments regarding the identification of al-Bīrūnī's source;[1] Sections 5.1.1 and 5.1.2 recapitulate and develop the findings of their study.

Thanks to textual evidence presented in this chapter, it is possible to definitively exclude the *Vivaraṇa*, the *Tattvavaiśāradī* and the *Rājamārtaṇḍa* as the Sanskrit sources of al-Bīrūnī's translation. I thus corroborate Maas' and Verdon's hypothesis that the *Kitāb Pātanǧal* was based on the *Pātañjalayogaśāstra*, or a text very much like it. Lastly, I address, whenever possible, the underlying causes for the discrepancies between the *Kitāb Pātanǧal* and the *Pātañjalayogaśāstra*.

5.1.1 *Before the Edition of the Manuscript*

This section illustrates the problems faced by scholars, such as Carl Edward Sachau and others, who researched the question of the Sanskrit source of the *Kitāb Pātanǧal* before the mid-20th century CE, that is, before Ritter edited the complete manuscript of the Arabic translation. In this context, Maas and Verdon have already pointed out another philological reason why Sachau was unsuccessful in his endeavour to identify al-Bīrūnī's source: Sachau compared and contrasted the extracts of the *Kitāb Pātanǧal* scattered throughout the *Taḥqīq mā li-l-Hind* with the English translation of the *Yogasūtra* and the *Rājamārtaṇḍa*, the only work related to the Yoga philosophical tradition available in English at the time.[2] The following paragraphs provide specific examples of Sachau's difficulties.

1 Maas & Verdon 2018: 291–315.

2 Maas & Verdon 2018: 291–293. Sachau apparently used the translation and edition by Rājendralāl Mitra (1883).

In the *Taḥqīq mā li-l-Hind*, al-Bīrūnī refers to the *Kitāb Pātanǧal* in order to provide a description of four paths leading to liberation (خلاص). In this passage, he also includes references to the *Kitāb Gītā* (*Bhagavadgītā*) and the *Viṣṇudharma* (*Viṣṇudharmottarapurāṇa*).[3] For Sachau, this description does not find any parallel in Yoga literature. The passage found in the *Taḥqīq mā li-l-Hind* and discussed by Sachau, however, paraphrases and intermingles the contents of the original KP 6, KP 11 and KP 57. This combination provides an additional reason for why Sachau did not find any connection with the Yoga texts he used in this place.[4]

Al-Bīrūnī characterizes the first path as "practical, that is, habituation" (عملى و هو التعويد),[5] while he describes the second as "intellectual, that is, a mental asceticism" (عقلى هو الزهد الفكرى). KP 6 elaborates on the two paths, where al-Bīrūnī defines the first one, practical as habituation (تعويد), and the second one, intellectual as renunciation (زهد). The third path mentioned in the *Taḥqīq mā li-l-Hind* is devotion (عبادة), which is referred to in KP 11. As seen in Chapter 4 of the present book, the first two paths can be identified in the *Pātañjalayogaśāstra* (I.12–16) as repeated practice (*abhyāsa*) and dispassion (*vairāgya*), which are two interconnected means leading to isolation (*kaivalya*), while the third path is profound contemplation on God (*īśvarapraṇidhāna*), which results in the cessation of mental dispositions and which PYŚ I.23 and PYŚ II.45 discuss.[6]

The fourth path which al-Bīrūnī describes in the *Taḥqīq mā li-l-Hind* is *rasāyan* (رساين), which is the Arabic transliteration of the Sanskrit *rasāyana*. In Sanskrit, this term refers to a substance and to a science: 1) a rejuvenating or restorative elixir, and 2) the branch of Indian medicine aiming at reaching immortality for human bodies thanks to such elixirs. Al-Bīrūnī defines it as the "procedures involving drugs and resembling Alchemy in the obtainment of what is by nature impossible" (هى تدابير بأدوية تجرى مجرى الكيمياء فى تحصيل الممتنعات بها).[7] He transliterates this Sanskrit term into Arabic in KP 57, which corresponds to PYŚ (1904) IV.1. Furthermore, in both the *Kitāb Pātanǧal* and the *Pātañjalayo-*

---

3 *Taḥqīq* (1958), pp. 58.5–61.7; Sachau 1910: I/76–80.

4 KP 6, 11, 57, pp. 171.14–172.10, 173.8–11 and 193.2–10; Pines & Gelblum: 1966: 316–319 and 1989: 267.

5 See p. 151, above, in Chapter 4, on the type of habituation which is prescribed by the *Kitāb Pātanǧal*.

6 PYŚ (1904), pp. 17–20, 25, and 110; Woods 1914: 34–38, 49 and 190. Devotion is also broached in KP 41 which refers to the content of PYŚ (1904) II.45. On non-theistic and theistic yogic concentrations, see Maas 2009.

7 *Taḥqīq* (1958), p. 61.5–6; Sachau 1910: I/80.

gaśāstra, five causes of obtaining supernatural powers (*siddhi*) are mentioned. According to *sūtra* IV.1, (medicinal) plants (*oṣadhi*) are one of them. The *Pātañjalayogaśāstra* explains that the supernatural power that originates from plants comes about first of all through *rasāyana*.[8] It is clear from this Sanskrit passage that plants, or *rasāyana* drugs or therapy, do not lead to isolation, but to a specific supernatural power.

In the corresponding passage of the *Kitāb Pātanǧal*, al-Bīrūnī specifies that *rasāyan* is one way to reach *siddha-hood* (زهادة الزاهد).[9] In this passage, he does not, however, mention the concept of liberation, or the yogic isolation, at all. It appears that he only equates *rasāyan* with a way to reach liberation in the *Taḥqīq mā li-l-Hind*. Despite this substantial discrepancy, it is possible, via the *Kitāb Pātanǧal*, to connect al-Bīrūnī's description of the fourth path to liberation in the *Taḥqīq mā li-l-Hind* to a specific portion of the *Pātañjalayogaśāstra*.[10] Due to these differences, Sachau could not connect these passages of the *Kitāb Pātanǧal* to their respective Sanskrit source-texts.

However, Sachau also establishes parallels between the Sanskrit works available to him and the *Kitāb Pātanǧal*.[11] He compares a quotation from the *Kitāb Pātanǧal* in the *Taḥqīq mā li-l-Hind* to the last *sūtra* of the *Pātañjalayogaśāstra*. The *sūtra* reads: "Isolation is the return to their original state of the constituents which are void of the conscious self's goal, or [it is] the power of consciousness (i.e., the conscious self) abiding in its own form."[12] The quotation from the *Kitāb Pātanǧal* found in the *Taḥqīq mā li-l-Hind* reads:

> Therefore, the questioner asks about the nature of liberation at the end of the *Kitāb Pātanǧal*. The respondent says: If you wish, you may say it is the cessation (تعطّل) of the three forces (القوى الثلث), and their return to the source from which they came; or if you wish, you may say it is the return of the knowing soul (نفس) to its own nature.[13]

8 See also Pines & Gelblum 1989: 283–284, n. 24.

9 I follow here the translation by Pines and Gelblum (1983: 285, n. 172, 1989: 267 and 282, n. 13).

10 In this context, it is worth recalling that al-Bīrūnī does not always differentiate between the ultimate state of liberation and the mental states that lead to it, although they are clearly distinguished in the *Pātañjalayogaśāstra*, as illustrated by his interpretation in KP 6, KP 11 and KP 57.

11 Sachau 1910: II/287.

12 PYŚ (1904) sū. IV.34, p. 207.2–3: *puruṣārthaśūnyānāṃ guṇānāṃ pratiprasavaḥ kaivalyaṃ svarūpapratiṣṭhā vā citiśaktir iti*. See Woods 1914: 347.

13 *Taḥqīq* (1958), p. 61.16–19; Sachau 1910: I/81.

This passage corresponds to KP 78:

> What is liberation? If you wish, you may say it is the cessation (تعطّل) of the action of the three primary forces (القوى الثلث الاول) and their return to the source from which they came; or if you wish, you may say it is the return of the soul (نفس) to its own nature.[14]

Sachau's identification of this passage with PYŚ (1904) IV.34 is correct. The first part of this passage defines liberation as "the return of the [three forces] to the source from which they came" (عودها الى المعدن الذى وفدت منه), a very close parallel to the first part of *sūtra* IV.34, which reads "the return to their original state of the constituents" (*guṇānāṃ pratiprasavaḥ*). The second part, "or if you wish, you may say it is the return of the soul to its own nature" (و ان شئت فقل هو رجوع النفس الى طباعها), corresponds to the second part of *sūtra* IV.34, namely, "or [it is] the power of consciousness (i.e., the conscious self) abiding in its own form" (*svarūpapratiṣṭhā vā citiśaktiḥ*). Thus, although Sachau did not identify the *Kitāb Pātanǧal* with any Sanskrit text known to him, he noticed striking parallels between this passage and a *sūtra* of the *Pātañjalayogaśāstra* which al-Bīrūnī almost literally translated into Arabic.

The difference in how al-Bīrūnī handled the two quotations from the *Kitāb Pātanǧal* in the *Taḥqīq mā li-l-Hind* is also noteworthy; in the example dealing with the four paths to liberation, the original passages in the *Kitāb Pātanǧal* underwent many transformations when al-Bīrūnī later on quoted them in the *Taḥqīq mā li-l-Hind*. In the second example on the definition of liberation, in contrast, he quotes almost literally from the *Kitāb Pātanǧal*. This difference is reflected in Sachau's interpretation of them. In the case of the first example, Sachau cannot identify corresponding passages in a Sanskrit text, whereas in the case of the second example he easily parallels the passage in the *Kitāb Pātanǧal* to a known Sanskrit passage.

Further, Garbe identifies the source of the *Kitāb Pātanǧal* with the *Rājamārtaṇḍa* by Bhoja.[15] Several elements based on circumstantial and textual evidence, however, tend to rule out the *Rājamārtaṇḍa* as a tenable candidate for being the source of the *Kitāb Pātanǧal*. First, despite the chronological prox-

14 Basing himself on the passage in the *Taḥqīq mā li-l-Hind*, Ritter adds in his edition the adjective "knowing" (عاقلة) as a qualification of the soul, which is unnecessary in my view. See KP 78, p. 198.19–22 and Pines & Gelblum 1989: 271. See also the translations of the two passages by Kozah (2016: 160).

15 Garbe (1894: 63, 1896: 41–42 and 1917: 91) first identified the source-text of the *Kitāb Pātanǧal* with the *Pātañjalayogaśāstra* and later on changed his mind.

imity between al-Bīrūnī and Bhoja, there is only one reference to the king in the *Taḥqīq mā li-l-Hind*,[16] and al-Bīrūnī most likely never visited his kingdom in Mālava There is no evidence for the political establishment of the Ghaznavids in Bhoja's kingdom, which would have facilitated a collaboration between officials of the two courts. In addition, al-Bīrūnī was very keen to find astronomical Sanskrit literature, as the abundance of references to it in the *Taḥqīq mā li-l-Hind* indicates. He does not, however, mention the medical and astronomical work *Rājamṛgāṅka* composed by Bhoja.[17]

Textual study also presents evidence that al-Bīrūnī did not use Bhoja's text. Maas and Verdon analyse two analogies, one agricultural and one mythological, provided in the *Kitāb Pātanǧal*. For Garbe, these passages are closely connected to the *Rājamārtaṇḍa*.[18] The agricultural analogy explains that the ripening of the accumulation of karma ceases if its root, the afflictions (*kleśa*), is stopped, in the same way as a rice grain does not sprout if its husk is removed (PYŚ (1904) II.13; KP 29). Maas and Verdon argue that al-Bīrūnī likely relied on a Sanskrit work other than the *Rājamārtaṇḍa* when rendering this example in the *Kitāb Pātanǧal*. First, they point out that this illustration may not have been an original and authentic part of the *Rājamārtaṇḍa*. They further connect this example in al-Bīrūnī's translation to a passage of the *Pātañjalayogaśāstra*, while also noting discrepancies in the use of this example in the *Pātañjalayogaśāstra* and the *Kitāb Pātanǧal*:

> The Sanskrit work explains how future consequences of the storage of karma can be prevented, whereas the Arabic work explains that the soul is covered by ignorance like a rice grain may be covered by its husk. In the *Kitāb Pātanǧal*, the husk has to be removed in order to prevent changes of the soul, whereas, according to the *Pātañjalayogaśāstra*, removing the husk prevents the ripening of karma.[19]

The two authors observe that al-Bīrūnī adapted the Sanskrit phrasing in his own manner and ignored the concept of karma referred to in the *Pātañjalayogaśāstra*. Al-Bīrūnī's interpretation can be indeed accounted for by his cultural

16 Al-Bīrūnī mentions king Bhoja in the *Taḥqīq mā li-l-Hind* in a passage that narrates a tale (حديث) about a piece of silver in the door of the governmental house in Dhāra, the capital of Bhoja's kingdom (*Taḥqīq* [1958], p. 152.4–6; Sachau 1910: I/191). See above Section 1.2 on al-Bīrūnī's visits to al-Hind.

17 Information on the *Rājamṛgāṅka* is given, for instance, by Sarma (1985a: 12).

18 See the detailed and full discussion in Maas & Verdon 2018: 293–301.

19 Maas & Verdon 2018: 296.

and intellectual background. In his version of the example, the soul's covering, that is, the husk in the analogy, has to be taken out in order to remove changes to the soul, to purify it (فاذا ازيل القشر عنه انقطعت تلك الحوادث و صفا للبقاء على حال).[20] In Islamic thought, when the soul is purified, it ascends to celestial spheres and gradually frees itself from gross matter. This conception was equally present among ancient Greek philosophers.[21] The idea of the soul being covered by a cloth existed in connection with purification, as interpreted by the Neo-Platonic philosopher Porphyry. This concept was also known to the philosopher al-Tawḥīdī (922/932–1023).[22] Charles Genequand notes that al-Tawḥīdī sometimes substitutes the Arabic term for "cloth" (ملبس) with the word "covering" or "scale" (قشر),[23] exactly the same term that al-Bīrūnī used in this analogy.[24]

It appears then that al-Bīrūnī interprets the *kleśa*s described in the *Pātañjalayogaśāstra* in the light of theories developed by earlier or contemporary philosophers. Whereas the consequences of the removal of the husk from the rice grain differ in the two works, the goal of the analogy is the same, that is, to illustrate the freeing of the soul—or the conscious self—from impurities that impede it from reaching a higher level of spirituality.

Further, the possibility of al-Bīrūnī having read other commentaries on the *Pātañjalayogaśāstra*, such as the *Vivaraṇa* or the *Tattvavaiśāradī*, still fails to explain this particular interpretation, as they do not substantially deviate from the explanation of the *Pātañjalayogaśāstra* with regard to this illustration.[25] Referring to David Pingree's opinion about al-Bīrūnī's level of knowledge of Sanskrit, Maas and Verdon also suggest that al-Bīrūnī's idiosyncratic interpretation may simply be owed to a limited knowledge of Sanskrit.[26] However, in the light of the above observations it is likely that al-Bīrūnī's interpretation constitutes a translational strategy, namely that of substitution.

The second analogy that led Garbe to believe that al-Bīrūnī used the *Rājamārtaṇḍa*, and which Maas and Verdon also analysed, narrates how Nandikeśvara (or Nandīśvara) and Nahuṣa, two mythological figures, metamorphosed on account of their deeds (PYŚ (1904) II.12; KP 28). In al-Bīrūnī's ver-

20 KP 29, p. 180.2–3; Pines & Gelblum 1977: 524; *Taḥqīq* (1958), p. 42.7–11; Sachau 1910: I/55.

21 Genequand 1996: 110.

22 Stern 2012.

23 Genequand 1996: 110–111.

24 In a different context, the *Quran* uses the image of the veils that cover the heart (قلب), not the soul (نفس) (Massignon 1954 [1922]: 108; Sūra 51.4).

25 *Vivaraṇa* (1952), pp. 146.18–147.20 and TVŚ, pp. 68.22–69.18. See also Woods 1914: 126 and Rukmani 2001: I/248–249.

26 Pingree 1983: 353; Maas & Verdon 2018: 297. See above p. 135.

sion, Nandikeśvara (نندكيشفر), who was a devotee of Śiva (مهاديو; *mahādywa*),[27] became an "angel" (ملك), a word corresponding to the Sanskrit *deva*, meaning "deity," whereas Nahuṣa (نهش), the evildoer, became a snake. Maas and Verdon observe that here the *Kitāb Pātanğal* parallels the *Pātañjalayogaśāstra*, in wording and content, to a greater extent than Bhoja's work. The latter, for instance, mentions Viśvāmitra and Urvaśī, whose names are absent from both the *Kitāb Pātanğal* and the *Pātañjalayogaśāstra*. Maas and Verdon also shed light on differences in the narration of the above myth between the *Kitāb Pātanğal* and the *Pātañjalayogaśāstra*, which, they convincingly argue, are most probably due to al-Bīrūnī's addition of contextual elements and his combination of two distinct myths.

Further, in the corresponding passage in the *Vivaraṇa*, its author employs several illustrations, referencing diverse figures, including Viśvāmitra, Ambā, Draupadī and Kumbhakarṇa, and eventually recounting the story of Nandīśvara and Nahuṣa.[28] The *Tattvavaiśāradī* gives an account of the story of Dhruva and refers to Nandīśvara, but not to Nahuṣa.[29] Thus, reading the *Pātañjalayogaśāstra*, or a work similar to it, was sufficient for al-Bīrūnī to expound the story of Nandikeśvara and Nahuṣa. These preliminary remarks lead to the observation that the *Pātañjalayogaśāstra* may be the Yoga text that best parallels the *Kitāb Pātanğal* if contrasted with the *Rājamārtaṇḍa*, the *Vivaraṇa* and the *Tattvavaiśāradī*.

Lastly, Maas and Verdon summarize Surendra Nath Dasgupta's conclusion that the *Kitāb Pātanğal* was not based on any known Yoga work, and that another Patañjali was the author of its source-text.[30] Without referring to Sachau's or Garbe's earlier analyses, Dasgupta points out that the commentary provided by al-Bīrūnī covers the same subject matter as the *sūtra*-part of the *Pātañjalayogaśāstra*—such as God, soul, bondage, salvation and karma. Yet, according to Dasgupta, the *Kitāb Pātanğal* differs from this Sanskrit work in the way in which it deals with these subjects. Dasgupta notes the following differences:

> (1) the conception of God has risen here to such an importance that he has become the only object of meditation, and absorption in him is the goal; (2) the importance of the yama and the niyama has been reduced to the minimum; (3) the value of the Yoga discipline as a sep-

27 Al-Bīrūnī makes use of the epithet *mahādeva* to refer to Śiva.

28 *Vivaraṇa* (1952), pp. 143.21–144.6. See also Rukmani 2001: I/241.

29 TVŚ, pp. 67.25–68.18. See also Woods 1914: 122.

30 Dasgupta 1930: 64; also referred to in Maas & Verdon 2018: 293.

> arate means of salvation apart from any connection with God as we find in the *Yoga sūtra* has been lost sight of; (4) liberation and Yoga are defined as absorption in God; (5) the introduction of Brahman; (6) the very significance of Yoga as control of mental states (*cittavṛttinirodha*) is lost sight of, and (7) rasāyana (alchemy) is introduced as one of the means of salvation.[31]

For Dasgupta, Vedāntic and Tantric ideas influenced the doctrine presented in the *Kitāb Pātanǧal*.[32] His discussion on these differences and the reasons underlying them are disputable, especially since he was only able to access excerpts of the *Kitāb Pātanǧal*. First, as I suggested in Chapter 2 of the present book, al-Bīrūnī was most probably unaware of Sanskrit works related to Vedānta.[33] Second, in Chapter 4 I have highlighted that al-Bīrūnī's descriptions of God, or Allah, mostly reflect his tendency to domesticate the Yoga concept of Īśvara.[34] Al-Bīrūnī's religious and philosophical backgrounds may actually constitute the reason for the differences stated by Dasgupta under numbers 1, 3 and 4 of the above quotation, rather than the influence of Vedāntic and Tantric ideas. In addition, once one is able to access the *Kitāb Pātanǧal* in its entirety, it is possible to see that al-Bīrūnī indeed addressed "Yoga as control of mental states" and dealt with the *yama*s and the *niyama*s.[35] Thus, like Sachau and Garbe, Dasgupta could not reach a conclusive and satisfactory answer to the question of al-Bīrūnī's source.[36]

5.1.2 *After the Edition of the Manuscript*

Pines and Gelblum were the first to study the complete manuscript of the *Kitāb Pātanǧal*. They published an annotated English translation of it in the form of four articles.[37] In these publications, they summarize the previous attempts made to identify the *Kitāb Pātanǧal*'s source and reach conclusions different from those of their predecessors. According to them, al-Bīrūnī based his Arabic translation on the *sūtra*s and an unknown commentary, which they consider having more in common with the *bhāṣya*-part of the *Pātañjalayogaśāstra* than

31 Dasgupta 1922: 235. Numbering and transliterations are by Dasgupta.
32 Dasgupta 1922: 235 and 1930: 63–64.
33 See fn. 126 of Chapter 2 above.
34 See above pp. 150–155.
35 On al-Bīrūnī's treatment of the *yama*s and *niyama*s, see below p. 174–175.
36 Jean Filliozat, who briefly refers to al-Bīrūnī's translations, appears to regard the *Kitāb Pātanǧal* as an independent work (Renou & Filliozat 1953: 46; Maas & Verdon 2018: 301).
37 Pines & Gelblum 1966, 1977, 1983 and 1989.

with the *Rājamārtaṇḍa*. Despite their comparison of the content of the *Kitāb Pātanğal* to numerous Sanskrit works and their abundant use of secondary literature, they were unable to identify the text used by al-Bīrūnī.

They thus put forward several remarks and hypotheses: 1) the commentary used by al-Bīrūnī is unknown and could either still be lying in an Indian library or simply be lost; 2) the commentary may have had theistic tendencies that would be characteristic of a later development of the classical Yoga system; 3) an analysis of similes, metaphors and of the *Kitāb Pātanğal*'s laudatory introduction would be conducive to identifying al-Bīrūnī's source; 4) al-Bīrūnī's choices in his interpretations depended upon his own cultural and religious background, as well as upon his intelligence and creativity, the *Kitāb Pātanğal* being thus a non-literal translation; 5) an investigation of these choices of interpretation is a *desideratum* in the further analysis of the relationship between the *Kitāb Pātanğal* and its main source.[38]

Pines and Gelblum provide a thorough and pertinent study that constitutes the necessary first step in such an analysis. However, I suggest three fundamental reasons for their having difficulty pinpointing a source. First, they consider the *sūtra*s and the *bhāṣya* of the *Pātañjalayogaśāstra* as dissociable entities. When they point out that the *sūtra*s are interwoven with a commentary in al-Bīrūnī's source-text,[39] they do not, as a first hypothesis, conceive the possibility that this commentary could in fact be the *bhāṣya*-part of the *Pātañjalayogaśāstra*. Searching for correspondences of al-Bīrūnī's text in other Sanskrit commentaries, they are unable to identify the commentary mentioned by al-Bīrūnī. The second drawback to their analysis, resulting from the first, is their assumption of the existence of an unknown commentary. The Arabic expression used by al-Bīrūnī to refer to "the commentary" or "the commentator" (تفسير كتاب پاتنجل or مفسّر) can be interpreted in two ways: "the commentary that comments upon the *Kitāb Pātanğal*" or "the commentary that is included in the *Kitāb Pātanğal*." If one adopts the latter interpretation, this wording suggests that the commentary referred to by al-Bīrūnī was included in his source-text and was not drawn from another Yoga text.[40] Further, the role which al-Bīrūnī's intelligence and creativity played in his interpretive choices is more important than the two scholars thought, as demonstrated above in Chapter 4. This, too, accounts for their struggling to identify his source.

38 Pines & Gelblum 1966: 302–308.

39 Pines & Gelblum 1966: 303. For references to the commentary in the *Kitāb Pātanğal*, see in its preface, as well as in KP 46 (KP, pp. 168.5, 185.16 and 188.3; Pines & Gelblum 1966: 310, 1983: 260 and 261), and Section 5.2.1 of the present book.

40 See above pp. 119–120.

More recently, Kozah have noted strong parallels between the *Kitāb Pātanǧal* and the *Pātañjalayogaśāstra*, which led him to assume that the latter is very close to either the source-text of al-Bīrūnī's translation or the oral tradition that had been transmitted to him.[41] Lastly, Maas and Verdon posit the hypothesis that the *Kitāb Pātanǧal*'s main Sanskrit source was the *Pātañjalayogaśāstra*.[42]

In the subsequent section, I adopt a different perspective from that of Pines and Gelblum, that is, that the commentary, which al-Bīrūnī refers to, could indeed be the one included in the *Pātañjalayogaśāstra*. I also analyse al-Bīrūnī's hermeneutics. This perspective will not only untangle the problems Pines and Gelblum faced, but also support the observations made by Kozah, Maas and Verdon.

5.2 The Commentary as an Integrated Part of the Kitāb Pātanǧal

The source of the *Kitāb Pātanǧal* included a fundamental text and a commentary which al-Bīrūnī merged with each other in his translation. In addition, it appears, as a general rule, that al-Bīrūnī integrated this commentary into his translation without differentiating it from the main primary text, and only in a few cases, two to be precise, he explicitly identifies passages drawn from the commentary. In these two cases, not only did he inform his readers that he quoted the commentary in full, but also distinguish it from the fundamental text. I analyse in the following section each of the two situations, in order to understand why al-Bīrūnī may have dealt with this source in two different ways and to determine if indeed the *Pātañjalayogaśāstra* was his main source.

5.2.1 *Explicit Integration*

Two passages of the *Kitāb Pātanǧal* explicitly refer to a commentator (hereafter the commentator).[43] Both occur in KP 46 corresponding to the content

41 Kozah 2016: 164.

42 Maas & Verdon 2018: 312–315.

43 Sachau posits that Balabhadra, an author who is often quoted in the *Taḥqīq mā li-l-Hind* primarily with regard to cosmography, may have composed the commentary translated in the *Kitāb Pātanǧal*. The discussion of some common points between portions of the *Kitāb Pātanǧal* and *bhāṣya*-parts of the *Pātañjalayogaśāstra* in the present chapter demonstrates that this position is untenable (Sachau 1910: II/264; on al-Bīrūnī's interpretation of Balabhadra's work, see Pingree 1983).

found in PYŚ (1904) III.21–35.[44] KP 46 enumerates several objects of continuous thought (فكرة), or reflection (تفكّر)—two terms that render the notion of mental concentration (*saṃyama*) in the source-text—and the results of the reflection on them. These results are referred to with the generic Arabic expression "wondrous acts" (افعال اعاجيب). Table 10 in Section 5.2.2 below presents fifteen objects of such reflection. The two instances, in which al-Bīrūnī explicitly mentions the commentator, occur after the passages that discuss the reflections upon the sun (passage no. 8; PYŚ (1904) III.26) and upon the navel (passage no. 11; PYŚ (1904) III.29), respectively.

The Arabic sentence "The commentator has in this place an explanatory discourse" (للمفسر فى الموضع كلام شرحى)[45] opens the first of these explicit quotations, while the words "Let us go back to the text" (فلنعد لى النص) conclude it.[46] The quotation deals with the cosmography developed in the *Pātañjalayogaśāstra*. After having introduced it, al-Bīrūnī comments on the way in which his source describes the topic and on the measurement units used in it, which he transposes into Arabic miles, a move that once again reveals his efforts to make his translation as intelligible as possible for his readership. Subsequently, al-Bīrūnī goes on with the translation of his source. He organizes seven broad categories of cosmic regions in the following order: 1) seven hells (*naraka*); 2) seven netherworlds (*pātāla*); 3) seven islands (*dvīpa*); 4) seven oceans (*samudra*); 5) the end of the world (*lokāloka*); 6) three regions above (?); 7) seven world-regions (*loka*, or *brahmaloka*).

Pines and Gelblum consider the content of the *Kitāb Pātanğal* in this place to diverge too much from the *bhāṣya*-part of the *Pātañjalayogaśāstra* in order to allow a correlation of the two texts.[47] Table 2 in Maas and Verdon, however, compares the above seven categories with those enumerated in the *Pātañjalayogaśāstra*.[48] This table shows discrepancies and similarities between the two accounts. For instance, al-Bīrūnī describes "three regions above" as containing the world of the fathers (بترلوك), the half of Brahmā's egg (برهماند) and a darkness (ظلمة) called *tama* (تم). This description does not find any parallel in the *Pātañjalayogaśāstra*. However, all other categories appear in both works. The defining features of the oceans, such as being salty or containing sugar cane water, are the same,[49] as are the names of the seven world-regions. The

44 KP 46, pp. 185–188; Pines & Gelblum 1983: 259–262.
45 KP 46, p. 185.16; Pines & Gelblum 1983: 260.
46 KP 46, p. 187.15; Pines & Gelblum 1983: 261.
47 Pines & Gelblum 1966: 304 and 1983: 258.
48 Maas & Verdon 2018: 306–312 and Table 2, 309–311.
49 Whereas the *Pātañjalayogaśāstra* uses the Sanskrit term *lavaṇa* ("salty") for describing

name and order of the enumerated seven hells and seven islands, however, do not entirely match across both works.

More importantly, Maas and Verdon note that the two accounts agree with regard to the number of hells and the position of the netherworlds above the hells, whereas other Brahminical works generally present a higher number of hells and locate the netherworlds at the bottom of the cosmos.[50] Another common point between the *Pātañjalayogaśāstra* and the *Kitāb Pātanǧal*, as opposed to other available cosmographical expositions, is the order of the description of the islands and oceans. Both works list them in two separate sequences, whereas other Brahminical Sanskrit literature enumerates each island and ocean consecutively.

Thus, despite some terminological and descriptive discrepancies, al-Bīrūnī's account coincides with that of the *Pātañjalayogaśāstra* in its global structure and representation of the cosmos. Further, no other known Sanskrit work related to Yoga is demonstrably closer to this passage of the *Kitāb Pātanǧal* than the *Pātañjalayogaśāstra*. None, as will become evident subsequently, can account for any of the differences highlighted in the preceding paragraphs. The author of the *Vivaraṇa* merely quotes from the *Pātañjalayogaśāstra* and does not comment upon the quoted passage, with the exception of these short sentences:

> Knowledge of the worlds [arises] from the [threefold] mental concentration upon the sun. Having concentrated upon the sun, he shall perceive [with his eyes] the whole extent of the worlds. The meaning of the commentary (*bhāṣya*), however, is understood, as it is established in all Purāṇas.[51]

The author of the *Vivaraṇa* considered this cosmographical description as part and parcel of a common knowledge shared by both the author of the *bhāṣya* and all Purāṇas. As for the *Tattvavaiśāradī*, it generally does not deviate from

---

the salty ocean, al-Bīrūnī gives the Arabic transliteration *kšāra* (كشار) of the term *kṣāra* (meaning "corrosive, caustic, acid," but also "saline").

50 See Kirfel 1920: 148–173. Al-Bīrūnī was aware of the variety of cosmographical views in Sanskrit literature, as shown by some of his comments in the *Taḥqīq mā li-l-Hind* indicate (*Taḥqīq* [1958], p. 185.11–15; Sachau 1910: I/228; *Taḥqīq* [1958], p. 186.9–11; Sachau 1910: I/229).

51 *Vivaraṇa* (1952) III.26, p. 287.16–17: *bhuvanajñānaṃ sūrye saṃyamāt* (sū. III.26). *sūrye saṃyamaṃ kṛtvā samastaṃ bhuvanaprastāraṃ pratyakṣīkurvīta. bhāṣyaṃ tu gatārthaṃ, sarvapurāṇaprasiddhatvāt*. See also Rukmani 2001: II/81.

the description provided in the *Pātañjalayogaśāstra*, except for its discussion of metaphysical notions, such as the original cause (*prakṛti*) and the essence of the intellect (*buddhisattva*), which neither the *Kitāb Pātanğal* nor the *Pātañjalayogaśāstra* mentions in this place.[52] Finally, the *Rājamārtaṇḍa* could not have inspired al-Bīrūnī here either, as it does not provide any cosmographical description at this place.[53]

The *Vivaraṇa*, the *Tattvavaiśāradī* and the *Rājamārtaṇḍa* do not mention the additional elements, such as the three regions above or the specific names of three of the hells, namely, *vajra*, *garbha* and *suvarṇa*, that however found their way into al-Bīrūnī's translation. The differences between the *Kitāb Pātanğal* and the *Pātañjalayogaśāstra*, on the other hand, could be explained by the fact that in some instances al-Bīrūnī deemed it necessary for the sake of his Muslim readership to supplement the cosmographical description, and, conversely, that in other instances he regarded some elements as irrelevant and therefore not worth expounding in detail. His knowledge drawn from other sources, such as the Purāṇas and his oral informants, may also have played a significant part in his approach to rendering this part of the Yoga work into Arabic.

The second passage in KP 46 that explicitly refers to the commentator corresponds to PYŚ (1904) III.29 and discusses medical notions strongly inspired by Āyurvedic medicine. It begins with "This too belongs to the commentator's discourse" (و هذا من كلام المفسر ايضا),[54] and ends with the sentence "At this point we return to the text" (و قد رجعنا الى النص).[55] In this passage, al-Bīrūnī describes the process by which food is transformed into matter, a process that is not explained in the *Pātañjalayogaśāstra*. Further, both the *Pātañjalayogaśāstra* and the *Kitāb Pātanğal* enumerate seven bodily constituents. Differences in these lists, however, can be observed: "(1) Al-Bīrūnī's list starts with the item chyle instead of skin and (2) it contains the bodily element fat instead of sinew as item no. 4."[56]

Pines and Gelblum conclude that the *Pātañjalayogaśāstra* could not have been the source of the *Kitāb Pātanğal* because of these two major discrepancies, that is, the addition of the description of food transformation and differences in the listed items. Maas and Verdon, on the other hand, explain the first

52 TVŚ, pp. 149.25–152.29. See also Woods 1914: 258 ff.
53 RM, pp. 38.26–39.3 on sū. III.26.
54 KP 46, p. 188.3; Pines & Gelblum 1983: 261.
55 KP 46, p. 188.11; Pines & Gelblum 1983: 262.
56 Maas & Verdon 2018: 305.

discrepancy as due to al-Bīrūnī's having been inspired by his oral informants or by his knowledge of medical Sanskrit literature. As for the second one, they argue, based on a critical edition of portions of the *Pātañjalayogaśāstra*, that Pines and Gelblum had access to an edition of the Sanskrit text which does not always render original readings of the *Pātañjalayogaśāstra* and that scribes may have introduced, during the evolution of the transmission of the text, these secondary readings.[57]

Maas and Verdon also note a possible peculiar understanding, on al-Bīrūnī's part, of the sentence "This arrangement is such that each preceding [item] among them is exterior [to the following one]" (*pūrvaṃ pūrvam eṣāṃ bāhyam ity eṣa vinyāsaḥ*), found in the *bhāṣya*-part of the *Pātañjalayogaśāstra*.[58] Al-Bīrūnī appears to have interpreted it as follows: "Whatever is farther from matter is more excellent" (كل ما هو ابعد عن المادة فهو افضل),[59] an interpretation most probably indebted to his socio-cultural background. The idea that impurity is to be connected with gross matter, whereas purity, or the good, should be associated with the immaterial, is a common conception not only among ancient Greek thinkers, but also Muslim philosophers, as shown by the above example of the covering of the soul.[60]

Al-Bīrūnī takes on a similar view in the following examples. He describes the higher level of contemplation (تصوّر), which corresponds to the non-cognitive absorption (*asaṃprajñāta samādhi*), as free from matter (المجرد عن المادة). With regard to the eight components (*aṅga*), he states that the last three, namely, *dhāraṇā*, *dhyāna* and *samādhi*, "are more remote from the senses, closer to the intellect (عقل) [than the other qualities] and border a representation of the known [object], as free from matter that pertains to the bonds of senses."[61] This peculiar interpretation is not found in any commentaries consulted for the present research, that is, the *Pātañjalayogaśāstra*, the *Vivaraṇa*, the *Rājamārtaṇḍa* and the *Tattvavaiśāradī*. Therefore, this peculiar interpretation may be owed to al-Bīrūnī's hermeneutics rather than to his having relied on a work different from the *Pātañjalayogaśāstra*.

While the exact reasons for the full insertion of the commentator's discourse in these two places are not clear, al-Bīrūnī expressly mentions one of them in his statements on the cosmographical digression:

---

57 Maas & Verdon 2018: 303–306.

58 PYŚ (1904) III. 29, p. 153.9.

59 KP 46, p. 188.8–9; Pines & Gelblum 1983: 261.

60 See above pp. 163–164.

61 On the way in which al-Bīrūnī deals with the eight components, see below pp. 175–177. On al-Bīrūnī's position vis-à-vis Ibn Sīnā on this question, see Kozah 2016: 79.

> The commentator has in this place an explanatory discourse that describes the world and the earths.[62] Its presentation in the right manner is useful, as [cosmography] is one of the widespread sciences among them (i.e., the Indians).[63]

Al-Bīrūnī thus considered it important to insert the commentator's words "in the right manner," simply because he regards the topic of cosmography as "one of the widespread sciences among" the Indians. A major part of the *Taḥqīq mā li-l-Hind* is devoted to it and the related scientific field of astronomy, with references *inter alia* to Brahmagupta, Āryabhaṭa and Varāhamihira, as well as to the Purāṇas. Section 2.1 of the present book has already highlighted al-Bīrūnī's knowledge of—and initial interest in—Indian astronomy and astronomical mathematics. Thus, his interest in cosmographical science and its importance for his informants may explain the explicit integration of the commentary in this place.

As for the medical discussion related to PYŚ (1904) III.29, al-Bīrūnī does not clarify why he quotes the commentary specifically here. It is possible that just as in the case of his cosmographical digression al-Bīrūnī considered medicine a popular and essential science among the Indians. Furthermore, as I highlighted in the introduction of this book, Sanskrit astronomical and medical treatises were among the first scientific writings that were translated into Arabic, that is, from the second half of the eighth century onward. Cosmography and medicine were fundamental disciplines for al-Bīrūnī's readership, or were considered by him to be so, which may have led him to include the whole commentary in these two places.

5.2.2 *Tacit Integration*

In addition to the two explicit integrations of the commentary into his translation of the Yoga text, al-Bīrūnī includes many portions of a gloss without however indicating these incorporations. He may have not expressed it because this tacit integration was less literal than the explicit ones, whereas the opacity of the *sūtras* compelled him to add explanations in order to comprehend their meanings. By analysing several passages of the *Kitāb Pātanğal* in which al-Bīrūnī makes use of the commentary, I show below that these portions of text resemble much the *bhāṣya*-parts of the *Pātañjalayogaśāstra* and at the same time contrast with the other extant Sanskrit commentaries.

---

62 The reading proposed by Pines & Gelblum is followed here. See above p. 117.

63 KP 46, p. 185.16–17; Pines & Gelblum 1983: 260.

TABLE 9 Comparison of the wordings of the chapter-colophons of the *Kitāb Pātanǧal* and *Pātañjalayogaśāstra*

Chapter-colophons of the *Kitāb Pātanǧal*	Chapter-colophons of the *Pātañjalayogaśāstra*
Here ends the first section of the *Kitāb Pātanǧal* on making the heart steadfastly fixed. (KP, p. 177.10; Pines & Gelblum 1966: 325). تمت القطعة الاولى من كتاب باتنجل فى اقرار القلب على مقر واحد	The first chapter in Patañjali's Yoga treatise, which is expressive of Sāṅkhya [focused] on absorption. *iti pātañjale yogaśāstre sāṅkhyapravacane samādhipādaḥ prathamaḥ.*
Here ends the second section on the guidance to the practice of what has been mentioned in the first section. (KP, p. 183.18; Pines & Gelblum 1977: 527). تمت القطعة الثانية فى ارشاد الى عمل ما كان تقدم فى القطعة الولى	The second chapter in Patañjali's Yoga treatise, which is expressive of Sāṅkhya is called instruction on the means. *iti pātañjale yogaśāstre sāṅkhyapravacane sādhananirdeśo nāma dvitīyaḥ pādaḥ.*
Here ends the third section pertaining to the reward and the way to obtain this reward. (KP, p. 192.22; Pines & Gelblum 1983: 265). تمت القطعة الثالثة المقصورة على ذكر الجزاء و كيفية المجازاة	The third chapter in Patañjali's Yoga treatise, which is expressive of Sāṅkhya [focuses] on supernatural powers. *iti pātañjale yogaśāstre sāṅkhyapravacane vibhūtipādas tṛtīyaḥ.*
Here ends the fourth section on liberation and union. As [this section] concludes, so does the book. (KP, p. 199.1; Pines & Gelblum 1989: 271). تمت القطعة الرابعة الخلاص والاتحاد و تم بتمامها الكتاب	The fourth chapter in Patañjali's Yoga treatise, which is expressive of Sāṅkhya [focused] on isolation. And [with this] the composition (i.e., work) ends. *iti pātañjale yogaśāstre sāṅkhyapravacane kaivalyapādaś caturthaḥ. samāptaś cāyaṃ granthaḥ.*

Table 9 displays the text of the chapter-colophons of the *Kitāb Pātanǧal* and the *Pātañjalayogaśāstra*[64] and reveals striking commonalities between them.[65] It also highlights how al-Bīrūnī adhered to the structure specific to the *Pātañjalayogaśāstra* in four chapters (*pāda*).

64 According to Maas 2006: xx–xxi.
65 This table is based on Maas & Verdon 2018: 288, Table 1.

The colophons of chapters 1, 2 and 4 of the two works are almost identical. Al-Bīrūnī rephrased in Arabic the concepts of absorption (*samādhi*), of instruction on the means (*sādhananirdeśa*) and of isolation (*kaivalya*), referred to in the *Pātañjalayogaśāstra*. The colophon of chapter 3 of the *Kitāb Pātanğal* does not literally correspond to that of chapter 3 of the *Pātañjalayogaśāstra*. The Arabic and Sanskrit texts, however, address the same topic, namely the results of the practices described in the previous chapters. Generally, a notable difference is the absence of the expression "which is expressive of Sāṅkhya" in the *Kitāb Pātanğal*.[66] On the whole, however, al-Bīrūnī conveys the meaning of the chapter headings of the *Pātañjalayogaśāstra*.

Before examining the structure of the two texts and their content, it is worth comparing the colophons with that of the available Yoga Sanskrit commentaries. Their colophons, as seen in the printed editions used for this study, all include the names of their authors. The *Vivaraṇa* is attributed to Śaṅkara, the disciple of Govinda, the *Tattvavaiśāradī* is ascribed to Vācaspatimiśra, and the *Rājamārtaṇḍa* to King Bhoja. However, none of these names appear in the *Kitāb Pātanğal* or in the *Taḥqīq mā li-l-Hind*.

In addition to the correspondences of the chapter-colophons, the content of the *Kitāb Pātanğal* is organized just as the *Pātañjalayogaśāstra* across their respective chapterization. A particularly striking concordance between the *Kitāb Pātanğal* and the *Pātañjalayogaśāstra* can be ascertained in terms of the organization of the content of their chapters 2 and 3, that is, of KP 41 on the one hand, and PYŚ (1904) II.29–55 and III.1–8 on the other.[67] The two passages deal with the eight components (*aṣṭāṅga*), which, together with their subdivisions and respective benefits, are described extensively. Al-Bīrūnī's enumeration of the eight components corresponds relatively well to that found in the *Pātañjalayogaśāstra*. He translates the term "component" (*aṅga*) by a word meaning "quality," "property" or "characteristic" (خصلة).

Al-Bīrūnī provides the following list: 1) refraining from evil (الكفّ عن الشرّ), which corresponds to ethical rules (*yama*) in the *Pātañjalayogaśāstra*; 2) holiness, outward and inward (القدس ظاهرا و باطنا), which can be paralleled to observances (*niyama*); 3) a state of rest (سكون), which may match posture (*āsana*);[68] 4) quieting the breath (تسكين التنفس), clearly the equivalent of breath-control (*prāṇāyāma*); 5) compression of the senses (قبض الحواسّ), which

66 On the interpretation of the compound *sāṅkhyapravacana*, see fn. 117 in Chapter 3 of the present book.

67 KP 41, p. 182.7–184.5; Pines & Gelblum 1977: 526–527 and 1983: 258–259; PYŚ (1904), pp. 101.7–122.3. See also Woods 1914: 177–208.

68 See Pines & Gelblum 1977: 547, n. 154.

renders the Sanskrit "withdrawal [of the senses]" (*pratyāhāra*); 6) quietude and tranquillity (السكينة الطمأنينة), which corresponds to fixation (*dhāraṇā*) [of the mind on a specific part of the body]; 7) maintenance of thought [about some object] (ادامة الفكرة), an interpretation of "meditation" (*dhyāna*); and finally 8) purity (اخلاص), which can be associated with absorption (*samādhi*).[69] A comparison between the enumerations provided in the *Kitāb Pātanğal* and the *Pātañjalayogaśāstra* reveals quasi-literal translations, paraphrases and conceptual adaptations on al-Bīrūnī's part.[70]

Furthermore, in this passage, al-Bīrūnī combines the various *sūtra*s with their *bhāṣya*s according to his own logic in one set of question and answer in KP 41, whereas the treatment of the eight components in the *Pātañjalayogaśāstra* is distributed over several *sūtra*s. Combining several passages of the Sanskrit text in a single set of question and answer is an approach that al-Bīrūnī frequently took.[71] However, in spite of such combinations, he apparently maintains the chapter structure and designation of his source. The *Pātañjalayogaśāstra* deals with the first five components at the end of chapter 2, and with the remaining three components at the beginning of chapter 3. Following this division, al-Bīrūnī distributes KP 41 across sections 2 and 3 of the *Kitāb Pātanğal*. Accordingly, he describes the last three items at the beginning of section 3 of his translation. He explains the reasons for this particular distribution:

> The last three qualities fall into the third section because they are separate from the first five [qualities], because they are more remote from the senses, closer to the intellect (عقل) [than the other qualities] and border a representation of the known [object], as free from matter that pertains to the bonds of senses.[72]

The corresponding passage of the *Pātañjalayogaśāstra* reads:

> The triad is an internal component, as compared with the preceding [components]. This triad, that is, fixation [of the mind], meditation and absorption, is an internal component of cognitive absorption, as compared with the preceding five means, namely [fulfilment of] ethical rules, etc. This [triad] is also an external component of [non-cognitive absorption that is] without seed. This triad of means that is also an internal

69 KP 41, p. 182.7–184.5; Pines & Gelblum 1977: 526–527 and 1983: 258–259.
70 See Kozah (2016: 116–119) on the eight components as presented by al-Bīrūnī.
71 For other examples of this strategy, see above Section 4.3.1.
72 KP 41, p. 184.3–5; Pines & Gelblum 1983: 258–259.

> component [is at the same time] an external component of yoga without seed (i.e., non-cognitive absorption). Why? Because it arises [even] when the [triad] does not arise.[73]

The two texts treat in a similar way the last three components separately from the earlier five, namely in chapter 3 instead of chapter 2. PYŚ (1904) III.4, defines them as (mental) concentration on a single thing (*trayam ekatra saṃyamaḥ*), which implies that the earlier five components were not conceived in these terms.[74] The difference between the last three components and the earlier five lies in that the triad is defined as an internal component (*antaraṅga*) of the first type of absorption, the cognitive one. However, it is also said to be an external component (*bahiraṅga*) of the second type of absorption, the non-cognitive one.

Al-Bīrūnī's definition of the last three qualities as "more remote from the senses and closer to the intellect" (ابعد عن الحس واقرب الى العقل) parallels the Sanskrit expression "an internal component, as compared with the earlier [components]" (*antaraṅgaṃ pūrvebhyaḥ*). His statement that they "border a representation (على شفا تصور) of the known [object], as free from matter that pertains to the bonds of senses" seems to refer to the conception that the three last components are internal to the first type of absorption and at the same time external to the second type. Lastly, this discussion also shows that al-Bīrūnī could not base his explanation of the two categories of the eight components only on the *sūtra*s. Therefore, he must have consulted a commentary, which in this case very much looks like the *bhāṣya*-part of the *Pātañjalayogaśāstra*.

Thus, although al-Bīrūnī modifies his source-text by grouping some *sūtras* and their *bhāṣya*-parts according to his own logic, he also expresses ideas that are found in the same general order as in the *Pātañjalayogaśāstra*, including its both layer of text. This structural similarity, as well as the evidence implied by the concordance between the chapter-colophons of the two works, indicates that al-Bīrūnī did not have to resort to a further commentary than that contained in the *Pātañjalayogaśāstra*. His Sanskrit source had structural features similar to those of the *Pātañjalayogaśāstra*, which enabled him to bestow related designations to the chapters of the *Kitāb Pātanǧal*.

73 PYŚ (1904) III.7–8, pp. 121.8–122.3: *trayam antaraṅgaṃ pūrvebhyaḥ* (sū. III.7). *tad etad dhāraṇādhyānasamādhitrayam antaraṅgaṃ saṃprajñātasya samādheḥ pūrvebhyo yamādibhyaḥ pañcabhyaḥ sādhanebhya iti. tad api bahiraṅgaṃ nirbījasya* (sū. III.8). *tad apy antaraṅgaṃ sādhanatrayaṃ nirbījasya yogasya bahiraṅgaṃ bhavati. kasmāt, tadabhāve bhāvād iti.*

74 PYŚ (1904), p. 120.1.

TABLE 10 Concordance of the passages on the different objects of concentration in the *Kitāb Pātanğal* and the *Pātañjalayogaśāstra*[75]

No.	KP 46	PYŚ III.21–35
1	Whoever wishes to be hidden from the eyes [of others] continuously applies his reflection to [his] body and to his representation of it [...] and [thus] persists in averting the sight [of other people] and in compressing sight [from the others]. Accordingly, he becomes invisible to [other] people. من اراد الاستتار عن العين ادام التفكر فى البدن وما تصور به[...] و دأب على غض البصر وقبض حاسة العين فانه يخفى عن الناس	When there is the absence of contact between the light of the sight [of others and one's own body and] when there is obstruction of [the body's] ability to be perceived disappearance [arises] from [one's] mental concentration [focused] on the [outer] form of [one's] body. *kāyarūpasaṃyamāt tadgrāhyaśaktistambhe cakṣuḥprakāśāsaṃprayoge 'ntardhānam* (sū. III.21)
2	Similarly, whenever he continuously applies his reflection to speech and its compression, his voice is hidden, and he is not heard [anymore]. كما انه اذا ادام التفكر فى الكلام وقبضه خفى صوته فلم يسمع	It is to be understood that, by this, the disappearance of sound and so on, has [also] been stated (i.e., explained). *etena śabdādyantardhānam uktaṃ veditavyam* (PYŚ (1904) III.21, p. 146.13–14).

75 The above translations from Arabic are based on the translation provided by Pines and Gelblum (1983: 259–262). The following are a few philological comments on some of the Sanskrit passages. Passage no. 3: By way of its maturation, the action discussed in this passage has impact on the life span and is twofold: with *upakrama* and without *upakrama* (*āyurvipākaṃ karma dvividham—sopakramaṃ nirupakramaṃ ca* PYŚ (1904) III.22). The word *upakrama* literally means "beginning" or "approach." The author of the *Vivaraṇa* ad loc. explains the twofold karma in terms of the speed with which it provides its result (*Vivaraṇa* [1952], pp. 282.13–15). Thus, I follow Filliozat's translation (2005: 276–277 and n. 537) for this passage. Passage no. 4: The printed editions of the *Vivaraṇa* available to me read *vetti* ("he knows"), instead of *paśyati* ("he sees") in the sentence referring to the forefather (*Vivaraṇa* [1952] III.22, p. 283: *pitṝn atītān akasmād vetti*; Rukmani 2001/II III.22, p. 74: *pitṝn atītān akasmāt vetti*). Passage no. 7: The term *pravṛtti* ("activity") refers to a practice leading to the stability of the mind. See PYŚ I.35. Passage no. 14: It is not necessary, in my view, to interpret the Arabic *sirra* ("secret"; سرّ) as a transliteration of the Sanskrit *siddha* (accomplished) as Pines and Gelblum do.

TABLE 10 Concordance of the passages on the different objects of concentration (*cont.*)

No.	KP 46	PYŚ III.21–35
3	Whoever wishes to grasp the circumstances of his death, continuously applies [his] reflection to [his] actions. من اراد الاحاطة بكيفية موته ادام التفكر فى الاعمال	Action is [twofold]: with impetus and without impetus. Knowledge of [one's] end [arises] from mental concentration [focused] on this [action], or from ill omens. *sopakramaṃ nirupakramaṃ ca karma tatsaṃyamād aparāntajñānam ariṣṭebhyo vā* (sū. III.22).
4	Whoever wishes to conceive heaven and hell, angels and spirits [who drive the damned into hell], and the dead among his ancestors, should continuously apply [his] reflection to them. من اراد ان يتصور له الجنة والنار والملايكة والزبانية والموتى من اسلافه فليدم التفكر فيهم	Likewise, [the ill omen] related to [other] beings [is as follows]: [He] sees the men of Yama. [Or] he sees the forefathers, without any reason. Likewise, [the ill omen] related to heaven: He sees heaven or the perfected [ones] without any reason. Or [he sees] everything reversed. Or by this, one knows that death has approached. *tathādhibhautikaṃ yamapuruṣān paśyati, pitṝn atītān akasmāt paśyati. tathādhidaivikaṃ svargam akasmāt siddhān vā paśyati. viparītaṃ vā sarvam iti anena vā jānāty aparāntam upasthitam iti* (PYŚ (1904) III.22, p. 147.15–17)
5	Whoever wants to strengthen his soul should continuously remember to rejoice in good and turn away from evil. من اراد تقوية نفسه فليدم تذكار السرور بالخير والاعراض عن الشرّ	The forces [of loving kindness, etc., arise] from [mental concentration focused] on loving kindness, etc. *maitryādiṣu balāni* (sū. III.23).

TABLE 10 Concordance of the passages on the different objects of concentration (*cont.*)

No.	KP 46	PYŚ III.21–35
6	Whoever wants to strengthen his body directs [his] reflection to [its] strength and its locations in the [body], as by this continuous [practice] he acquires a strength that does not fall short of that of an elephant. من اراد تقوية بدنه صرف الفكرة الى القوة ومواضعها منه فانه يكتسب بادامة ذلك قوة لا تخلف عن قوة الفيل	The strengths of an elephant and so on [arises from mental concentration focused] on [these] strengths. *baleṣu hastibalādīni* (sū. III.24).
7	Therefore, if he directs his reflection to the light of senses after having subdued and compressed them, he is rewarded by the knowledge of subtle things, [both] present and absent. لهذا اذا صرف فكرته الى نور الحواس بعد قمعها وقبضها كوفىء بمعرفة الدقايق الحاضرة الغايبة	Knowledge of the subtle, the concealed and the remote [arises] from placing the activity [of the mind called] "luminous" [on these objects]. *pravṛttyālokanyāsāt sūkṣmavyavahitaviprakṛṣṭajñānam* (sū. III. 25).
8	Whoever directs it to the sun is rewarded by the comprehension of everything that is in the worlds, and he [can] see them (i.e., the worlds). من صرفها الى الشمس كوفىء بالاحاطة بجميع ما فى العوالم ابصرها	Knowledge of the worlds [arises] from mental concentration [focused] on the sun. *bhuvanajñānaṃ sūrye saṃyamāt* (sū. III.26).
9	Whoever directs his reflection to the moon gains knowledge of the arrangement of the stars, their positions and their activities. من صرف فكرته الى القمر احاط علما بترتيب الكواكب واوضاعها وافعالها	Knowledge of the arrangement of the stars [arises from mental concentration focused] on the moon. *candre tārāvyūhajñānam* (sū. III.27).

TABLE 10 Concordance of the passages on the different objects of concentration (*cont.*)

No.	KP 46	PYŚ III.21–35
10	Whoever directs it (i.e., reflection) to the pole star, which is in a complex of fourteen stars […], knows the movements of the stars. من صرفها الى القطب ـ و هو فى جملة اربعة عشر كوكبا ـ عرف حركات الكواكب […]	Knowledge of their movements [arises from mental concentration focused] on the pole star. *dhruve tadgatijñānam* (sū. III.28).
11	Whoever wishes to know his body should continuously reflect on the navel. من اراد معرفة بدنه فليدم التفكر فى السرّة	Knowledge of the arrangement of the body [arises from mental concentration focused] on the navel's circle. *nābhicakre kāyavyūhajñānam* (sū. III.29).
12	Whoever wishes to remove from oneself the harm of hunger and thirst should direct his reflection on the space [connecting] the chest and the larynx, [that is] the channel [through which] the air [passes] as breath. من اراد نفى اذى الجوع والعطش عنه فليصرف فكرته الى فضاء الصدر الحلقوم مجرى الريح بالتنفّس	Cessation of hunger and thirst [arises from mental concentration focused] on the well in the throat (i.e., part of the larynx below the vocal cords). *kaṇṭhakūpe kṣutpipāsānivṛttiḥ* (sū. III.30).
13	Whoever wishes to dispense with motion should reflect on the tortoise, namely the twisted veins above the navel that resemble it. من اراد الاستغناء عن الحركة فليتفكر فى السلحفاة و هى عروق ملتوية فوق السرة شبّهت بها	[Mental] stability [arises from mental concentration focused] on the tortoise canal. Below [the above-mentioned] well, there is a canal resembling a tortoise in the chest. *kūrmanāḍyāṃ sthairyam* (sū). *kūpād adha urasi kūrmākārā nāḍī* (PYŚ (1904) III.31, p. 153.14–15).

TABLE 10 Concordance of the passages on the different objects of concentration (*cont.*)

No.	KP 46	PYŚ III.21–35
14	Whoever wishes to see the secret of the ascetics who [...] inhabit *bhuvarloka* should direct his reflection to the light of the orifice located in the bone at the crown of the head. من اراد ان يعاين سرّ الزهّاد الذين[...] سكنوا بهوبرلوك فليصرف الفكرة الى نور الثقبة التى على عظم اليافوخ	The sight of the perfected [ones arises from mental concentration focused] on the light in the head. *mūrdhajyotiṣi siddhadarśanam* (sū. III.32).
15	Whoever wishes knowledge, let his reflection be [focused] on the heart, which is its source and dwelling place. من اراد العلم فليكن فكرته فى القلب الذى هو ينبوعه ومسكنه	[The state] of mental consciousness [arises from mental concentration focused] on the heart. *hṛdaye cittasaṃvit* (sū. III.34).

A second example of al-Bīrūnī having integrated parts of the commentary he used into his translation, without however stating it, is found in KP 46. In this passage, al-Bīrūnī enumerates fifteen objects of continuous thought or reflection (تفكر; فكرة). If these objects are continuously reflected upon, one obtains peculiar powers or specific knowledge connected with them. Al-Bīrūnī explains that a person, most probably an ascetic, can achieve these powers "by thoughts and determinations, as he finds his recompense and reward wherever he applies his thought and [wherever] he directs his determination, although any reward without liberation is neither complete nor pure good" (بالافكار والعزايم فانه يجد مكافأته وثوابه حيث انزل فكرته و صرف اليه عزيمته وان كان كل ثواب دون الخلاص ليس بتام ولا خير محض).[76]

The *Pātañjalayogaśāstra*, in a passage extending from III.21 to 35, also lists these fifteen objects,[77] as Table 10 shows. Except for two *sūtras*, the correspondences highlighted cover all *sūtras* and parts of respective *bhāṣya* of *Pātañjalayo-*

76 KP 46, p. 185.2–3; Pines & Gelblum 1983: 259.
77 PYŚ (1904), pp. 146.9–155.9.

gaśāstra III.21 to 35. Al-Bīrūnī appears to have omitted PYŚ (1904) III.33 and greatly reworked PYŚ (1904) III.35 in the last part of KP 46. Nevertheless, the otherwise high degree of correspondences between the two texts cannot be a mere coincidence. Every object of concentration enumerated in the *Kitāb Pātanǧal* finds its analogue in the *sūtra*-part of the *Pātañjalayogaśāstra*, except numbers 2 and 4 which are addressed in the *bhāṣya*. The *sūtra*-part of PYŚ (1904) III.21 addresses the first object of the *Kitāb Pātanǧal* in passage no. 1. The *Kitāb Pātanǧal* describes the second one, related to sound (passage no. 2), as an object distinct from the first one, which the *Pātañjalayogaśāstra* discusses in its *bhāṣya*-part of *sūtra* III.21.

Further, despite a different wording, passages no. 3 and 4 of the above table show parallels between the *Kitāb Pātanǧal* and *sūtra* III.22 and the end of the *bhāṣya* on III.22, respectively.[78] First, there are a few technical notions which are present in the Sanskrit text but absent from the Arabic translation: the twofold action, the omens of imminent death and Yama. Al-Bīrūnī summarized and simplified the content of the *sūtra* and some portions of the *bhāṣya*. The *bhāṣya*-part of the *Pātañjalayogaśāstra* describes three types of omens of imminent death, as related to oneself (*ādhyātmika*), related to other beings (*ādhibhautika*) and related to heaven (*ādhidaivika*).[79] Concentration on each of these omens leads to different results. Al-Bīrūnī appears to have totally overlooked the first type of omens, while merging the other two as leading to the same result (passage 4). The reference in the *Kitāb Pātanǧal* to hell, spirits and the dead among the forefathers invokes the men of Yama and the deceased forefathers mentioned in the corresponding section of the *bhāṣya*. The mention of heaven and angels in the *Kitāb Pātanǧal* reminds one of mention of heaven (*svarga*) in section of the *bhāṣya* that describes the kind of omen related to heaven.

Third, al-Bīrūnī appears to have organized the structure of the content slightly differently from his possible Sanskrit source, as he has described as two different categories the results of the reflection upon one's action and upon deceased persons/heaven in these two passsages.

The authors of the *Vivaraṇa, Tattvavaiśāradī* and *Rājamārtaṇḍa* do not provide glosses on the concept of the omens of imminent death in a way that could account for these differences between the *Kitāb Pātanǧal* and the *Pātañ-*

78 Pines and Gelblum already noted the correspondence with the *bhāṣya* (1983: 274, n. 7).

79 The same terminology *ādhyātmika, ādhibhautika* and *ādhidaivika* is found in classical Sāṅkhya-Yoga in order to refer to the triple suffering (*duḥkha*) (See for instance kā. 1 and PYŚ I.31, pp. 49.3). However, al-Bīrūnī never refers to the three kinds of suffering.

jalayogaśāstra. The discrepancies between these two latter works may be simply due to al-Bīrūnī's idiosyncratic understanding of his source-text and to transformations he may have done to it in order to transfer those ideas to his readership. For instance, by not rendering the two qualifications *sopakrama* and *nirupakrama* ascribed to the action, he may have avoided an explanation of the Indian theory of karmic retribution and an interpretation of the difficult distinction of two types of karma.[80]

Furthermore, in passage no. 12 dealing with reflecting on a part of the larynx, al-Bīrūnī explains this spot as "the space [connecting] the chest and the larynx" and defines it as the "the channel [through which] the air [passes] as breath." While the *sūtra* simply refers to this place as "the well of the throat" (*kaṇṭhakūpa*), the *bhāṣya*-part of the *Pātañjalayogaśāstra* defines it in the following words: "Below the tongue, there is a cord; below the cord, there is the throat; below this there is a well" (*jihvāyā adhastāt tantuḥ, tantor adhastāt kaṇṭhaḥ, tato 'dhastāt kūpaḥ*). The description of al-Bīrūnī reflects this definition and is thus perhaps based on it, despite a different phrasing.

In passage no. 10, which shows a parallel between the *Kitāb Pātanǧal* and *sūtra* III.28, al-Bīrūnī specifies that the pole star is in a complex of fourteen stars (هو فى جملة اربعة عشر كوكبًا). This explanation is not found in the corresponding passages of the *Pātañjalayogaśāstra* or the other Yoga commentaries that could have been available to him.[81] In the *Taḥqīq mā li-l-Hind*, however, al-Bīrūnī quotes the *Viṣṇudharma* (بشن دهرم) and writes that "he (i.e., the author of the *Viṣṇudharma*) placed fourteen of these [stars] around the pole star" (وضع منها حول القطب أربعة عشر).[82] In this case, al-Bīrūnī may have regarded it necessary to add some information to the text he translated, perhaps drawn from the *Viṣṇudharma*.

On the whole, KP 46 appears to present translations of several *sūtra*s of the *Pātañjalayogaśāstra*, and at the same time to include some of its *bhāṣya*-parts, namely in passages 2, 4 and 13. If so, al-Bīrūnī incorporated these portions of the *bhāṣya*, without however explicitly saying it.

Another example of his tacit integration of the commentary into his translation relates to the introductory questions found in the *bhāṣya*-part of the *Pātañjalayogaśāstra*. Al-Bīrūnī adopted these questions in the *Kitāb Pātanǧal*, as the following table shows:

80 On possible other omissions in al-Bīrūnī's translations, see above Section 4.4.1.
81 TVŚ, p. 153.19–23, RM, p. 39.9–13 and *Vivaraṇa* (1952), pp. 287.22–288.2; Woods 1914: 260.
82 *Taḥqīq* (1958), p. 199.13; Sachau 1910: I/242.

TABLE 11 Correlation of some questions from the *Kitāb Pātanğal* and the *Pātañjalayogaśāstra*[a]

Kitāb Pātanğal	*Pātañjalayogaśāstra*
KP 6: How is it possible to quell the soul and to compress its faculties away from external things? فكيف يمكن قمع النفس وقبض قواها عن الخارجات؟ (KP, p. 171.14; Pines & Gelblum 1966: 316).	Now what is the means for the cessation of these [mental dispositions]? *athāsāṃ nirodhe ka upāya iti* (intro. to sū. I.12; PYŚ, p. 21.1).
KP 19: What are these [obstacles] which prevent the soul from attaining its own self? فما هذه الموانع التى تمنع النفس عن القبال على ذاتها؟ (KP, p. 175.11; Pines & Gelblum 1966: 322).	But what are these obstacles and (*vā*) how many are they? *atha ke 'ntarāyāḥ, kiyanto veti* (intro. to sū. I.30; PYŚ, p. 46.1).
KP 26: What are these afflictions which burden the heart? وما هذه الاثقال التى تؤود القلب؟ (KP, p. 177.21; Pines & Gelblum 1977: 522).	Now what are those afflictions and (*vā*) how many are they? *atha ke kleśāḥ kiyanto veti* (intro. to sū. II.3; PYŚ (1904), p. 59.1).
KP 66: If both [merit and demerit] are non-existent in [the ascetic's] past and future, and liberation is an existent, how can an existent come from two non-existents? اذا عُدما معا فى ماضيه وفى مستقبله و للخلاص ايس فكيف يحصل انس من ليسين؟ (KP, p. 196.1–2. Pines & Gelblum 1989: 269; 294, notes 81 and 82).	There is no production of what is non-existent nor destruction of what is existent. Considering this, how past impressions, which occur as substance, disappear?[b] *nāsty asataḥ saṃbhavaḥ, na cāsti sato vināśa iti dravyatvena saṃbhavantyaḥ kathaṃ nivartiṣyante vāsanā iti* (intro. to sū. IV. 12; PYŚ (1904), p. 186.1–2).

a See also the question in the *bhāṣya* introducing *sūtra* I.24 rendered by al-Bīrūnī in KP 12 (KP 12, p. 173.12; Pines & Gelblum 1966: 319; see above, pp. 152–153 and Maas 2013: 59). This passage also occurs in the *Taḥqīq mā li-l-Hind* as a quotation from the *Kitāb Pātanğal*. Such correlations between the introductory questions of the *bhāṣya*-part of the *Pātañjalayogaśāstra* and the *Kitāb Pātanğal* are also found in KP 2 and 3 corresponding to PYŚ I.3 (Maas & Verdon 2018: 317–318), in KP 7 and in KP 12.

b This passage has been translated by Bronkhorst (2011b: 58).

Furthermore, KP 23, partly corresponding to PYŚ I.41, describes the ascetic's psychic faculty (قوة نفسية), in a level (رتبة) preceding liberation. The Arabic expression "psychic faculty" stands here for the Sanskrit *cetas*, psyche or mind, of the *Pātañjalayogaśāstra*. This level referred to in the *Kitāb Pātanğal* is seemingly that of *samāpatti* (contemplative state). Al-Bīrūnī compares the psychic faculty to a crystal, which, while it reflects the external world, is yet not similar to it. The passage reads as follows:

> KP 23: What is his state, when he (i.e., the ascetic) achieves this level before liberation?
>
> His psychic faculty overcomes his body, the bodily obstacles disappear, and he masters his soul. If he wishes, he makes it as small and subtle as a dust particle, and if he wishes, he makes it as large and wide as the air. It is like a crystal in which its surroundings are seen. Thus, objects are [seemingly] in it, although they are external to it. In the same way, he (i.e., the ascetic) comprises that which encompasses him, so that when knowing and known [objects] are united in him who is the knower, then intellection, the one who intellects and what is intellectualized upon become one thing in him.[83]

In the *Pātañjalayogaśāstra*, a similar analogy is used to describe the mind when it has ceased its activities. It reads:

> Then, what is the form or the object of the contemplative state (*samāpatti*), when the psyche (*cetas*) obtained a stability?
>
> In order to teach this, the *sūtra* proceeds: the contemplative state, including the identity with what is located in (*tatsthatadañjanatā*) the perceiver, the perception and the perceptible,[84] [presents itself] to the

83 KP 23, p. 176.7–12; Pines & Gelblum 1966: 323–324. See the translation of this passage by Kozah (2016: 112).

84 The two sentences, "knowing and known [objects] are united in him who is the knower" (اتحد العلم والمعلومات به و هو العالم) and "intellection, the one who intellects and what is intellectualized upon become one thing in him" (صار العقل والعاقل والمعقول فيه شيأ واحدا), interestingly refer to the *sūtra*-part of this passage and its definition of *samāpatti* as being "the identity with what is located in the perceiver, the perception and the perceptible" (*grahītṛgrahaṇagrāhyeṣu tatsthatadañjanatā*). The three concepts perceiver (*grahītṛ*), act of perceiving (*grahaṇa*) and perceptible (*grāhya*) are translated into Arabic by al-Bīrūnī as knower (عالم) or the one who intellects (عاقل), as act of knowing (علم) or of intellecting (عقل) and as known (معلوم) or intellected (معقول) object. The syntax used in both languages presents parallels, as both versions offer three forms belonging to the same verbal

> [mind], when the latter's activities have ceased, [and therefore it has become] like a beautiful gem. [...]. He (i.e., the author) offers an example with [the phrasing] "like a beautiful gem": just as a crystal tinted by different colours because of the variety of its environment, irradiates [differently] depending upon the colour and the form of its environment, likewise the mind (*citta*) tinted by the support of the perceptible, fallen into the state of the perceptible, irradiates [differently] depending upon the colour and the form of the perceptible. In the same way, tinted by a subtle element, fallen into the state of the subtle element, it takes on the [same] appearance and form as the subtle element. In the same way, tinted by a gross support, fallen into the state of the gross [support], it takes on the [same] appearance and form as the gross [support].[85]

The analogy of the gem is only contextualized in the *bhāṣya*-part of the *Pātañjalayogaśāstra*, which specifies: "just as a crystal tinted by different colours because of the variety of its environment, irradiates [differently] depending upon the colour and the form of its environment" (*yathā sphaṭika upāśrayabhedāt tattadrūpoparakta upāśrayarūpākāreṇa nirbhāsate*). The *bhāṣya* thus provides a synonym for gem (*maṇi*), the lexical field of which is vast. According to the Monier-Williams, it can be translated in a flurry of ways such as jewel, gem, pearl, any ornament or amulet, globule, crystal, a magnet, but also glans, penis, clitoris, the hump (of a camel), thyroid cartilage, the name of different mythological figures and so on. The *bhāṣya* specifies the meaning that has to be understood in this context by employing as a synonym the Sanskrit masculine term *sphaṭika*, one which cannot be understood differently than as crystal or quartz.

---

root: an agent noun and an action noun for the first two items in both languages, and, for the third item, an adjective verbal of obligation with a passive sense, called gerundive, in Sanskrit and a passive participle in Arabic. This elegant syntactic correspondence may reflect the importance of these notions in philosophical debates of both Islamic and Indian thoughts. Similarly, in KP 33, which roughly corresponds to PYŚ II.17, the Sanskrit terms *draṣṭṛ* (the one who sees) and *dṛśya* (to be seen) occurring in the *bhāṣya*-part of the text are translated by the Arabic terms knower (عالم) and known (معلوم) respectively.

85 PYŚ I.41, pp. 65.1–67.10: *atha labdhasthitikasya cetasaḥ kiṃrūpā, kiṃviṣayā samāpattir iti? tan nididarśayiṣayedaṃ sūtraṃ pravavṛte. kṣīṇavṛtter abhijātasyeva maṇer grahītṛgrahaṇagrāhyeṣu tatsthatadañjanatā samāpattiḥ* (sū. I.41) [...] *abhijātasyeva maṇer iti dṛṣṭāntopādānam. yathā sphaṭika upāśrayabhedāt tadrūpoparakta upāśrayarūpākāreṇa nirbhāsate, tathā grāhyālambanoparaktaṃ cittaṃ grāhyasamāpannaṃ grāhyasvarūpākāreṇa nirbhāsate. tathā bhūtasūkṣmoparaktaṃ bhūtasūkṣmasamāpannaṃ bhūtasūkṣmarūpābhāsaṃ bhavati. tathā sthūlālambanoparaktaṃ sthūlasamāpannaṃ sthūlarūpābhāsaṃ bhavati.*

The second part of the passage drawn from the *Kitāb Pātanǧal*, that is: "Thus, objects are [seemingly] in it, although they are external to it. In the same way, he (i.e., the ascetic) comprises that which encompasses him, so that when knowing and known [objects] are united in him who is the knower, then intellection, the one who intellects and what is intellectualized upon become one thing in him" appears to be a simplified and summarized version of the content found in the *Pātañjalayogaśāstra*, which reads: "likewise the mind (*citta*) tinted by the support of the perceptible, fallen into the state of the perceptible, irradiates [differently] depending upon the colour and the form of the perceptible." The Sanskrit text then provides a few additional examples. Even if al-Bīrūnī's wording is very concise as compared the *bhāṣya*-part of the *Pātañjalayogaśāstra* in this place, he must have based his account on it.

Instances of such silent insertion are numerous. The aforementioned analogies of Nandīśvara and Nahuṣa, as well as that of the husked or unhusked rice grains, equally stand as examples of the *bhāṣya*'s influence on al-Bīrūnī's works. Only the *bhāṣya*-parts, and not their respective *sūtra*s, indeed, refer to these analogies.[86]

It is worth digressing from the main topic of the present section in order to highlight an additional way in which al-Bīrūnī organized the content of his source. The first quotation below is a passage drawn from the *Taḥqīq mā li-l-Hind* which refers to the commentator in the *Kitāb Pātanǧal*. The second one is the corresponding passage in *Kitāb Pātanǧal* found in KP 46. The quotations deal with cosmography and the Mount Meru. The first quotation reads:

> For instance, the commentator of the *Kitāb Pātanǧal* exaggerates [by transforming] the square [shape of Meru] into an oblong shape.[87] He fixes [the length of] one of its four sides at fifteen *koṭi*s *yojana*s (كورتى جوژن), which is 150'000'000 [*yojana*s], and that of the other [side] at five *koṭi*s, namely a third of the former. With regard to its four sides, on the east are the mountain Mālava (مالو) and the ocean, and between them the kingdoms called Bahaḍrāsa (بهدراس). On the north are the mountains Nīra (نير), Šīta (شيت) and Šrangādar (شرنگادر), and the ocean, and between them the kingdoms Ramīku (رميك), Harinmāyān (هرنماى) and Kur (كر). On the west are the mountain Gandamādan (گندمادن) and the ocean, and between them the kingdom Kītumāla (كيتمال). On the south are the mountains Mrābta (مرابت), Nišada (نشد), Hīmakūta (هيمكوت), Himagiru

86 PYŚ (1904), pp. 67.7–68.10.

87 On this passage see Pines & Gelblum 1983: 278, n. 131.

> (همگر) and the ocean, and between them the kingdoms Baharaṯa Barša (بهارث برش), Kīnpuruśa (كينپرش) and Haribarša (هرپرش).[88]

Whereas al-Bīrūnī provides the names of several mountains and kingdoms that surround Mount Meru in this quotation, the parallel passage in the *Kitāb Pātanğal*, does not specify those names and is much shorter:

> In the middle of the island which we inhabit is Mount Meru, the habitation of the angels. One of its four sides is five *koṭis* (كورتى). On its four sides are mountains, kingdoms, rivers and seas; there is no use in enumerating them because they are unknown, nor in naming them because the [names] are in the Indian [language].[89]

Thus, in this passage, al-Bīrūnī explicitly justifies his choice to not enumerate the names of the different mountains, kingdoms, etc. His statement suggests that he knew these names, despite their absence in this passage, and his knowledge of them is indeed confirmed by the parallel passage in the *Taḥqīq mā li-l-Hind* quoted above. In addition, these names are found in the *bhāṣya*-part of PYŚ (1904) III.26. The *Pātañjalayogaśāstra* reads:

> North of the Sumeru are three mountains, Nīla, Śveta and Śṛṅga, which span two thousand *yojana*s. Between these mountains, are three regions (*varṣa*), [spanning] nine thousand *yojana*s each called Ramaṇaka, Hiraṇmaya and Northern Kurus. To the south [of Mount Sumeru], are the mountains Niṣadha, Hemakūṭa and Himaśaila, [covering] two thousand *yojana*s. Between these [mountains, are] three regions [stretching over] nine thousand *yojana*s each, called Harivarṣa, Kiṃpuruṣa and Bhārata. To the East of Sumeru, [lies] Bhadrāśva, bounded by Mālyavat [mountains]. To its West, [is the country of] Ketumāla, bounded by the Gandhamādana [mountains]. In the middle, [is] the region [called] Ilāvṛta.[90]

---

88 *Taḥqīq* (1958), p. 205.14–206.3; Sachau 1910: I/248–249.

89 KP 46, p. 187.4–7; Pines & Gelblum 1983: 261.

90 PYŚ (1904) III.26, pp. 149.20–150.2: *tasya nīlaśvetaśṛṅgavanta udīcīnās trayaḥ parvatā dvisāhasrāyāmāḥ. tadantareṣu trīṇi varṣāṇi nava nava yojanasāhasrāṇi ramaṇakaṃ hiraṇmayam uttarāḥ kurava iti. niṣadhahemakūṭahimaśailā dakṣiṇato dvisāhasrāyāmāḥ. tadantareṣu trīṇi varṣāṇi nava nava yojanasāhasrāṇi harivarṣaṃ kiṃpuruṣaṃ bhāratam iti. sumeroḥ prācīnā bhadrāśvamālyavatsīmānaḥ pratīcīnāḥ ketumālā gandhamādanasīmānaḥ. madhye varṣam ilāvṛtam.*

In addition to the differences between the two Arabic quotations, there are a few discrepancies between them, on the one hand, and the *Pātañjalayogaśāstra*, on the other. Al-Bīrūnī's two Arabic passage do not provide the same size of the different regions than the *Pātañjalayogaśāstra* and make mention of oceans (and rivers), which are absent from *Pātañjalayogaśāstra*.

Notwithstanding these discrepancies, the two enumerations correspond quite well: to the east of Mount Meru can be found the mountains called Mālwa (مالو), that is, Mālyavat; between Mālyavat and Mount Meru is situated the kingdom of Bahadrāsa (بهدراس), or Bhadrāśva. Al-Bīrūnī's Nīra (نير), Šīta (شيت) and Šrangādar (شرنگادر) located to its north stand for the Sanskrit Nīla, Śveta and Ṣṛṅga. The kingdoms of Ramīku (رميك), Harinmāyān (هرنماى) and Kur (كر), respectively corresponding to Ramaṇaka, Hiraṇmaya and Kuru, are situated between these mountains and Mount Meru. To its west are the mountains Gandamādan (گندمادن), that is, Gandhamādana, and between Mount Meru and these mountains is situated the kingdom of Kītumāla (كيتمال), standing for Ketumāla. To its south can be found Mrābta (مرابت), the probable rendering of Ilāvṛta, Nišada (نشد), namely Niṣadha, Hīmakūta (هيمكوت), or Hemakūṭa, and Himagiru (همگر), which may stand for the Sanskrit *himaśailā*. At the end of this passage, al-Bīrūnī enumerated the names of the following kingdoms: Baharaṯa Barša which transposes Bhāratavārṣa, Kīnpuruśa the rendering of Kiṃpuruṣa, and Haribarša that corresponds to Harivarṣa. All names found in the *bhāṣya*-part of the *Pātañjalayogaśāstra* thus appear in al-Bīrūnī's description.

Thus, al-Bīrūnī used a section of his Sanskrit source, presumably the *Pātañjalayogaśāstra*, in the *Taḥqīq mā li-l-Hind*, but omitted details of the corresponding section in his full translation of it, that is, the *Kitāb Pātanǧal*. In addition, he appears to have given a summary of the *Pātañjalayogaśāstra* in the *Kitāb Pātanǧal*. This short digression indicates an additional way in which al-Bīrūnī dealt with this Sanskrit source, beside his two manners of incorporating the commentary of his source-text into his translation.

5.3 A Problematic Laudatory Passage

Al-Bīrūnī was thus greatly inspired by the *Pātañjalayogaśāstra* when he wrote the *Kitāb Pātanǧal*. Nevertheless, an essential point can impede the definite identification of the *Pātañjalayogaśāstra* as his source. The *Kitāb Pātanǧal* contains a problematic laudatory passage that occurs after al-Bīrūnī's own preface to his translation and before the beginning of section 1 of the *Kitāb Pātanǧal* itself.[91] This laudatory passage begins after the sentence "This is the beginning

91 The question of the laudatory passage was first broached in Maas & Verdon 2018: 314–315.

of the *Kitāb Pātanǧal*, its text [being] interwoven with its commentary" (و هذا هو ابتداء كتاب باتنجل مركّبًا نصه بشرحه),[92] which obviously originates from al-Bīrūnī's own hand. The whole passage reads as follows:

> I prostrate [myself] before Him above whom there is nothing, and I glorify Him who is the beginning of things and to whom they shall return, the knower of all beings. Then, I exalt those below Him, the angels and the spiritual beings, with a humble soul and a pure intention, and I call upon them to help me in my discourse—which I wish to keep short—according to the method of Hiraṇyagarbha (هيرنيكرب).
>
> The ancients have been deeply engaged in the study of the things through which the four objectives may be achieved. These [objectives] are religion and conduct [of life], possessions and comfort, [enjoyable] way of living and pleasure, liberation and permanence. [The ancients examined these objectives] in such a way that [they] scarcely left room for discourse for the later [authors]. However, my discourse excels in clearing up the ambiguities which they stated. It is confined to the means leading to the perfection of the soul through liberation from these bonds and to the attainment of eternal bliss.
>
> Accordingly, I say: with regard to things which are hidden from apprehension [by senses], the attribute [of not being apprehended] can only be ascribed to them on [various] conditions: [their] essential smallness, like atoms which are hindered from the senses because of [their] minuteness; [their] remoteness as the distance prevents apprehension when it extends beyond its limit; an obstacle which conceals, like a wall which prevents apprehension of what is placed behind it, like bones covered by flesh and skin, and like [liquid] mixtures inside the body, as they cannot be perceived because of veils [which are] between us and them; their being distant from the present time, either [because of their being] in the past, like the past generations and the tribes which have perished, or [because of their being] in the future, like things [which are] expected [to happen] in the time to come; the deviating from the methods of knowledge through which comprehension becomes perfected, like the case of necromancy whereby the state of hidden things is discovered. It is known that the perfection of ascertainment can necessarily only exist through direct perception. This [direct perception as a means of knowledge] is eliminated in the case of hidden things, because what

92 KP, p. 168.5; Pines & Gelblum 1966: 310.

is absent can only be inferred from something present, and that which can be attained through arguments is not in the same [category] as that which is known through direct perception. Similarly, logical demonstration removes doubts as [does] direct perception. As long as ambiguities affect the soul, the latter is occupied with confusion, not being free [to reach] what [procures] its liberation from this entanglement, its deliverance from adversity and fetter, and its eternal sojourn where there is neither death nor birth.

Most of the intentions of the expounders of books are either [directed] to the creation of a discourse peculiar to them or to the guidance toward a goal which they pursue. The goals are determined according to the knower, and the knowledge is divided into two parts: the higher of them leading to liberation, because it procures the absolute good; and the lower of them in comparison to the other one [referring to] the remaining [aforementioned four] objectives which rank lower than [liberation]. I shall try to [see] if my discourse has for those who hear it a status [similar] to the convincing [means of knowledge of] direct perception, in comparison to the arguments which the ancient [authors] presented on this hidden subject.[93]

Such an introduction does not occur in the *Pātañjalayogaśāstra* and does not correspond to the introductory passages found in other Sanskrit commentaries. It contains a benedictory verse to a superior being, probably Allah/Īśvara, to the angels (ملايكة), translating the Sanskrit *deva*, and to spiritual beings (روحانيين). The author of this passage recognizes Hiraṇyagarbha's method as authoritative and as a source of inspiration. Whereas this name does not appear in the *Pātañjalayogaśāstra* itself, Hiraṇyagarbha is explicitly acknowledged as playing a part in the transmission of Yoga teachings in several of this text's commentaries, as seen in Section 3.3.3 of this book. Al-Bīrūnī does not refer to Hiraṇyagarbha anywhere else. However, it is possible that his informants assisted him and suggested he pay homage to Hiraṇyagarbha in the *Kitāb Pātanǧal*.[94]

The other elements present in this passage are not discussed in the extant commentaries al-Bīrūnī could have used. These are: the four human goals (المطالب الاربعة; *puruṣārtha*), namely religion and conduct of life (الدين والسيرة; *dharma*), possessions and comfort (المال والنعمة; *artha*), [enjoyable] way of living and pleasure (العيش واللذه; *kāma*) and liberation and permanence (الخلاص

93 KP, pp. 168.6–169.9; Pines & Gelblum 1966: 310–313.

94 On the significance of al-Bīrūnī's informants, see chapter 1, section 2.4 and chapter 4, section 4.3 of the present book.

والديمومة; *mokṣa*); the reasons why things are hidden from perception (الاشياء التى تغيب عن الادراك); and the means of valid knowledge (برهان; *pramāṇa*).[95] The four human goals are fundamental beliefs in Brahminical thought, while reasons why things are hidden from perception and means of valid knowledge are common themes in Indian thought. For instance, Sāṅkhya (kā. 7) examines the former concepts,[96] while both Sāṅkhya and Yoga (kā. 4; PYŚ I.7) discuss the latter ones. Therefore, Indian informants and/or Sanskrit works may have influenced al-Bīrūnī to add these topics in the laudatory introduction to his translation.

Lastly, al-Bīrūnī makes use of the first person in this introduction. He also employs the first person for his preface to the *Kitāb Pātanğal* directly preceding this passage. It, however, appears unlikely that, in the case of the laudatory introduction, the first person should refer to al-Bīrūnī himself. First, the statement "this is the beginning of the *Kitāb Pātanğal*, its text [being] interwoven with its commentary" introducing the passage strongly suggests that the translation per se starts at this point in the text. Moreover, whereas the beginning of the passage praising the superior being, the angels and spiritual beings could reflect al-Bīrūnī's own beliefs, other elements such as Hiraṇyagarbha, the four human goals, the reasons for the non-perceptions of things and the means of valid knowledge, are clearly related to Indian thought. It would then be very surprising for al-Bīrūnī to officially avow such a position for himself.

It is thus likely, as Maas and Verdon suggested, that al-Bīrūnī, inspired by his own knowledge of Indian philosophy and religion, decided to include this passage to his translation of his own initiative and/or under the guidance of his Indian informants. Such introductions including a laudation to divine beings and further describing the author's motives as well as the work's subject matter are shared common practices in the Arab literary tradition and in Indian scholastic tradition.[97] If the identification of the *Pātañjalayogaśāstra* with the main source of the *Kitāb Pātanğal* is correct, al-Bīrūnī would have included in his translation elements that his source and/or his informants considered essential topics so as to provide a complete Arabic text on Yoga for his Muslim audience.

95 See Gelblum's comments on this introductory passage in Larson & Bhattacharya (2008: 263).

96 These reasons for why somebody does not perceive objects provided by the Sāṅkhya system and by the *Kitāb Pātanğal*, however, differ in number and in kind.

97 Funayama 1995: 181; Minkowski 2008; Maas & Verdon 2018: 314–315.

5.4 Concluding Remarks

Thus, the present chapter aimed to show first that al-Bīrūnī needed to study a commentary in order to understand the aphoristic text related to Yoga philosophy, which he translated into Arabic. Second, it demonstrated that this commentary existed in the *bhāṣya*-part of the *Pātañjalayogaśāstra* or in a work very much similar to it. It was possible to reach a conclusion by studying the whole of the *Kitāb Pātanǧal*, instead of excerpts found in the *Taḥqīq mā li-l-Hind*, and by taking on the new perspective presented in Chapter 4, that is, analysing in depth al-Bīrūnī's hermeneutics, intellectual background, contacts with Indians, knowledge of Indian literature and possible motives.

Thanks to this approach, this chapter thoroughly examined different ways in which al-Bīrūnī transformed his source. Al-Bīrūnī in many cases rephrased the commentary he used and integrated it in the *Kitāb Pātanǧal*, either implicitly or explicitly. In general, he silently combined the two layers of the Sanskrit text in his Arabic translation, while often summarizing the commentary part. On two occasions, however, he expressed that he integrated the Sanskrit commentary into his translation. The reasons why he handled his source differently only in these two cases remain uncertain. However, the topics of these incorporated passages relate to astronomy and medicine. Since al-Bīrūnī considered these subjects particularly essential, he may have opted for the full retransmission of these passages in his translation.

In addition, al-Bīrūnī extensively rephrased and defined technical concepts of the original Sanskrit *Pātañjalayogaśāstra*. These modifications are particularly evident from the way he translated the titles of the chapter-colophons and from his interpretation of the passage extending from PYŚ (1904) III.21 to III.35 (Section 5.2.2). As he reshaped his source into a dialogue, some of the questions in the Arabic text almost directly reflect questions introducing the *sūtras*. Al-Bīrūnī also used quotations of the *Kitāb Pātanǧal* in his *Taḥqīq mā li-l-Hind*, but the quoted passages did not always correspond exactly to the *Kitāb Pātanǧal*. For instance, Section 5.2.2 also highlighted that he provided in the *Taḥqīq mā li-l-Hind* a translation of a passage, originally drawn from the *Kitāb Pātanǧal*, which may have been more faithful to the original Sanskrit than the text in the *Kitāb Pātanǧal* itself.

One passage of the *Kitāb Pātanǧal* remains, however, enigmatic, the initial laudatory introduction, because it did not find any correspondence in the *Pātañjalayogaśāstra* nor in any other Sanskrit commentary on this work. The topics dealt with in this passage are however widely known subjects, related to Indian thought and religion. Thus, Indian informants and other Sanskrit literature known to al-Bīrūnī could well have inspired him to add such an introduction to his translation. In addition, one or several of his informants provided him with the necessary supplementary information to understand his Sanskrit source in the form of an oral commentary.

Further, investigating similes and metaphors occurring in the *Kitāb Pātanğal* may constitute an additional way of determining its source. However, as seen in Chapters 4 and 6 of the present book, analogies were also highly subject to al-Bīrūnī's adaptations, namely substitutions, additions and omissions.

On the whole, I foregrounded the importance of the role of al-Bīrūnī's informants, providing him with an oral commentary, of his Sanskrit sources, his concerns for his readership and his motives in his transmission of Indian thought to his peers. Lastly, I argued that al-Bīrūnī's use of a different extant Sanskrit commentary than the *Pātañjalayogaśāstra*, such as the *Vivaraṇa*, the *Tattvavaiśāradī* and the *Rājamārtaṇḍa* could not explain differences between the *Pātañjalayogaśāstra* and the *Kitāb Pātanğal*. Al-Bīrūnī thus did not need to use a written commentary besides the *Pātañjalayogaśāstra*, or a work similar to it, in order to compose the *Kitāb Pātanğal*.

CHAPTER 6

On the Kitāb Sānk and Its Sources

6.1 Scholarship Review

In the preceding chapter, I employed an effective method to understand the relationship between al-Bīrūnī's *Kitāb Pātanǧal* and its possible Sanskrit source. The present chapter adopts a similar approach in order to puzzle out the origin of the *Kitāb Sānk*. As in the preceding chapters, I do cross-examination from the perspective of al-Bīrūnī's interpretative choices and motives. As I describe below, however, due to the fragmentary character of the *Kitāb Sānk*, the study requires a slightly different method.

As in the case of the *Kitāb Pātanǧal*, scholars, such as Carl Edward Sachau (1910), Richard Garbe (1894, 1896 and 1917) and Junjiro Takakusu (1904) have all attempted to identify al-Bīrūnī's source for the *Kitāb Sānk*. New literary material has since been unearthed and, in general, academic insight into the Sanskrit textual tradition has significantly increased. These scholars compared the extracts from the *Kitāb Sānk* in the *Taḥqīq mā li-l-Hind* to the texts of the **Suvarṇasaptati, Gauḍapādabhāṣya, Tattvakaumudī* and *Sāṅkhyapravacana.* They had no access to other commentaries on the *Sāṅkhyakārikā*, such as the *Yuktidīpikā*, the *Sāṅkhyasaptativṛtti*, the *Sāṅkhyavṛtti*,[1] the *Māṭharavṛtti*[2] and the *Jayamaṅgalā*.[3] No one after them re-examined the question of the possible Sanskrit source of the *Kitāb Sānk*. In the present chapter, after outlining the observations and arguments presented by the above scholars, I analyse relevant passages belonging to the Arabic *Kitāb Sānk* and to the *Sāṅkhyakārikā* and its commentaries, the **Suvarṇasaptati, Gauḍapādabhāṣya, Tattvakaumudī, Yuktidīpikā, Sāṅkhyasaptativṛtti, Sāṅkhyavṛtti, Māṭharavṛtti* and *Jayamaṅgalā.*

Sachau was the first scholar to discuss the relationship between the *Kitāb Sānk* and Sanskrit literature on Sāṅkhya. The basis of his analysis is a comparison between the *Kitāb Sānk* and three Sanskrit works: the *Sāṅkhyapravacana* by Vijñānabhikṣu, the *Sāṅkhyakārikā* by Īśvarakṛṣṇa and the *Gauḍapādabhāṣya*

1 The *Sāṅkhyasaptativṛtti* and the *Sāṅkhyavṛtti* were both edited by Esther A. Solomon (1973a and 1973b).

2 The *Māṭharavṛtti* was discovered in 1917 (Keith 1924: 551).

3 The *Jayamaṅgalā* was edited for the first time in 1926 (Sarma 1926). See also Sarma 1985.

by Gauḍapāda.[4] The date of the composition of the *Sāṅkhyapravacana* follows that of the *Kitāb Sānk* by several centuries, as Vijñānabhikṣu lived in the sixteenth century CE.[5] Unsurprisingly, thus, Sachau found little in common between the text of this work and the *Kitāb Sānk*. The comparison between al-Bīrūnī's translation and the two other treatises yields, in his opinion, more fruitful results. Sachau noticed that both the *Sāṅkhyakārikā* and the *Kitāb Sānk* "teach *moksha* by means of knowledge."[6]

He then concludes that Īśvarakṛṣṇa's "words show that he copied from a book like the *Sâṁkhya* of Alberuni" and that Gauḍapāda "seems to have taken his information from a work near akin to, or identical with, that *Sâṁkhya* book which was used by Alberuni."[7] He also noticed that the analogies are presented in a more extensive way in al-Bīrūnī's work than in Gauḍapāda's. Sachau's preliminary observations suggest that al-Bīrūnī not only translated the *Sāṅkhyakārikā*, but also one of its commentaries, which will become evident thanks to the analyses of portions of the *Kitāb Sānk* presented below in Section 6.3.

Garbe was the second scholar to address the question of the source of al-Bīrūnī's *Kitāb Sānk* He noticed striking similarities between the latter and the *Gauḍapādabhāṣya*. For this reason, in the first edition of his *Die Sâṃkhya-Philosophie* published in 1894, he concludes that the source of the *Kitāb Sānk* is the *Gauḍapādabhāṣya*.[8] However, this identification raises two problems. First, as aforementioned, al-Bīrūnī enumerates beside the *Kitāb Sānk*, a book composed by "Gauḍa the ascetic" (گور الزاهد).[9] According to Garbe, Gauḍa's book should be identified with the *Gauḍapādabhāṣya*, and at the same time with the source of the *Kitāb Sānk*. However, al-Bīrūnī lists the book by Gauḍa and the *Kitāb Sānk* one after the other, as if he considered them as two distinct works yet belonging to a group of related literature. If indeed the Gauḍa mentioned by al-Bīrūnī is the author of the *Gauḍapādabhāṣya*, then this latter commentary cannot be the source of the *Kitāb Sānk*. This is also the view of Takakusu (1904a: 26 and 35).

Garbe's conclusion, however, is mostly due the fact that at his time, among the very few extant commentaries on the *Sāṅkhyakārikā*, the text of the *Gauḍ-*

4 Sachau: 1910: II/266–268.
5 Sachau used the edition by Ballantyne (1885). About Vijñānabhikṣu's date see Larson & Bhattacharya 1987: 375–412 and 2008: 295–333; Maas 2006: xviii; Nicholson 2010: 6.
6 Sachau 1910: II/267.
7 Sachau 1910: II/267.
8 Garbe 1894: 7 and 66.
9 See above p. 91.

apādabhāṣya was the only one chronologically fit for such identification and provided the closest parallels to the relevant passages.[10] The availability of other commentaries on the *Sāṅkhyakārikā* that were composed prior to the *Taḥqīq mā li-l-Hind*, however, solves this first problem. Second, among these commentaries, some of them resemble the *Kitāb Sānk* more than the *Gauḍapādabhāṣya*, for instance the *Sāṅkhyavṛtti* and the **Suvarṇasaptati*. Furthermore, I present below further textual evidence that suggests that the *Gauḍapādabhāṣya* is unlikely to have been the main source of the *Kitāb Sānk*.

In 1904, Takakusu brought to light, and translated into French, the Chinese translation by Paramārtha of a Sanskrit commentary on the *Sāṅkhyakārikā*. In his study, Takakusu compared the **Suvarṇasaptati* to the *Gauḍapādabhāṣya* and to the *Kitāb Sānk*.[11] One of Takakusu's aims was to determine the Sanskrit text upon which the Chinese **Suvarṇasaptati* was based. In summary, Takakusu observed that the *Gauḍapādabhāṣya* is shorter than the **Suvarṇasaptati* and the *Kitāb Sānk*.[12] He thus came to the conclusion that Paramārtha and al-Bīrūnī used the same commentary as the source of their respective translations, and that the *Gauḍapādabhāṣya*—which he did not consider as the source of the Chinese or Arabic translations—is also indebted to this initial work.[13] In the second edition of *Die Sâṃkhya-Philosophie* published in 1917, Garbe changed his initial position, following Takakusu's analysis, and identified the source of the *Kitāb Sānk* as identical to the source of the Chinese **Suvarṇasaptati*.[14]

Esther A. Solomon, after having edited the *Sāṅkhyavṛtti* and *Sāṅkhyasaptativṛtti*, examined the relationship of these two texts to other commentaries, including the *Kitāb Sānk*. She first observed that three commentaries, namely the *Yuktidīpikā*, the *Jayamaṅgalā* and the *Tattvakaumudī*, generally diverge from the other extant commentaries.[15] She also highlighted a striking resemblance between the *Māṭharavṛtti* and the *Sāṅkhyasaptativṛtti*. Her analysis led her to conclude that the *Sāṅkhyavṛtti* must be not only the earliest extant commentary on the *Sāṅkhyakārikā*, but also the Sanskrit source of Para-

10 Garbe 1894: 63–64 and 66.

11 Takakusu 1904a and 1904b.

12 Takakusu 1904a: 27 and 33–34.

13 Takakusu 1904a: 2–4, 25–26 and 35.

14 Garbe 1917: 91–93. Filliozat, who broaches the question of the *Kitāb Sānk*'s source in passing, writes: "Takakusu demonstrated that al-Bīrūnī's source is Paramārtha" (Fr. Takakusu a démontré que la source d'al-Bīrūnī est Paramārtha; Renou & Filliozat 1953: 37).

15 Solomon 1974: 1.

mārtha's Chinese translation and al-Bīrūnī's Arabic work.[16] Her editions of the *Sāṅkhyavṛtti* and *Sāṅkhyasaptativṛtti* represent a significant contribution to the scholarly research on Sāṅkhya. However, as Wilhelm Halbfass observed in two reviews of her aforementioned works, deeper analysis needs to be done in order to confirm or refute her claims.[17]

In the following sections, I highlight the intimate connection between the *Sāṅkhyavṛtti*, the **Suvarṇasaptati* and al-Bīrūnī's *Kitāb Sānk* as already noticed by Solomon, as well as the close relationship between the *Māṭharavṛtti* and the *Sāṅkhyasaptativṛtti*. I also identify some differences between the *Sāṅkhyavṛtti*, the **Suvarṇasaptati* and the *Kitāb Sānk*. In my view, thus, the *Sāṅkhyavṛtti* was probably not the direct source of the two translations, but rather a text very much like it.

6.2 Methodological Considerations

This section discusses general questions related to methodology, as well as to the selection and analysis of some of these passages presented in Section 6.3. While the method applied here is the same as the one employed in Chapters 4 and 5, in contrast with the *Kitāb Pātanǧal*, however, it is nearly impossible to establish for sure whether al-Bīrūnī omitted some parts of his Sanskrit source in the *Kitāb Sānk*, as no manuscript of this work is available. I also highlighted in Chapter 5 that al-Bīrūnī dealt differently with his complete Arabic translation of the *Kitāb Pātanǧal* from its quotations found in the *Taḥqīq mā li-l-Hind*.[18] Thus, observations pertaining to excerpts of the *Kitāb Sānk* found in the *Taḥqīq mā li-l-Hind* may not be true for a hypothetical complete text.

Furthermore, one of the works used in this study, the **Suvarṇasaptati*, is a Chinese translation of the *Sāṅkhyakārikā* and a commentary. As such, its translator Paramārtha also made changes to it.[19] Only comparing the **Suvarṇasaptati* and the *Kitāb Sānk*, which however appear to present many common points cannot lead to a definitive conclusion. Despite these problems, it is possible to examine the relationship between the *Kitāb Sānk* on the one hand, and the Sanskrit (and Chinese) commentarial tradition on the *Sāṅkhyakārikā* on the other.

16 Solomon 1973b: 5–7, 1974: 100 and 106.
17 Halbfass 1976 and 1977.
18 See above pp. 188–190.
19 Toru Funayama (2010: 168–177) discusses this process in Paramārtha's other translations.

In addition, many passages of the *Kitāb Sānk* that are analysed below narrate analogies, which are either found in the *kārikās* and contextualized in the commentaries on them, or referred to in the commentaries themselves. The *Yuktidīpikā*, the *Tattvakaumudī* and the *Jayamaṅgalā* do not contextualize these analogies or do it in a different way from the other commentaries on the *Sāṅkhyakārikā* and from the *Kitāb Sānk*. In general, these three Sanskrit commentaries do not offer the same narrative form as the aforementioned group of five commentaries and the *Kitāb Sānk*. Even so, I take the *Yuktidīpikā*, the *Tattvakaumudī* and the *Jayamaṅgalā* into account, as it will enable to definitively exclude each of the three commentaries from being the source of the *Kitāb Sānk*.

Lastly, Table 12 lists the topics of all passages of the *Taḥqīq mā li-l-Hind* that refer to the *Kitāb Sānk* and which I provide in full in the Appendix of the present book. Takakusu compared most of them to the text of the *Gauḍapādabhāṣya* and **Suvarṇasaptati*. In doing so, he identified the *kārikās* to which these passages relate.[20] In Table 12, I reproduce Takakusu's results, while summarizing my comments upon, or my revisions of them. I add to Takakusu's list one passage (no. XXI), which contains a mere reference to the *Kitāb Sānk*.

Furthermore, among al-Bīrūnī's many references to the *Kitāb Sānk* in the *Taḥqīq mā li-l-Hind*, I designate some of them as "indirect references" and others as "direct references". The former are not explicitly quotations from the *Kitāb Sānk*, but their content relates in some way to Sāṅkhya, while the latter are quotations from the Arabic translation by al-Bīrūnī.

Takakusu established preliminary connections between the above passages and the content of specific *kārikās*. Therefore, his study offers an invaluable starting point for an analysis of these passages. Even so, it is marred by a particular flaw, because his list also includes passages that are mere references to Sāṅkhya topics, rather than explicit quotations from the *Kitāb Sānk*. Moreover, some of the indirect references that can be traced back to commentaries on the *Sāṅkhyakārikā*, such as the *Gauḍapādabhāṣya* and **Suvarṇasaptati*, concern topics that Indian thought has widely developed. These topics, such as the analogy of the driver of a chariot and the five vital breaths, are thus not confined to Sāṅkhya philosophy.

In Section 6.3, I do not take into the account seven excerpts that al-Bīrūnī does not explicitly connect to the *Kitāb Sānk* or that deal with general topics of Indian thought, because they may have been drawn from another source than

20 Takakusu 1904a: 27–32.

the *Kitāb Sānk*, be it by way of oral transmission or in the form of Sanskrit texts. These seven extracts are passages II,[21] III, IV,[22] V,[23] XIII, XIV[24] and XV of Table 12.

Further, a particular section of the *Taḥqīq mā li-l-Hind* that is explicitly marked as indebted to the *Kitāb Sānk* lists five analogies that deal with the relationship between matter, action and soul, in connection with the three primary forces (القوى الثلاث الاول), that is, the Sanskrit constituents (*guṇa*). They are passages VI, VII, VIII, IX and X in Table 12. These passages succeed each other in the text of the *Taḥqīq mā li-l-Hind* as if they constituted a logical whole in al-Bīrūnī's eyes. Their separation from each other, which admittedly assists their analysis, is thus artificial.

Only the first analogy (passage VI) is explicitly drawn from the *Kitāb Sānk*, as it directly follows a general statement that is introduced by the sentence "and as for the *Kitāb Sānk*, it relates action to matter" (و أما في كتاب سانك فانّه ينسب الفعل إلى المادّة).[25] The second of these analogies (passage VII) most probably consists of a quotation from the *Kitāb Sānk*, because it starts with the expression "it relates action to the soul" (ينسب الفعل إلى النفس),[26] the subject of the sentence, the *Kitāb Sānk*, being implied here. The three subsequent analogies in passages VIII, IX and X may belong to the source of the *Kitāb Sānk*. They are introduced with the verbal form "they say" (قالوا) in the *Taḥqīq mā li-l-Hind*, which likely refers to philosophers who accompanied al-Bīrūnī in his translation of the *Kitāb Sānk*, as no other referent can be linked to this verbal form. They can be connected to some commentaries on kā. 16, 13 and 17, respectively.

Among the analogies, only the ones in passages VI (the soul as a passer-by) and VIII (the rainwater which varies in taste) provide fruitful results for clarifying the relationship between the *Kitāb Sānk* and the Sanskrit commentaries. Observations arising from the analysis of passage VIII, however, need to be con-

21 Due to the very obscure origin of this passage and the fact that it refers to the general Sāṅkhya theory of evolution, I refrain from connecting it with a specific *kārikā*. For a detailed discussion of this passage, see above pp. 85–88.

22 Takakusu (1904a: 28) entitles this excerpt "comparison of nature (i.e., *prakṛti*) to a female dancer" (Fr. Comparaison de la nature à une danseuse). Since al-Bīrūnī compares the soul to a female dancer in this passage, the passage is referred to in this book with the title "the soul as a female dancer."

23 According to Takakusu (1904a: 3), the analogy of the blind and the lame is peculiar to the *Sāṅkhyakārikā* and its commentarial tradition. See also Apte 1957: Appendix E: 66 and Jacob 2004: 34.

24 Takakusu (1904a: 31) links this passage to the **Suvarṇasaptati* and the *Gauḍapādabhāṣya* glossing kā. 49. If this excerpt is based on the *Kitāb Sānk*, its content, however, rather parallels that of kā. 43 and commentaries.

25 *Taḥqīq* (1958), p. 36.16–17; Sachau 1910: 1/48.

26 *Taḥqīq* (1958), p. 37.5; Sachau 1910: 1/48.

TABLE 12 Passages related to the *Kitāb Sānk* and/or to Sāṅkhya concepts found in the *Taḥqīq mā li-l-Hind*[27]

No.	*Kitāb Sānk*	Corresponding *kārikās* (according to Takakusu)	Comments and revisions by the author
I	Six views on action and agent.	Kā. 27 and 61.	Kā. 61; Number 2 of Table 6, in Chapter 3.
II	Enumeration of the twenty-five principles (*tattva*).	Kā. 3.	Indirect reference, possibly based on oral transmission supplementing other sources. Takakusu counts twenty-four *tattva*s although al-Bīrūnī enumerates twenty-five elements.
III	The five vital breaths.	Kā. 29.	Indirect reference.
IV	The soul as a female dancer.	Kā. 42, 65, 66 and 59.	Indirect reference; kā. 59 and 65 (~ kā. 61, 64 and 66).
V	The blind and the lame.	Kā. 21	Indirect reference.
VI	The soul as a passer-by.	Kā. 19.	Direct reference; Number 3 of Table 6, in Chapter 3.
VII	The innocent man among thieves.	Kā. 20.	Direct reference; Number 3 of Table 6, in Chapter 3.
VIII	The rainwater which varies in taste.	Kā. 16.	Indirect reference.
IX	The production of light from oil, wick and fire.	Kā. 13.	Indirect reference.
X	The driver of a chariot.	Kā. 17.	Indirect reference.
XI	Rewards of attaining heaven being no special gain.	Kā. 1–2.	Direct reference; Number 4 of Table 6, in Chapter 3.
XII	Births depending on virtue and vice.	Kā. 39.	Direct reference (~ kā. 39–43): Number 5 of Table 6, in Chapter 3.
XIII	The eight powers.	Kā. 23.	Direct reference to the *Kitāb Pātanǧal*, not to the *Kitāb Sānk*.

27 Takakusu 1904a: 27–35.

TABLE 12 Passages related to the *Kitāb Sānk* and/or to Sāṅkhya concepts (*cont.*)

No.	*Kitāb Sānk*	Corresponding *kārikās* (according to Takakusu)	Comments and revisions by the author
XIV	The three types of knower.	Kā. 49.	Indirect reference to the *Kitāb Sānk*; reference to Kapila (~ kā. 43).
XV	Nine rules how to conduct one's life.	Kā. 23.	Indirect reference.
XVI	The limitations of humankind.	?	Direct reference; Number 6 of Table 6, in Chapter 3.
XVII	The movement of a weaver's wheel.	Kā. 67.	Direct reference; Number 7 of Table 6, in Chapter 3.
XVIII	States other than liberation.	Kā. 50.	Direct reference; kā. 44 and 45; Number 8 of Table 6, in Chapter 3.
XIX	Four levels of knowledge.	Kā. 46.	Continuation of Number XVIII; Number 8 of Table 6, in Chapter 3.
XX	Categories of beings.	Kā. 39, 44 and 53.	Direct reference to kā. 53; to kā. 44 (in passing); Number 9 of Table 6, in Chapter 3.
XXI	Criticism of a list of spiritual beings.	Not mentioned by Takakusu.	Direct reference, but no elaboration.

sidered with caution, since it is uncertain whether it really belongs to the *Kitāb Sānk* or not, although connected to it as explained above. Further, I also exclude passages VII, IX and X from the analysis, because these excerpts, somehow related to the *Kitāb Sānk*, however, underwent much adaptation.

The remaining passages listed in the above table (passages I, XI, XII, XVI, XVII, XVIII, XIX, XX and XXI) either consist of explicit references to, or quotations from, the *Kitāb Sānk*. Among them, passages XI,[28] XVI,[29] XVII[30] and XIX are not studied here. The first three of them have been subject to major trans-

28 This passage is not a literal translation of kā. 1–2 and one of the extant commentaries, but its content relates to the Sāṅkhya idea that the traditional teachings of the Vedas and Āyurveda are not sufficient to permanently remove suffering of existence.

29 This excerpt could not be linked to any particular portion of the *Sāṅkhyakārikā* and its commentaries.

30 This analogy appears at least in works by Vedānta and Sāṅkhya works (Apte 1957: Appendix E: 61; Jacob 2004: 27).

formations by al-Bīrūnī and are therefore not helpful in determining the relationship between the *Kitāb Sānk* and a specific commentary on the *Sāṅkhyakārikā*. The last excerpt (no. XXI) is only a reference in passing to the *Kitāb Sānk* that does not contain any substantial content.

Three passages (VII, XII and XX) lead to the exclusion of specific commentaries as the source of the *Kitāb Sānk*.[31] Thus, I consider jointly these excerpts in Section 6.3.1. Similarly, passage VI, which directly connects to the *Kitāb Sānk*, also allows us to exclude some commentaries from being the source of the *Kitāb Sānk*. It does not, however, enable us to connect it with one or two specific Sanskrit commentaries on the *Sāṅkhyakārikā*. Nevertheless, its analysis well illustrates the problems one encounters when comparing the extracts of the *Kitāb Sānk* with the extant Sanskrit commentaries on the *Sāṅkhyakārikā*. Therefore, I analyse it in Section 6.3.2.

Three passages (I, XVIII and XIX) allow us to draw parallels between the content of the *Kitāb Sānk* and that of specific Sanskrit commentaries. I thus examine them separately in Sections 6.3.3 to 6.3.5.

6.3 Passages of the Kitāb Sānk and the Tradition of the Sāṅkhyakārikā

6.3.1 *The Innocent Man among Thieves, Birth Depending on Virtues and Vices, and Categories of Beings*

Analysis of three passages of the *Kitāb Sānk* tends to exclude the *Yuktidīpikā*, the *Jayamaṅgalā*, the *Tattvakaumudī* and the *Gauḍapādabhāṣya* from being the sources of the *Kitāb Sānk*. The analysis of these excerpts, however, does not enable further parallels to be drawn between the *Kitāb Sānk* and the remaining commentaries under consideration.

The first of them is the analogy of the innocent man among thieves referenced under Number VII. It runs as follows:

> It (i.e., the *Kitāb Sānk*) ascribes action to soul, even if the [soul] is exempt from it (تبرّئها),[32] in the manner of a man who happens [to get in] a company of a group [of people] whom he does not know. They are thieves returning from a village which they have attacked and destroyed. He only marched with them a short distance, until the pursuers caught them. He is

31 Passage XII, describing two conditions of a future life resulting from one's action, has been briefly discussed above in Sections 4.3.2 and 4.4.1.

32 This reading is that of the two Arabic editions used in the present book, while the manuscript has تبروه.

> locked up among the group. The innocent [man] (البرىء) is taken in their group and, as [if he were] in their situation: [the pursuers] hit him, as they hit the [group], without him having taken part in their actions.[33]

Some commentaries on kā. 20 narrate the illustration.[34] The *Yuktidīpikā*, the *Jayamaṅgalā* and the *Tattvakaumudī* do not, on their part, mention this illustration on kā. 20, and thus may not have constituted the work used by al-Bīrūnī for this passage.[35] Whereas the *Kitāb Sānk* explains the illustration in a relatively detailed manner, the *Gauḍapādabhāṣya* offers an extremely concise reading: "just like [someone who is] not a thief, when caught together with thieves, is considered as a thief" (*yathā acauraś cauraiḥ saha gṛhītaś caura ity avagamyate*).[36]

A further analysis of this quotation and the remaining commentaries that are the **Suvarṇasaptati*, the *Sāṅkhyavṛtti*, the *Sāṅkhyasaptativṛtti* and the *Māṭharavṛtti*, does not make it possible to pinpoint one specific source for al-Bīrūnī's translation. They all narrate the story in a more developed manner than the *Gauḍapādabhāṣya* does, while presenting few discrepancies in their respective descriptions. Because of the brevity of the *Gauḍapādabhāṣya* in contrast to the description provided by the *Kitāb Sānk* and to that of the other commentaries, the *Gauḍapādabhāṣya* is unlikely to have been al-Bīrūnī's unique source for this passage.

The next excerpt, passage XII corresponding to the content of kā. 39, consists of a description of two types of births which depend on virtue and vice. Al-Bīrūnī quoting from the *Kitāb Sānk* explained that "the one who deserves ascension and reward becomes like one of the angels mixing with the community of spiritual [beings]" (أمّا من استحقّ الاعتلاء و الثواب فانّه يصير كأحد الملائكة مخالطا للمجامع الروحانيّة), whereas "the one who deserves to descend because of [his] sins and crimes becomes an animal or a plant" (أمّا من استحقّ السفول بالأوزار و الآثام فانّه يصير حيوانا أو نباتا).[37] The **Suvarṇasaptati*, the *Gauḍapādabhāṣya*, the *Sāṅkhyasaptativṛtti* and the *Māṭharavṛtti* commenting on kā. 39 all provide similar passages to this quotation.[38] These commentaries explain that the

33 *Taḥqīq* (1887), p. 24.6–10; *Taḥqīq* (1958), p. 37.5–9; Sachau 1910: I/48–49.
34 Sachau 1910: II/275. Takakusu 1904a: 29.
35 YD, pp. 181.1–183.26, JM, p. 86.9–19 and TK, p. 126.6–8. See Kumar & Bhargava 1992: 163–169.
36 GPBh, p. 23.7–8 on kā.20.
37 *Taḥqīq* (1958), pp. 48.16–49.2; Sachau 1910: I/64.
38 Takakusu 1904b: 1025, GPBh, p. 38.5–7, V1, p. 55.18–19 and MV, p. 42.6–9 on kā.39. Folios in the *Sāṅkhyavṛtti* are missing here (V2, p. 53).

subtle body (*sūkṣmaśarīra*) can transmigrate into two forms, either as animal and plant or as a deity. Although the sequence is inverted as compared to al-Bīrūnī's account, the message is the same. In contrast, the *Yuktidīpikā*, the *Jayamaṅgalā* and the *Tattvakaumudī* do not describe the two different conditions of life, nor do they mention the types of beings into which the subtle body could transmigrate on kā. 39.[39]

The third of these extracts (passage no. XX), equally consisting in a quotation from the *Kitāb Sānk*, enumerates several categories of beings. The passage reads:

> First, we tell what is in the *Kitāb Sānk* about this [topic].
>
> The hermit said: "How many categories of living bodies are there?"
>
> The sage replied: "There are three categories of them. The spiritual [ones] in the higher [world], man in the middle [world] and the animals in the lower [one]. As for their species, they are fourteen, eight of which belong to the spiritual [beings]: Brāhma, Indra, Prajāpati, Saumya, Gāndharva, Yakṣa, Rākṣasa and Piśāca (براهم وإندر وپرجاپت وسومى وگاندهرب وجكش وراكشس وپيشاچ). Among the [species], five belong to the animals: cattle, wild beasts, birds, reptiles and plants, I mean the trees. The human species is [only] one."
>
> The author of this book enumerates them in another place of the [*Kitāb Sānk*] with other names: Brāhma, Indra, Prajāpati, Gāndharva, Yakṣa, Rākṣasa, Pitra and Piśāca (براهم، اندر، پرجاپت، گاندهرب، جكش، راكشس، پتر، پيشچ).[40]

This quotation enumerates three categories of beings (أجناسها ثلاثة): spiritual [beings] (الروحانيّون), man (الناس) and animals (الحيوانات). These three categories include fourteen species (وأمّا انواعها فهى أربعة عشر) distributed as follows: spiritual beings are eight (الروحانيّين ثمانية), animals are five (الحيوانات خمسة) and man is one (الانس نوع واحد). Al-Bīrūnī also provided the names of eight types of spiritual beings and of the five kinds of animals. In this quotation, he complained that the names of the eight types of spiritual beings are given twice in the *Kitāb Sānk* with two different orders.

---

39 YD, p. 228.2–230.12, JM, p. 99.4–16 and TK, p. 146.2–12 on kā 39.

40 *Taḥqīq* (1887), p. 43.14–20; *Taḥqīq* (1958), p. 67.11–19; Sachau 1910: I/89–90. I propose the most probable corresponding Sanskrit terms according to the Arabic transliterations of the names found in the *Taḥqīq mā li-l-Hind*, even when they are found in their adjectival forms (e.g., Saumya and Gāndharva instead of Soma and Gandharva).

Sachau links this quotation to kā. 44, 53, while Takakusu adds its connection with kā. 39.[41] The two lists of spiritual beings occur in comments on kā. 44 and 53, respectively. Whereas the commentaries on kā. 39 mention three types of births, as gods, humans and animals, they remain too concise on this topic for al-Bīrūnī to have drawn all of his material from kā. 39. Furthermore, the commentaries on kā. 44 only list the divine beings and the world of animals, foregoing any mention of human beings. The perspective adopted in kā. 44 is also different from that of kā. 53. The two different worlds, divine and animal, are referred to in kā. 44 because they are consequences of the binary notions of righteousness (*dharma*) and unrighteousness (*adharma*). In contrast, the contents of the commentaries on kā. 53 and this quotation share a comparable way of numbering the different species. The Arabic extract is thus most probably a rendering of kā. 53 and commentary, while referring in passing to the second list of spiritual beings which occurs in the commentaries on kā. 44.

The *Yuktidīpikā*, the *Jayamaṅgalā* and the *Tattvakaumudī* enumerate the divine beings only when commenting on kā. 53. They could not thus give al-Bīrūnī the opportunity to comment on two different lists of these beings. The **Suvarṇasaptati*, the *Sāṅkhyasaptativṛtti*, the *Māṭharavṛtti* and the *Gauḍapādabhāṣya* provide two lists of divine beings on both kā. 44 and 53.[42] However, when confronted to al-Bīrūnī's remark that two lists are given with a different sequence in the source of the *Kitāb Sānk*, only the **Suvarṇasaptati* lists these categories in a different order in each places. The enumerations in the **Suvarṇasaptati* do not, however, reflect the sequence in which the items are listed in the *Kitāb Sānk*. Changes in the sequence of the listed names may also originate from an evolution in the textual transmission of the works. The above observations suggest that the source of the *Kitāb Sānk* was probably not the *Gauḍapādabhāṣya*, the *Yuktidīpikā*, the *Jayamaṅgalā*, or the *Tattvakaumudī*.

6.3.2 *The Soul as a Passer-by*

Chapter 4 of the *Taḥqīq mā li-l-Hind*, entitled "On the cause of action and on the attachment of the soul to matter" (فى سبب الفعل و تعلّق النفس بالمادّة),[43] references, both implicitly and explicitly, passages of the *Kitāb Sānk* that are intimately connected to metaphysical Sāṅkhya concepts. The next extract (no. VI in the Appendix) directly relates to the *Kitāb Sānk*, as it is a quotation drawn from it, and relates to its understanding of action (فعل):

41 Sachau 1910: II/290; Takakusu 1904a: 34.

42 Leaves of the *Sāṅkhyavṛtti* manuscript are missing in these two places and the commentary is partially available, from kā. 43 to kā. 50 and from kā. 51 to kā. 55. See V2, pp. 57–58.

43 *Taḥqīq* (1958), pp. 34.4–38.2; Sachau 1910: I/45–49.

a) As for the *Kitāb Sānk*, it ascribes action to matter because the shape which becomes visible is much different from the [matter] due to the three primary forces and to their dominance of one or two [over the other(s)], that are angelic, human and animal. These are the forces which belong to the [matter] not to the soul.

b) The soul shall know the actions of [matter], with the position of a spectator, in the manner of a passer-by who sits down in a village to rest. Everyone among the villagers is busy with his own particular work. The [passer-by] looks at them and considers their conditions, disliking some, liking others, and learning from them. Thus, he is busy without having himself any share in their [business] and without being the cause for the consequence of their [business].[44]

The first part a) of this reference constitutes a relatively accurate summary of the role of the three constituents (*guṇa*) in Sāṅkhya metaphysics and thus could refer to the content of several *kārikās*.[45] Therefore, this section analyses in depth only the second part b) of this reference that includes an illustration described in commentaries on kā. 19, as was rightly noted by Sachau and Takakusu.[46] The analogy aims to exemplify the relationship between the three primary forces (القوى الثلاث الاول), rendering the three constituents (*guṇa*), the soul (نفس), namely the conscious self (*puruṣa*), and action/activity. This *kārikā* and its commentaries attempt to define the conscious self:

> It is established that the conscious self is a witness, separated, neutral, seeing and non-agent because of being opposed [to the three *guṇas*].[47]

Several commentaries of the *Sāṅkhyakārikā* make use of an analogy staging an inactive spectator who observes working villagers without however intermingling with them in order to illustrate one or several of the five qualities which this *kārikā* ascribes to the conscious self. The *kārikā* itself does, however, not refer to this analogy. First, this fact confirms the observation that the *Kitāb Sānk* is a

44 *Taḥqīq* (1887), pp. 24.1–6; *Taḥqīq* (1958), pp. 36.16–37.4; Sachau 1910: I/48. Passages Numbers VI to X follow upon each other.

45 See for instance kā. 11 to 13, 16, 19, 27 and 54. Sachau suggests that this first part is connected with kā. 12 and 25 (Sachau 1910: II/274–275). However, the topic of kā. 25 is different from that of this Arabic passage.

46 Sachau 1910: II/275; Takakusu 1904a: 29.

47 SK 19: *tasmāc ca viparyāsāt siddhaṃ sākṣitvam asya puruṣasya | kaivalyaṃ mādhyasthyaṃ draṣṭṛtvam akartṛbhāvaś ca*. The opposition between the conscious self and the three constituents is made explicit in kā. 17 and 18, as well as in the comments on kā. 19.

translation of a treatise comprising a basic text resembling the *Sāṅkhyakārikā* and a commentary. The *Yuktidīpikā*, the *Jayamaṅgalā* and the *Tattvakaumudī* do not make use of this illustration at all on kā. 19.[48]

The analysis of the different versions of this passage in the remaining five commentaries under consideration and in the *Kitāb Sānk* does not, however, indicate an evident correspondence between the Arabic translation and any one single Sanskrit commentary. It rather illustrates the complexity of examining the relationship between the *Kitāb Sānk* and the Sanskrit textual tradition of classical Sāṅkhya.

First, the ways in which the commentaries make us of the analogy differ from each other. The *Kitāb Sānk* uses this analogy to illustrate only the observing quality (نظّارة) of the soul, which may refer to two qualities attributed to the conscious self in kā. 19, that of witnessing (*sākṣitva*) or/and that of seeing (*draṣṭṛtva*). The **Suvarṇasaptati* and the *Gauḍapādabhāṣya* make use of this analogy in order to explain the quality of being neutral (*mādhyasthya*) of the conscious self, and therefore are least likely to constitute al-Bīrūnī's source for this passage. As for the *Sāṅkhyavṛtti*, it positions the analogy at the end of its comment on kā. 19, thus referring to the qualities of being separated (*kaivalya*) and neutral (*mādhyasthyaṃ*) of the conscious self. The *Sāṅkhyasaptativṛtti* and the *Māṭharavṛtti* could have inspired the *Kitāb Sānk* here, as they both narrate this analogy at the beginning of their discussion to evidently explain the quality of witnessing (*sākṣitva*) of the conscious self. However, al-Bīrūnī's choice to explain the observing quality of the soul through this illustration may also be due to adaptations he made when interpreting the Sanskrit text.

Second, al-Bīrūnī's quotation labelled the person involved in the events a passer-by (أحد السابلة). Yet, this exact qualifying term cannot be found in any other commentary under scrutiny here. The *Sāṅkhyasaptativṛtti* and the *Māṭharavṛtti* both use the term religious mendicant (*bhikṣu*), while the **Suvarṇasaptati* refers to this person as an ascetic mendicant (Fr. *ascète mendiant*). The *Gauḍapādabhāṣya* and the *Sāṅkhyavṛtti* employ the term wandering religious mendicant (*parivrājaka*). The idea of wandering, or travelling, is associated with the Sanskrit *parivrājaka* used in the *Gauḍapādabhāṣya* and the *Sāṅkhyavṛtti*, rather than with *bhikṣu* used by the other commentaries. Therefore, the traveller al-Bīrūnī depicted is perhaps a free translation of *parivrājaka*.

Third, the narrative takes place in a village (القرية) in al-Bīrūnī's version, as it does in both the *Gauḍapādabhāṣya* (*grāmīṇeṣu*) and the *Sāṅkhyavṛtti* (*grāme*).

---

48 YD, pp. 176.18–180.32, JM, pp. 85.20–86.4 and TK, p. 126.1–3. See also Kumar & Bhargava 1992: 148–160.

In contrast, the *Sāṅkhyasaptativṛtti* and the *Māṭharavṛtti* locate the story in a city (*nagara*). The **Suvarṇasaptati* does not specify in which place the event occurs. Although these elements constitute minor hints, they may be indicative of the identification of al-Bīrūnī's source with a commentary similar to the *Gauḍapādabhāṣya* or the *Sāṅkhyavṛtti*.

The last element that may be worth analysing here is the way the commentaries describe the activities of the villages. Although the *Kitāb Sānk* does not specify the types of activities, the *Sāṅkhyasaptativṛtti* enumerates various activities the citizens are involved in, such as worshipping, studying, ploughing and trading,[49] and the *Sāṅkhyavṛtti* describes the villagers' activities by saying that "some villagers are farming and some are not" (*te grāmyā lokāḥ kṣetrakarmaṇi pravartante nivartante ca*).[50] On the other hand, the **Suvarṇasaptati*, the *Gauḍapādabhāṣya* and the *Māṭharavṛtti* do not specify what type of activities are meant, paralleling al-Bīrūnī's version. It is also possible that he summed up the content of his source here.

The above observations may be summarized in the following way. The manner in which the different works used the analogy suggests similarities between the *Kitāb Sānk*, the *Māṭharavṛtti* and the *Sāṅkhyasaptativṛtti*. The resemblance of the description of the person involved in the illustration, as well as the place of the event, relate the *Kitāb Sānk* to the *Sāṅkhyavṛtti* and the *Gauḍapādabhāṣya*. The description of the villagers' activities indicates resemblance between the Arabic text and the *Gauḍapādabhāṣya*, the *Māṭharavṛtti* and the **Suvarṇasaptati*. The many uncertainties as to what elements originate from al-Bīrūnī's adaptations and to what other detail directly comes from the Sanskrit source he used do not allow for drawing conclusions based on this passage. However, this excerpt at least indicates that the *Yuktidīpikā*, the *Jayamaṅgalā* and the *Tattvakaumudī* were probably not al-Bīrūnī's source, as these three commentaries do not mention the illustration at all in this place.

6.3.3 *Six Views on Action and Agent*

The passage entitled "six views on action and agent" occurs in a chapter of the *Taḥqīq mā li-l-Hind* entitled "On their belief in Allah—glorified be He" (ذكر اعتقادهم فى الله سبحانه) and consists of an explicit quotation from the *Kitāb Sānk* (passage I of the Appendix).[51] Before the passage itself, al-Bīrūnī makes use

49 V1, p. 34.14–15.

50 The emended reading proposed by Solomon (V2, p. 31) is accepted here.

51 *Taḥqīq* (1958), pp. 22.9–23.10; Sachau 1910: I/30–31. Unlike Sachau, I decide to keep the original Arabic term Allah, so as to highlight the terminology used by al-Bīrūnī in this title.

of the *Kitāb Pātanğal* and the *Kitāb Gītā* in order to discuss the Indian conception of God. A general statement by al-Bīrūnī on different Indian opinions about action (فعل) introduces the quotation from the *Kitāb Sānk*. In this passage, al-Bīrūnī spells out six views on the nature of action and agent, and their relationship: 1) Allah is the agent; 2) union of action and agent occurs by nature; 3) the agent is *puruṣa* (پورش) according to the Vedas;[52] 4) the agent is time; 5) action is only the reward of the preceding act; 6) matter is the cause and the agent. The sixth and last view is that held by the wise man expounding the position of the *Kitāb Sānk*, that is, of Sāṅkhya.[53]

This passage is based on kā. 61 and one of its commentaries, which expose different opinions regarding the source of the phenomenal world. The **Suvarṇasaptati*, the *Gauḍapādabhāṣya*, the *Sāṅkhyasaptativṛtti* and the *Māṭharavṛtti* on kā. 61 all provide the first four opinions conveyed in the *Kitāb Sānk*: God (*īśvara*), effected by nature (*svabhāva*), conscious self (*puruṣa*) and time (*kāla*). As for the *Sāṅkhyavṛtti*, some folios are missing in its comment on kā. 61, and the available text for this passage begins with the discussion on the opinion that the world is produced by nature.[54] All commentaries similarly refute these views and acknowledge that the original cause (*prakṛti*), or the primary matter (*pradhāna*), is the cause of the phenomenal world (See Table 13).[55]

Takakusu links this passage to the **Suvarṇasaptati* and the *Gauḍapādabhāṣya* glossing kā. 27 and 61.[56] This quotation, however, appears to be a relatively free translation of a commentary on kā. 61, but not on kā. 27. The available Sanskrit commentaries on kā. 27 first discuss the role of the mind (*manas*) as a special sense organ endowed with discernment (*saṃkalpaka*). In the Sanskrit works, the question arises of how the phenomenal world, which is multiple, originates from a unique cause, namely *prakṛti*. According to the Sāṅkhya view, as explained on kā. 27, the multiplicity of the phenomenal world exists because of the different combinations of the three constituents (*guṇa*) in the world.

The constituents, *sattva*, *rajas* and *tamas*, are present in every principle (*tattva*) from the unmanifest subtle original cause (*prakṛti*) down to the mani-

---

52 On *puruṣa* interpreted as the soul by al-Bīrūnī, see above p. 85.

53 See also the discussion of this passage by Kozah (2016: 67–70).

54 The commentary is missing for the passage extending from kā. 55 up to the middle of the comment on kā. 61 approximately. See V2, pp. 58–59.

55 Takakusu 1904b: 1050–1051, GPBh, pp. 54.6–55.6, V1, pp. 72.14–73.16, V2, pp. 59.13–60.17 and MV, pp. 55.28–56.22 on kā. 61.

56 Takakusu 1904a: 27–28.

fest gross elements (*mahābhūta*). The three constituents bear different specific qualities, and the qualities of a *tattva* depend on the mutual combinations of the constituents. Although originating from one unique cause, the phenomenal world can be multiple "because of the specific modifications of the constituents" (*guṇapariṇāmaviśeṣāt*) in each principle. The content of kā. 27 and its commentaries thus significantly diverges from the topic of the quotation drawn from the *Kitāb Sānk*. For this reason, I suggest that al-Bīrūnī did not base this passage on kā. 27.

Further, four commentaries, the **Suvarṇasaptati*, the *Gauḍapādabhāṣya*, the *Sāṅkhyasaptativṛtti* and the *Māṭharavṛtt* on kā. 61, connect to the content of the *Kitāb Sānk*. The *Jayamaṅgalā* and the *Tattvakaumudī*, on their part, do not expound different opinions at all on this *kārikā*,[57] while there is a lacuna in this passage in the *Yuktidīpikā*.[58] This commentary discusses the origin of the world when glossing kā. 15. It does not, however, present the discussion in the same form as the *Kitāb Sānk* does in this passage. Moreover, the *Yuktidīpikā* has a different list: atoms (*paramāṇu*), the conscious self (*puruṣa*), God (*īśvara*), action (*karman*), fate (*daiva*), time (*kāla*), chance (*yadṛcchā*) and non-existence (*abhāva*).[59] This quotation from the *Kitāb Sānk* could not thus be based on the *Jayamaṅgalā* or the *Tattvakaumudī*, and probably not on the *Yuktidīpikā*.

On the other hand, the passage from the *Kitāb Sānk* bears striking resemblance to the *Gauḍapādabhāṣya*, the *Sāṅkhyasaptativṛtti*, the *Sāṅkhyavṛtti* and the *Māṭharavṛtti* with regard to the form in which they present the different positions. The source of the *Kitāb Sānk* and these commentaries indeed introduce the different views in the form of reported speech: "some people say" (قوم قال) and "others say" (آخرون قال) in the *Kitāb Sānk*; "some say" (*kecid* […] *bruvate*); "other [say]" (*apare*), "some masters say" (*kecid ācāryāḥ bruvate*) or "masters consider" (*ācāryāḥ manyante*) in these four Sanskrit commentaries. This similarity of form is an additional sign that the source of the *Kitāb Sānk* resembles these texts, more than others.

In order to determine which commentaries among these remaining possible candidates could have been the main Sanskrit source of the *Kitāb Sānk*, the following table displays the different opinions and elements relevant to the analysis, as they appear in each of these commentaries on kā. 61:

57 JM, p. 113.14–22 and TK, pp. 166.28–168.10 on kā.61.
58 Noted by Wezler and Motegi (YD, p. 265, n. 1).
59 See Bronkhorst (1983: 149–155) on this passage referring to the edition by Ram Chandra Pandeya (1967: 68.20–74.15). See also YD, pp. 154.13–162.12.

TABLE 13 Correspondences in the six views listed on kā. 61

Views	*Kitāb Sānk*	**Suvarṇasaptati*	*Gauḍapād-abhāṣya*	*Sāṅkhyavṛtti*	*Sāṅkhyasaptativṛtti Māṭharavṛtti*
1)	God (Allah)	God (Fr. *Īçvara*)	God (īśvara)	missing	God (*īśvara*)
	omission?	quotation from the *Mahābhārata*	quotation from the *Mahābhārata*[a]	missing	quotation from the *Mahābhārata*
2)	by nature (بالطباع)	spontaneity (Fr. *Spontanéité*)	by nature (*svabhāva*)	missing	*puruṣa*
	substitution?	unidentified *śloka*	unidentified *śloka*	missing	reference to the Vedas
3)	soul (نفس)	soul (Fr. *Âme*)	*puruṣa*	by nature (*svabhāva*)	by nature (*svabhāva*)
	reference to the Vedas	reference to the Vedas	no correspondence	unidentified *śloka*	unidentified *śloka*
4)	time (زمان)	time (Fr. *Temps*)	time (*kāla*)	time (*kāla*)	time (*kāla*)
	analogy of the time and the rope	reference to the *Mahābhārata*	quotation from the *Mahābhārata*[b]	quotation from the *Mahābhārata*	quotation from the *Mahābhārata*
5)	preceding action (العمل المتقدّم)	no correspondence	no correspondence	no correspondence	no correspondence
6)	matter (مادّة)	nature (Fr. *Nature*)	original cause (*prakṛti*)	primary matter (*pradhāna*)	primary matter (*pradhāna*)

a The same quotation from the *Mahābhārata* occurs in the **Suvarṇasaptati*, the *Sāṅkhyasaptativṛtti*, the *Māṭharavṛtti* and the *Gauḍapādabhāṣya*. The editors of the *Sāṅkhyasaptativṛtti*, the *Māṭharavṛtti*, however, connects it to *Mahābhārata* 3.31.27, that of the *Gauḍapādabhāṣya* to verse 3.30.88 of the Epic, while Takakusu connects this quotation with that given in the *Gauḍapādabhāṣya*.

b The *Sāṅkhyavṛtti* relates this quotation to *Mahābhārata* 11.2.24, the *Sāṅkhyasaptativṛtti* and *the Māṭharavṛtti* to *Mahābhārata* 3.13.70.57. The text in the printed edition of the *Gauḍapādabhāṣya* does not indicate any verse number, while Takakusu refers to the *Gauḍapādabhāṣya*'s reading without however giving any verse number.

The *Kitāb Sānk*, the *Gauḍapādabhāṣya* and the **Suvarṇasaptati* enumerate the different views in the same sequence. The *Māṭharavṛtti*, the *Sāṅkhyasaptativṛtti*—and most probably the *Sāṅkhyavṛtti*—list the positions in an inverted order as compared with the list provided in the *Kitāb Sānk*. They present the view advocating the conscious self as the cause before the position according to which the world is produced by nature. This first observation may lead to the conclusion that al-Bīrūnī's translation is based on the original source of the **Suvarṇasaptati*, or on the *Gauḍapādabhāṣya*. However, the order in which the elements are enumerated alone does not allow for connecting the *Kitāb Sānk* with a Sanskrit source. As seen in Chapters 4 and 5, discrepancies in the sequence of lists can be also due to the evolution of the manuscripts' transmission of the text, or to al-Bīrūnī's adaptation of his source-text.[60] Thus, they do not constitute, in my view, very strong reasons for determining al-Bīrūnī's sources or to reject some Sanskrit works to be his source.

The absence of the reference to the Vedas in the *Gauḍapādabhāṣya* is a more tenable argument to rule out the possibility of this Sanskrit commentary being the *Kitāb Sānk*'s source. It is doubtful that al-Bīrūnī added this reference on his own initiative since such reference occurs in other Sāṅkhya commentaries that are the **Suvarṇasaptati*, the *Sāṅkhyasaptativṛtti* and the *Māṭharavṛtti*.

The *Sāṅkhyasaptativṛtti* and the *Māṭharavṛtti*, respectively, refer to the Vedas as follows:

> The followers of the Vedas say that the cause is *puruṣa*.[61]

> Nevertheless, the followers of the Vedas consider the cause in this manner: "*puruṣa* is certainly everything." Therefore, they consider the cause as *puruṣa*.[62]

The references to the Vedas in the extant portions of the Sanskrit commentaries serving a similar purpose as in the *Kitāb Sānk* is no accident. One of these Sanskrit commentaries, or one similar to them, must have constituted the source for al-Bīrūnī's translation. As aforementioned, folios of the manuscript of the *Sāṅkhyavṛtti* are missing in this place. It is therefore not possible to draw

60 See above pp. 171–172 and 205–206.

61 V1, p. 72.17: *vedavādi⟨no⟩ br⟨u⟩vate puruṣaḥ kāraṇam iti.*

62 MV, p. 56.1–2: *vedāvadinaḥ punar itthaṃ kāraṇam āhuḥ. "puruṣa evedaṃ sarvam" ity ataḥ puruṣaṃ kāraṇam āhuḥ.* According to the editor of the *Māṭharavṛtti*, this quotation belongs to the *Śvetāśvatara Upaniṣad* (III.15).

parallels or highlight discrepancies between this Sanskrit commentary and the *Kitāb Sānk* for the first part of this passage. The *Sāṅkhyavṛtti*, however, matches with the *Sāṅkhyasaptativṛtti* in its available portion of text for this passage, and it may have thus originally contained the reference to the Vedas.

There are also differences between the *Kitāb Sānk* and the Sanskrit commentaries. Some of them do not, however, necessarily entail that al-Bīrūnī drew from a different Sanskrit text, as he may have had recourse in this passage to translational strategies, such as omissions, substitutions and possibly an addition.[63] First, al-Bīrūnī inserted the whole discussion of the different opinions in a dialogue between an ascetic (ناسك) and a wise man (حكيم) who propounds them. This form, which is absent from all Sanskrit commentaries on the *Sāṅkhyakārikā* in this place, was in all likelihood supplemented by al-Bīrūnī who systematized his translation in a dialogue, in the same way as he did in the *Kitāb Pātanğal*.[64]

Furthermore, the quotation from the *Kitāb Sānk* also appears to be a simplified version of a Sanskrit work on kā. 61. Two quotations from other texts which occur in the Sanskrit commentaries do not have parallels in the *Kitāb Sānk*. The *Sāṅkhyasaptativṛtti*, the *Māṭharavṛtti*, the *Gauḍapādabhāṣya* and the **Suvarṇasaptati* quote from the *Mahābhārata* in order to illustrate the view that Īśvara is the cause of the world.[65] It is possible that al-Bīrūnī deliberately omitted this quotation in his translation.[66] Other technical explanations, which are absent from the *Kitāb Sānk*, are, however, found in the Sanskrit commentaries. For instance, the original cause (*prakṛti*) is said to be extremely delicate (*sukumāratara*) in kā. 61. Therefore, when the conscious self (*puruṣa*) perceives it as a different entity from itself, the cause disappears from sight since it cannot be seen twice due to her delicacy. The separation between the two brings about the dissolution of the world and the isolation (*kaivalya*) of the conscious self. This process is referred to in kā. 61 and explained in the Sanskrit commentaries before they enumerate the different opinions. The *kārikā* reads:

63 See above Section 4.4 on translational strategies.

64 On the dialogue form, see above Section 4.3.1.

65 Only the second quotation from the *Mahābhārata* is found in the *Sāṅkhyavṛtti*, as the earlier portion of text is missing.

66 Al-Bīrūnī perhaps replaces the quotation from the *Mahābhārata* about Īśvara being the cause of the world by the following analogy: "just like what is alive and powerful sets in motion what is dead and impotent" (كما يحرّك الحىّ القادر الموات العاجز) (*Taḥqīq* [1958], pp. 22.15–16). This analogy, however, differs from the possible original quotation to such an extent that it is difficult to draw any conclusion.

> "Nothing is more delicate than the original cause." This is my thought. "She, who [realizes]: 'I have been seen,' does not show herself anymore to the conscious self."[67]

The *Kitāb Sānk* does not qualify the cause of the world—matter (مادّة) in Arabic—as extremely delicate, nor does it mention the specific interaction between matter and soul in this quotation. Neither it broaches the topic of the three categories constituting the world, that are the manifest (*vyakta*), the unmanifest (*avyakta*) and the knower (*jña*), which is the *puruṣa*, when refuting the opinion according to which time is the cause. These categories are, however, described in the available Sanskrit commentaries.[68]

Lastly, al-Bīrūnī may have omitted the word *sāṅkhya*, referring to the followers of Sāṅkhya, term, however, occurring in the *Gauḍapādabhāṣya*, the *Sāṅkhyasaptativṛtti*, the *Sāṅkhyavṛtti* and the *Māṭharavṛtti*, and thus possibly found in his source-text. These instances are a few examples of technical explanations or terms that al-Bīrūnī may have left out in his translation.

In addition, some other differences between the extant Sanskrit commentaries and the *Kitāb Sānk* can be identified with some confidence as adaptations by al-Bīrūnī who had in mind the difficulties for his readership in understanding some technical concepts and explanations. For instance, al-Bīrūnī probably made use of two substitutions in this passage. The *Gauḍapādabhāṣya*, the *Sāṅkhyasaptativṛtti*, the *Māṭharavṛtti* and the **Suvarṇasaptati* quote from an unknown work, when explaining the view that the world is produced by nature. It reads as follows:

> This [natural condition], which makes swans white, parrots green, peacocks multi-coloured, also produces our condition.[69]

Al-Bīrūnī perhaps deemed the Sanskrit illustration beyond understanding for his readership and decided to omit it or to substitute it with another explanation. In the corresponding passage of the *Kitāb Sānk*, the following explanation

67 Kā. 61 in GPBh, p. 53: *prakṛteḥ sukumārataraṃ na kiñcid astīti me matir bhavati. yā dṛṣṭāsmīti punar na darśanam upaiti puruṣasya.*

68 See above p. 146.

69 V1, p. 72.20–21: *yena śuklīkṛtā haṁsāḥ śukāś ca haritīkṛtāḥ. mayūrāś citritā yena sa no vṛttiṃ vidhāsyati.* The *Māṭharavṛtti* has the exact same reading as the *Sāṅkhyasaptativṛtti* (MV, p. 56.4–5). The *Gauḍapādabhāṣya* refers to the same strophe, although differently and in an incomplete manner (GPBh, p. 54.10). The corresponding passage in *Sāṅkhyavṛtti* is missing. Only the *Māṭharavṛtti* provides the reference for this quotation. See also Takakusu 1904b: 1050.

is found: "such is a common phenomenon (i.e., of being produced by nature) in everything which grows and decays" (فهكذا جرت العادة فى كل ناش بال).[70] This case of substitution is only hypothetical, as the semantic connection between the Sanskrit quotation and the Arabic text remains relatively loose. However, an additional example of al-Bīrūnī substituting a Sanskrit quotation for an idiomatic Arabic expression parallels this instance. It appears at the second quotation from the *Mahābhārata* occurring in the Sanskrit commentaries. They quote:

> Time ripens beings; time destroys the world; time is awake among the sleeping ones; indeed, time is insurmountable.[71]

Rather than literally translating this strophe, al-Bīrūnī made use of an analogy absent from the Sanskrit commentaries under review: "the agent is time, because the world is tied to it (i.e., time), [in the same way] as the sheep's tie [is] with a tight rope, since its movement depends upon the pulling or the loosening of the [rope]" (الفاعل هو الزمان فانّ العالم مربوط به رباط الشة بحنل مشدود بها حتى تكون حركتها بحسب انجذابه و استرخائه).[72] This analogy most probably consists of an idiomatic expression recalling the power of time, however, drawn from al-Bīrūnī's own background. His readership would have thus been more acquainted with it than with the quotation from the *Mahābhārata*. Indeed, although the Sanskrit and Arabic illustrations are different, the message is similar: time has control over the world.

Lastly, al-Bīrūnī also provides a fifth opinion absent from the Sanskrit texts, that "action is nothing but the recompense for a preceding deed" (قال آخرون: ليس الفعل سوى المكافاة على العمل المتقدّم).[73] The *Yuktidīpikā* is the only commentary that conveys this position, although not on kā. 61 but on kā. 15. In light of observations made so far in the present chapter, it is, however, unlikely that al-Bīrūnī drew his information from the *Yuktidīpikā*. This view, referring to karmic retribution, may have been added by al-Bīrūnī under the influence of his Indian informants or simply on his own initiative, if this viewpoint was a widespread one among the Indians he encountered.

---

70 *Taḥqīq* (1958), pp. 22.16–17.

71 V2, p. 60.8–9: *kālaḥ pacati bhūtāni kālaḥ saṃkṣipate jagat. kālaḥ supteṣu jāgartti kālo hi duratikramaḥ*. The quotation appears in a similar form, completely or incompletely, in the other Sanskrit commentaries (Takakusu 1904b: 1051; GPBh, p. 55.1–2; MV, p. 56.14–15; V1, p. 73.9).

72 *Taḥqīq* (1958), pp. 22.18–23.1.

73 *Taḥqīq* (1958), pp. 23.1–2.

Thus, the analysis of passage I in comparison to the available Sanskrit works on the *Sāṅkhyakārikā* reveals that the *Yuktidīpikā*, the *Jayamaṅgalā* and the *Tattvakaumudī* wouldnot be the source of the *Kitāb Sānk*. It also highlights, despite parallelisms, a certain number of formal and substantial discrepancies between the *Kitāb Sānk* and its remaining possible Sanskrit sources.[74] Many of these discrepancies, however, could be due to al-Bīrūnī's hermeneutics and cannot be merely explained by al-Bīrūnī's use of a different work than those available to us. As with the *Kitāb Pātanǧal*, al-Bīrūnī transformed his Sanskrit source-text when composing the *Kitāb Sānk* so as to adjust its content for his readership.

On the whole, only one element of this passage constitutes, in my view, a valid argument to determine an affinity between the *Kitāb Sānk* and the Sanskrit commentaries, namely the mention of the Vedas. This reference is found in the *Kitāb Sānk*, *Sāṅkhyasaptativṛtti*, *Māṭharavṛtti* and **Suvarṇasaptati*—and possibly the *Sāṅkhyavṛtti*—under the position according to which the world is produced by the soul or *puruṣa*. It was most probably not one of al-Bīrūnī's creations, but it illustrates that the source of the *Kitāb Sānk* at least parallels the *Sāṅkhyasaptativṛtti*, the *Māṭharavṛtti* and the source of the **Suvarṇasaptati*. Moreover, the **Suvarṇasaptati* shares an additional common point with the *Kitāb Sānk* in this passage, that it presents the different positions in the same order than the Arabic text. This similarity can constitute an indication of their possible connection, if supplemented with other pieces of evidence, such as the reference to the Vedas in this passage, and other common points highlighted in the present section. Lastly, this passage also evidences that al-Bīrūnī drew from a commentary on the *Sāṅkhyakārikā* and not solely from the *kārikās*.

6.3.4 *States Different from Liberation and Four Levels of Knowledge*

The next passages analysed in the present book are found in chapter 7 of the *Taḥqīq mā li-l-Hind*, which is devoted to the concept of liberation (في كيفيّة الخلاص من الدنيا وصفة الطريق المؤدى إليه). This analysis jointly considers two passages indexed under Numbers XVIII and XIX, because they follow each other in the *Taḥqīq mā l-Hind* and are connected to consecutive *kārikās*, as seen below.[75] The first Arabic passage provides a general discussion on three stages of human condition, which constitute steps toward liberation but do not lead

74 See also some words used by a-Bīrūnī to express the whole debate, i.e., agent (فاعل) instead of cause (*kāraṇa*) in some places or in terms of union between agent and action.

75 *Taḥqīq* (1958), pp. 63.7–64.8; Sachau 1910: 1/83–85.

to it. Due to the many probable adaptations on al-Bīrūnī's part on this passage, its analysis does not point to one specific Sanskrit commentary as a possible source. However, since it demonstrates well these adaptations, and as the first part of passage XIX analysed below, it deserves consideration. The second passage is an illustration of four levels of knowledge—of intellect (*buddhi*) in Sāṅkhya.

Takakusu connects the first part of this passage to kā. 50, which deals with nine reasons for not reaching isolation, which the Sanskrit commentaries call satisfactions (*tuṣṭi*).[76] In the Arabic passage, al-Bīrūnī does not describe nine reasons for not reaching liberation. He enumerates three situations that can be connected with some states (*bhāva*) of the intellect (*buddhi*). His translation of this passage thus fits better the contents of the commentaries on kā. 44 and 45 than with those commenting upon kā. 50. The commentaries on kā. 44 and 45 discuss the eight states (*bhāva*) inherent to the intellect (*buddhi*): righteousness (*dharma*), unrighteousness (*adharma*), knowledge (*jñāna*), ignorance (*ajñāna*) all enumerated in kā. 44, and dispassion (*vairāgya*), passion (*rāga*), mastery (*aiśvarya*) and lack of mastery (*anaiśvarya*) listed in kā. 45.

The connection of this Arabic passage with a commentary on kā. 44 and 45 is also supported by the fact that the subsequent passage drawn from the *Kitāb Sānk* (XIX) is unequivocally taken from kā. 46 and its related commentaries. If one accepts that the first part is indebted to kā. 44–45, it is then possible to trace the source for this passage to the three consecutive kā. 44, 45 and 46 and their commentaries, rather than to posit that al-Bīrūnī drew from two separate passages of his Sanskrit source.

Al-Bīrūnī's translation of kā. 44–45 and their related commentaries consists, however, of a summary and rephrasing of them. In order to contextualize al-Bīrūnī's translations, it is necessary to first quote kā. 44 and 45:

> Upward movement is [a result of] righteousness, downward movement [arises] from unrighteousness. Emancipation [takes place] with knowledge, while attachment [to this world] is caused by [its] reverse. The dissolution in the producers (*prakṛti*) [arises] from dispassion, transmigration from passion that is related to *rajas*. The absence of obstacles [originates] from mastery, its opposite from the reverse.[77]

76 Takakusu 1904a: 31.

77 SK 44–45: *dharmeṇa gamanam ūrdhvaṃ gamanam adhastād bhavaty adharmeṇa | jñānena cāpavargo viparyayād iṣyate bandhaḥ* ||44|| *vairāgyāt prakṛtilayaḥ saṃsāro bhavati rājasād rāgāt | aiśvaryād avighāto viparyayāt tadviparyāsaḥ* ||45||.

Al-Bīrūnī provides definitions to some states (*bhāva*) described in kā. 44–45 and rewords the content of the original Sanskrit text. He appears to have avoided translating the abstract Sanskrit concepts of *bhāva* and *buddhi*, preferring to depict human behaviours that can illustrate the *bhāvas*. For instance, he defines the Sanskrit righteousness (*dharma*) mentioned in kā. 44 as follows: "he who enters upon the world with an excellent conduct [of life], who is generous with what he possesses among [any goods, his] reward in the world is the obtention of [his] wishes and desires and [his] return in the [world] with happiness, happy in body, soul and the condition [of life]" (المقبل على الدنيا مع حسن السيرة الجواد بما يملك منها مكافى فى الدنيا بنيل الأمانىّ والارادة و التردّد فيها على السعادة مغبوطا فى البدن والنفس والحال).[78]

The second part of al-Bīrūnī's translation can be traced to kā. 45 and commentary. The passage reads: "the ascetic in the world [living] without knowledge obtains elevation and reward, but he is not liberated because [the other] means [to reach liberation] are lacking" (الزاهد فى الدنيا من غير علم يفوز بالاعتلاء والثواب ولا يتخلّص لعوز الآلة).[79] The whole Arabic expression is a rendering of the portions of Sanskrit commentaries dealing with dispassion (*vairāgya*). First, al-Bīrūnī uses the Arabic term ascetic (زاهد; *zāhid*) derived from the verbal root زهد (*z-h-d*) meaning to abstain or to renounce, notably from the worldly pleasures. The scholar translates dispassion (*vairāgya*) in the *Kitāb Pātanğal* (KP 6, corresponding to PYŚ I.12–16) with the same Arabic verbal root. Second, the *Kitāb Sānk*, the **Suvarṇasaptati*, the *Gauḍapādabhāṣya*, the *Sāṅkhyasaptativṛtti*, the *Māṭharavṛtti*, the *Jayamaṅgalā* the *Tattvakaumudī* and the *Yuktidīpikā* all mention on kā. 45 that dispassion is insufficient to reach isolation if it occurs without knowledge (*jñāna*). Similarly, al-Bīrūnī states that the ascetic who live without knowledge obtains elevation and reward, no liberation.

The last portion of the Arabic passage constitutes al-Bīrūnī's explanation of the concept of mastery (*aiśvarya*). It runs as follows: "the content one, self-sufficient, when he is a master of the eight aforementioned conditions (i.e., powers), he is fooled by them, he carries out [his activities] easily and successfully and he considers them as liberation, then he remains in these [powers]" (القانع المستغنى إذا اقتدر على الثمانية الحال المذكورة و اغترّبها و تنجّح و ظنّها الخلاص بقى عندها).[80] Al-Bīrūnī describes the eight conditions in the preceding lines of the same chapter of the *Taḥqīq mā li-l-Hind*.[81] All commentaries on kā. 45, except the

78 *Taḥqīq* (1958), p. 63.8–10; Sachau 1910: I/83.

79 *Taḥqīq* (1958), p. 63.11–12; Sachau 1910: I/83.

80 *Taḥqīq* (1958), p. 63.12–13; Sachau 1910: I/83–84.

81 *Taḥqīq* (1958), pp. 52.11–17 and 56.13–16; Sachau 1910: I/69 and 74. The concept of eight powers (*aṣṭa siddhi*) overlaps in Yoga and Sāṅkhya philosophies.

Jayamaṅgalā and the *Tattvakaumudī*, refer to these eight powers of mastery (*pūrvam aiśvaryam aṣṭavi⟨dha⟩m aṇimādi*).[82] Thus the Arabic passage is based on a Sanskrit work commenting upon kā. 45, which probably also referred to the eight powers resulting from mastery.

As aforementioned, al-Bīrūnī made several adaptations in his translation of kā. 44–45 and commentary. For instance, in contrast to the *Sāṅkhyakārikā*, his text does not refer to the binary notions opposed to righteousness, knowledge, dispassion, mastery, which are unrighteousness (*adharma*), ignorance (*ajñāna*), passion (*rāga*) and lack of mastery (*anaiśvarya*). Further, he defines the concepts of righteousness, dispassion and mastery in his own words, rather than literally translating these terms. He also fails to discuss the state corresponding to knowledge, whereas all Sanskrit commentaries explain this notion separately. If this constitutes an omission by al-Bīrūnī, the reasons behind it are not completely clear.[83]

The **Suvarṇasaptati*, the *Sāṅkhyasaptativṛtti*, the *Gauḍapādabhāṣya* and the *Māṭharavṛtti* that are the Sanskrit works resembling much al-Bīrūnī's *Kitāb Sānk* so far, define this type of knowledge as that of the twenty-five principles (*tattva*) constituting the world. The commentary in the *Sāṅkhyavṛtti* is only partially available from *kārikā* 41 to 45, since folios of its manuscript are missing in this place. The definition of the type of knowledge leading to isolation is technical, specific to Sāṅkhya and completely unknown to Muslim readers of the time. If al-Bīrūnī used a commentary akin to these four commentaries, he may have decided to forego such technical discussion at this particular point of the narrative in order to adjust the content of the *Kitāb Sānk* to his readership.

Al-Bīrūnī may also have omitted other technical concepts that were probably present in his source for this passage. When explaining righteousness (*dharma*) the Sanskrit commentaries specify that upward movement signifies reaching the land of the gods, whereas downward movement leads to the land of animals. Al-Bīrūnī made no mention of these two lands at all here. However, as seen in Section 6.3.1, he addressed this question when explaining that births depend on the way in which one conducts in life (Appendix, passage XII), and thus perhaps deemed redundant to repeat the concept here. The result of dispassion (*vairāgya*), is, according to the commentaries, the dissolution into the producers, that are the primary matter, the intellect, the "I" consciousness and the five subtle elements (*pradhānabuddhyahaṃkāratanmātra*).[84] This again is not rendered by al-Bīrūnī.

82 V1, p. 60.18.

83 Section 4.4.1 discusses this possible omission. See above p. 147.

84 See for instance GPB, p. 42.10–15, **Suvarṇasaptati*, p. 1031 and V1, p. 60.7–13.

The above elements, namely the specific type of knowledge leading to liberation, the lands of gods and animals and the dissolution into eight principles are all very technical Indian or Sāṅkhya conceptions. It appears that al-Bīrūnī does not, or sparingly, cover these elements, perhaps because he regarded them as too culturally loaded to be transmitted effectively to his readership. Because of the high degree of differences between the *Kitāb Sānk* and the possible corresponding Sanskrit passages, the analysis of the first part of this passage (XVIII) makes it impossible to retrace its specific Sanskrit source.

The second part of this passage (XIX) is more conducive to link the *Kitāb Sānk* to Sanskrit commentaries on the *Sāṅkhyakārikā*. It considers four levels of knowledge (معرفة) exemplified by four different disciples who are asked to ascertain the identity of an object they see from far. The same story illustrating four types of intellect (*buddhi*) appears in some commentaries on kā. 46. The terms to designate these levels of knowledge in Sanskrit and Arabic are respectively: error (*viparyaya*) and ignorance (جهل), inability (*aśakti*) and disability (عجز), satisfaction (*tuṣṭi*) and indolence (تراخى). Al-Bīrūnī paraphrases the fourth, accomplishment (*siddhi*) in Sanskrit, at the end of the illustration by the sentence: "Thus, the [student] obtained knowledge by his effort" (وقد فاز من يديه بالمعرفة).[85]

Some commentaries on kā. 30, notably the *Gauḍapādabhāṣya*, refer to the same illustration, as Sachau and Garbe noticed.[86] However, only commentaries on kā. 46 explain it at length. In addition, in the commentaries on kā. 30, the example does not illustrate four divisions of intellect, but rather aims at expounding the role and functioning of the intellect in relation to other *tattva*s in determining external objects. It is thus reasonable to connect this passage to kā. 46 and its commentaries in accordance with Takakusu.[87] The **Suvarṇasaptati*, the *Gauḍapādabhāṣya*, the *Sāṅkhyavṛtti*, the *Sāṅkhyasaptativṛtti* and the *Māṭharavṛtti* record this illustration, explain it and contextualize it, in a way similar as al-Bīrūnī does. The *Yuktidīpikā*, the *Jayamaṅgalā* and the *Tattvakaumudī*, however, do not reference this analogy on kā. 46.[88] Their lack of mention of this analogy in this place stands as an additional indication that al-Bīrūnī did not use any of these three commentaries to compose the *Kitāb Sānk*.

85 It must be noted that *siddhi* in this context does not have the same meaning as the supernatural powers described in the *Pātañjalayogaśāstra*.

86 Sachau 1910: II/288; Garbe 1894: 64–65.

87 Takakusu also provided a detailed analysis of the variants of this illustration found in the **Suvarṇasaptati*, the *Gauḍapādabhāṣya* and the *Kitāb Sānk* (Takakusu 1904a: 31–34).

88 YD, pp. 239.5–240.7, JM, p. 103.10–17 and TK, p. 152.15–25. See also Kumar & Bhargava 1992: 321–323.

The **Suvarṇasaptati*, the *Sāṅkhyavṛtti* and the *Kitāb Sānk* share the most common points with the *Kitāb Sānk* in the way in which they narrate the analogy, as opposed to the *Gauḍapādabhāṣya*, the *Sāṅkhyasaptativṛtti* and the *Māṭharavṛtti*. Therefore, the version of the *Sāṅkhyavṛtti* is presented below:

> ⟨It is said that a teacher set out at dawn for a town with four young religious students⟩. One student ⟨said⟩ to [his] teacher: "This, [which] is seen on this path, is it a pole or a thief?" This student had a doubt about the pole. ⟨The teacher said to the second student: "Let [me] know what this is." He (i.e., the student) observes [the pole] from afar⟩. Then, he said to the teacher: "I am not able to approach [it]."[89] Thus, inability arose to him. The teacher ⟨said⟩ to the third student: "Let [me] know what this is." The third student, having looked at [it], said to the teacher: "What is the use of determining it? Let us approach it at sun⟨rise⟩, with the caravan." Having [thus] spoken, and having not ascertained [what it is], he fell asleep at dawn. Thus, satisfaction arose to the third student. Again, the teacher asked the fourth student: "Let [me] know what this is." This one, having looked at this pole, sees a plant climbing on the pole, and a bird on top of it. Thereafter, having gone [there], having touched the pole with his foot, he returned back to the teacher and said: "This is a pole." ⟨This is accomplishment⟩.[90]

The emendations and additions of Solomon are not all certain.[91] However, even if these emendations were omitted, the passage still resembles that of the *Kitāb Sānk* in several respects, which I reproduce in full here:

---

89 The text is corrupt here.

90 V2, pp. 56.18–57.9: ⟨*kaścit kila upādhyāyaḥ anudite sūrye caturbhir baṭubhiḥ saha nagaram abhiprasthitaḥ*⟩. *kaścid baṭuḥ upādhyāyaṃ* ⟨*bravīti*⟩ *eṣo 'tra pathi dṛśyate kiṃ sthāṇuḥ syāt coraḥ syād iti. tasya baṭoḥ sthāṇau saṃśayaḥ.* ⟨*upādhyāyena dvitīyo baṭuḥ uktaḥ jñāyatāṃ ko 'yam iti, durāt nirīkṣate?*⟩ *tataḥ upādhyāya uktaḥ nāhaṃ śakto 'dhigantum iti. evaṃ asyāśaktir utpannā. upādhyāyena tṛtīyo baṭuḥ* ⟨*uktaḥ*⟩ *jñāyatām ko 'yam iti. sa tṛtīyo baṭuḥ nirīkṣya upādhyāyaṃ bravīti kim anenāvacchinnena, sūrya* ⟨*udite*⟩ *sārthena saha yāsyāmaḥ iti. uktvā ajñātveṣattame prasuptaḥ. evaṃ tṛtīyasya baṭoḥ tuṣṭir utpannā upādhyāyo bhūyaś caturthaṃ baṭuṃ bravīti jñāyatāṃ ko'*⟨*ya*⟩*m iti. sa nirīkṣya tasmin sthāṇau vallīṃ paśyati sthāṇunārūḍhāṃ tatrārūḍhaṃ śakunam* [*ca*]. *tato gatvā pādena sthāṇuṃ spṛṣṭvā punar āgata upādhyāyaṃ bravīti sthāṇur ayam iti.* ⟨*eṣā siddhiḥ*⟩. This passage has been also translated and published in Verdon 2019c.

91 Solomon's emendations are generally accepted in this book.

And to give an example for those who differ in [their] degrees of intellection [i.e., *buddhi*]: a man journeyed at night with his disciples for some business. Then, an upright figure, whose exact knowledge was prevented by the darkness of the night, befell them in the way. The man turned toward his disciples and asked them about it, one after the other.

The first [disciple] said: "I don't know what it is."

The second said: "I don't know it and I am unable to know it."

The third said: "It is useless to learn about it [now], as the rising of the day will reveal it. If it is frightful, it [will] go away by the morning and if it is different, its case will become clear to us."

All three [disciples] failed to [obtain] the knowledge [about the figure]: the first of them because of ignorance (بالجهل), the second because of disability (بالعجز) and damage of organ (آفة فى الآلة), the third because of indolence (بالتراخى) and of satisfaction in ignorance (الرضاء بالجهل).

As for the fourth [disciple], he did not find any answer before ascertaining [the identity of the figure]. He went to see it. While approaching it, he saw a squash (يقطين) intertwined over it. He knew that a living man would not remain still in this standing position of [his own] free will, until such a thing gets entangled around him, and was sure that a lifeless [object] was erected. Then, he could not trust whether it was a hiding place for garbage [or not]. Therefore, he came close to it, kicked it with his foot until it fell to the ground. The doubt [then] disappeared on the case of the [figure] and he returned to his master with absolute certainty. Thus, the [student] obtained knowledge by his effort.[92]

The *Sāṅkhyavṛtti* presents the illustration in the form of a dialogue between a teacher (*upādhyāya*) and four young boys (*baṭu*) in the same way as the *Kitāb Sānk* does. Similarly, the **Suvarṇasaptati* specifies that the discussion occurs between a Brahmin (Fr. *brahmane*) and his four disciples (Fr. *disciple*).[93] The *Gauḍapādabhāṣya*, the *Sāṅkhyasaptativṛtti* and the *Māṭharavṛtti* remain relatively concise and do not for instance provide the illustration as a story involving an erudite and four young people, but only narrate it by way of impersonal pronouns, such as "somebody" (*kaścit*) or "he" (*sa*). It may be argued that the specific form of this quotation from the *Kitāb Sānk* was due to al-Bīrūnī's own creativity, as he reshaped the Sanskrit text into a dialogue. In this case, the dialogue is between a master and four young people, or his disciples, rather than between a wise man and a hermit, as is the case of al-Bīrūnī's creation of the dia-

92 *Taḥqīq* (1887), p. 41.7–15; *Taḥqīq* (1958), pp. 63.14–64.8; Sachau 1910: 1/84–85.

93 Takakusu 1904a: 1033.

logue form.[94] Another common point between the *Kitāb Sānk*, the **Suvarṇasaptati* and the *Sāṅkhyavṛtti* lies in the details they provide. For instance, the three texts specify that the fourth disciple touches the object with his foot in order to ascertain the identity of the object. Thus, although being a free translation, the *Kitāb Sānk* describes the situation in a way that can be paralleled to the **Suvarṇasaptati* and the *Sāṅkhyavṛtti*.

The *Sāṅkhyavṛtti* and the **Suvarṇasaptati*, however, also differ from al-Bīrūnī's version, for instance in their structures. In the *Kitāb Sānk*, the teacher himself asks the disciples for the identification of the object. In the *Sāṅkhyavṛtti* and the **Suvarṇasaptati*, the interaction first starts with a question by the first disciple. Only after this initial question, the teacher requests each of the remaining pupils to establish the identity of the object. These two latter commentaries also mention a caravan (Fr. *caravane*; *sārtha*) when the third disciple tries to identify the object, an element that is absent from the *Kitāb Sānk*.

Conversely, the Arabic translation supplements the story with descriptions that do not appear in any of the Sanskrit commentaries under scrutiny. For instance, they lack two of al-Bīrūnī's explanations about the fourth disciple and type of knowledge, namely "[h]e knew that a living man would not remain still in this standing position of [his own] free will, until such a thing gets entangled around him, and was sure that a lifeless [object] was erected." (علم أنّ الانسان الحيّ المختار لا يبقى في مقضعه قائما إلى أن يحصل عليه ذلك الالتفات و تحقّق أنّه موات منصوب) and "he could not trust whether it was a hiding place for garbage [or not]" (لم يأمن أن يكون مخبئا لمزبلة شيء).[95] These omissions and additions might be due to al-Bīrūnī's creativity and interpretation, as much as to him having used a different Sanskrit source than the commentaries under consideration. However, the similarities between the *Kitāb Sānk*, the *Sāṅkhyavṛtti* and the **Suvarṇasaptati* are too important to be explained by a mere coincidence. Expanded with the other pieces of evidence provided in the present study, this common point may suggest a strong connection between these three works.

6.3.5 *The Rainwater Which Varies in Taste*

The last excerpt under review (passage VIII in the Appendix) consists of one analogies found in Chapter 4 of the *Taḥqīq mā li-l-Hind*. This analogy, as passage VI above (The soul as a passer-by), aims to illustrate a property of the soul (النفس) in the *Taḥqīq mā li-l-Hind*:

94 See passages I, XVII and XX of the Appendix.
95 *Taḥqīq* (1958), p. 64.4–7; Sachau 1910: I/84.

> They (i.e., the Indians or the adherents of the *Kitāb Sānk*) narrate the example of the soul, which resembles the rainwater falling down from the sky, being as it is and of one nature (على حاله و كيفيّة واحدة). When it is collected in receptacles placed for it in different materials, such as gold, silver, glass, ceramic, clay and saline earth (سبخة), it differs in appearance, taste and smell due to these. In this way, the soul does not affect the matter (i.e., *prakṛti*) except for life [which it gives to matter] because of the proximity [between the two].[96]

The Sanskrit phrasing *salilavat*, meaning "just like water," in kā. 16 refers to this analogy. There, the illustration is used to characterize the unmanifest (*avyakta*), and the way in which it emanates in various manifest forms in the phenomenal world. The fact that al-Bīrūnī's version compares the soul with water, and not the unmanifest original cause—which he would generally translate as matter or primary matter—is rather due to a misunderstanding, or adaptation, on his part, than him using an unknown Sanskrit text.

Except the *Yuktidīpikā*, all commentary comment upon this analogy.[97] The *Jayamaṅgalā* does not, however, provide much explanation on it here, while the *Tattvakaumudī* does not contextualize it in the same way as the *Kitāb Sānk* or other Sanskrit commentaries do.[98] Therefore, the *Yuktidīpikā*, the *Jayamaṅgalā* and the *Tattvakaumudī* could not have been the sources of the *Kitāb Sānk* for this passage and are thus not considered in the following analysis. Among the other commentaries, the *Gauḍapādabhāṣya* does not mention the role of the receptacle.[99] The comments upon this *kārikā* in the *Sāṅkhyasaptativṛtti* and the *Māṭharavṛtti* do not diverge much from each other. They both explain that water, when reaching the soil, has different tastes, depending upon the receptacles in which it falls, but they do not specify the types of these receptacles, as al-Bīrūnī does.[100] The **Suvarṇasaptati* specifies different tastes of water, stating the following: "It (i.e., water) has various tastes, depending upon its receptacles. If it is in a golden vase, its taste is very sweet; if it is in the earth (i.e., soil), its taste varies depending upon the quality of the earth."[101] The reading of the first compound of this passage in the *Sāṅkhyavṛtti* is uncertain.

96 *Taḥqīq* (1887), p. 24.10–13; *Taḥqīq* (1958), p. 37.9–13; Sachau 1910: I/49.

97 YD, p. 165.2–6. See also Kumar & Bhargava 1992: 119.

98 JM, p. 83.8–13 and TK, p. 118.27–30.

99 GPBh, p. 19.2–24 on kā. 16.

100 V1, p. 30.9–12 and MV, p. 21.7–8 on kā. 16.

101 From the French: "Elle devient d'un gout varié selon les différents réceptacles. Si elle est dans un vase d'or, son gout est très doux; si elle est dans la terre, son goût diffère selon la qualité de la terre." Takakusu 1904b: 1001.

However, Solomon proposes two possible emendations. Her two proposals are the following:

1) Water, received from the sky in a receptacle [made for the purpose of] retaining water, is transformed into sweetness.[102]
2) Water, received from the sky in a golden receptacle, is transformed into sweetness.[103]

The second proposal (*ākāśādondhāraṇa*) appears possible, as it is close to that of the **Suvarṇasaptati*. If one is willing to accept this emendation, then the *Kitāb Sānk* bears some similarities with the **Suvarṇasaptati* and the *Sāṅkhyavṛtti*. At any rate, al-Bīrūnī may have had a version of this analogy resembling the one found in these two commentaries. Even so, he may have added elements in his enumeration of receptacles, as he did not solely mention gold as a type of receptacle. In spite of this, the version of the explanation of the analogy drawn from the *Kitāb Sānk* resemble much that found in the **Suvarṇasaptati* and the *Sāṅkhyavṛtti*, in contrast with the other Sanskrit commentaries on kā. 16.

As a summary of this section and in order to demonstrate the intricate relationship of the *Kitāb Sānk* with the Sanskrit commentarial tradition on the *Sāṅkhyakārikā*, I reproduce the main results of the above discussion in Table 14.

6.4 Concluding Remarks

Thus, this chapter demonstrated that the *Kitāb Sānk* constitutes a free translation of the *Sāṅkhyakārikā* and one of its commentaries. Having reshaped his source-text into a dialogue form, al-Bīrūnī substantially transformed it and adapted his translation to meet the needs of his Muslim eleventh-century readership. Three specific types of substantial transformations emerged from the observation made in the present chapter: omission, addition and substitution. Al-Bīrūnī frequently made these specific adaptations of content when dealing with technical and/or abstract ideas elaborated by Indian thought, as he did with the source of the *Kitāb Pātanǧal*. A comparison between his Arabic translations and his possible Sanskrit sources without considering his hermeneutics and creativity is thus insufficient to comprehend his work.

---

102 V2, p. 28.5–6: *⟨ākāśād udandhāraṇa⟩bhājanena parigṛhītam ambhaḥ madhurabhāvena pariṇa⟨mate⟩.*

103 V2, p. 28.5–6: *⟨ākāśāt suvarṇa⟩bhājanena parigṛhītam ambhaḥ madhurabhāvena pariṇa⟨mate⟩.*

TABLE 14 Summary of the analytical comparisons of the passages drawn from the *Kitāb Sānk* with the Sanskrit commentaries on the *Sāṅkhyakārikā*

Passages	The possible source of the *Kitāb Sānk*	Arguments
No. VII: the innocent man among thieves (Section 6.3.1), kā. 20.	Possible exclusion of YD, JM and TK.	Absence of explanation of the analogy on kā. 20.
	Possible exclusion of GPBh.	Too brief.
No. XII: birth depending on virtues and vices (Section 6.3.1), kā. 39.	Possible exclusion of YD, JM and TK.	No concordance of description on kā. 39.
No. XX: categories of beings (Section 6.3.1), kā. 53.	Exclusion of YD, JM and TK.	Only one list of items
	Parallel with V1, MV and GPBh	Two lists, but same order
	Parallel with the **Suvarṇasaptati.*	Two lists in different orders, but not same as the *Kitāb Sānk.*
	V2 missing	
Passage VI: the soul as a passer-by (Section 6.3.2), kā. 19.	Possible exclusion of YD, JM and TK.	Absence of the analogy on kā. 19.
	Possible difference with the **Suvarṇasaptati*, GPBh and V2.	Qualities of the conscious self explained by the analogy.
	Possible parallel with V1 and MV.	Terminology related to the passer-by
	Possible parallel with GPBh and V2.	Terminology related to the passer-by.
	Possible parallel with GPBh and V2.	Terminology related to the place.
	Possible parallel with the **Suvarṇasaptati*, GPBh, MV (possibly V2).	Description of the activities of the villagers; possible simplification of the narrative on al-Bīrūnī's part.
Passage I: six views on action and agent (Section 6.3.3), kā. 61.	Exclusion of YD, JM and TK.	No list or a different list on a different *kārikā.*
	Parallel with the **Suvarṇasaptati* and GPBh.	Sequence order.
	Parallel with the **Suvarṇasaptati*, V1, MV and perhaps V2.	Reference to the Vedas.
Passage XIX: four levels of knowledge (Section 6.3.4), kā. 46	Exclusion of YD, JM and TK.	Absence of the analogy on kā. 46.
	Parallel with the **Suvarṇasaptati* and V2.	Same form and elaboration of the analogy.
Passage VIII: the rainwater which varies intaste (Section 6.3.5), kā. 16.	Possible exclusion of YD and GPBh.	No or very brief contextualization of the analogy on kā. 16.
	Moderate parallel with V1 and MV.	Missing detail in the description.
	Parallel with the **Suvarṇasaptati* and possibly V2.	Same elaboration and contextualization of the analogy.

The major discrepancies between the content of the *Kitāb Sānk*, as transmitted in the *Taḥqīq mā li-l-Hind*, and that of the *Yuktidīpikā*, the *Jayamaṅgalā* and the *Tattvakaumudī* allowed for excluding these three Sanskrit commentaries from being considered as the sources of the *Kitāb Sānk*. On the other hand, the Arabic translation matches the content of the commentaries belonging to the group of five, the **Suvarṇasaptati*, the *Gauḍapādabhāṣya*, the *Māṭharavṛtti*, the *Sāṅkhyavṛtti* and the *Sāṅkhyasaptativṛtti*, in a striking manner. Its source is therefore affiliated in some way to this group.

Among this group, the *Māṭharavṛtti*, the *Sāṅkhyasaptativṛtti*—which display parallel characteristics and structures—and the *Gauḍapādabhāṣya*, share several commonalities with the *Kitāb Sānk*, but were most likely not its source. They bear fewer resemblances to the *Kitāb Sānk* than the **Suvarṇasaptati* and the *Sāṅkhyavṛtti*. This trait, in addition to their general condensed style of composition, minimizes the possibility of these three commentaries of being the sources of the Arabic translation. The *Kitāb Sānk*, however, when compared to the *Māṭharavṛtti*, the *Sāṅkhyasaptativṛtti* and the *Gauḍapādabhāṣya* shows a higher degree of similarity to the *Gauḍapādabhāṣya*, than to the other two.

The **Suvarṇasaptati* and the *Sāṅkhyavṛtti* have affinities not only with one another, but also with the *Kitāb Sānk*. The **Suvarṇasaptati* parallels the *Kitāb Sānk* relatively well in both style and content. Yet, as aforementioned, it is a Chinese translation and, thus, it remains problematic to equate its source with that of the *Kitāb Sānk*. A comparison between these two translations only led to the hypothesis that their respective Sanskrit sources were similar to each other, without any possible further conclusion.

As for the *Sāṅkhyavṛtti*, its style and content are generally relatively similar to that of the Arabic text. This is particularly evident from the passages on the four levels of knowledge (Section 6.3.4). Their titles also correspond by containing *sānk* and *sāṅkhya* respectively. In general, however, the *Kitāb Sānk* provides more detailed descriptions than the *Sāṅkhyavṛtti*, although these may be explained by al-Bīrūnī's creativity. As many passages of the *Sāṅkhyavṛtti* corresponding to the quotations from the *Kitāb Sānk* are missing, or uncertain, due to the impaired condition of its manuscript, it is difficult to further define the exact relationship between this commentary and the Arabic translation.

In any case, some passages also indicated that the *Kitāb Sānk* may not have been based on the *Sāṅkhyavṛtti*. Table 13 above for instance shows that the sequence in which views on agent and action are listed probably does not match in the two works, while the reference to the Vedas found in the *Kitāb Sānk* cannot be ascertained in the *Sāṅkhyavṛtti*. The **Suvarṇasaptati* and the *Gauḍapādabhāṣya*, however, share these two common points with the *Kitāb Sānk*. Another element which may be shared by the **Suvarṇasaptati* and the *Kitāb Sānk* is the specification of the material of receptacle in which the rainwater falls and which alters its taste (Section 6.3.5), if the Arabic passage was indeed drawn from a text and not from an oral commentary in this place.

Therefore, I argue, based on the above analyses, that the **Suvarṇasaptati* and the *Kitāb Sānk* were based on a similar Sanskrit commentary on the

Sāṅkhyakārikā. This Sanskrit source was, however, perhaps not the *Sāṅkhya-vṛtti*, despite the three works sharing numerous common points. Thus, al-Bīrūnī could have accessed this source in order to compose the *Kitāb Sānk*, while being assisted by Indian thinkers, and possibly supplementing his information with the book by Gauḍa, namely the *Gauḍapādabhāṣya*, which he mentioned in the *Taḥqīq mā li-l-Hind*. Although these interpretations remain hypothetical and will remain so in the absence of the complete text of the *Kitāb Sānk*, they help understand the relationship of the Arabic translation with the **Suvarṇasaptati*, the *Sāṅkhyavṛtti* and the *Gauḍapādabhāṣya*.

Conclusion

The first part of this book (Chapters 1 and 2) focused on al-Bīrūnī's socio-historical and intellectual environment. It enabled a contextualization of the way in which he became acquainted with Indian science. It also shed light on the different locales in which al-Bīrūnī dwelt, highlighting the prosperous conditions of these towns in terms of trade and development of sciences. Residing in these flourishing centres, al-Bīrūnī usually benefited from the support of a ruler. These circumstances were conducive for him not only to devote himself to his research, but also to engage with scholars from different cultural and intellectual milieus.

Considering the specific geographical distribution of the various sites in which he lived, both within and beyond al-Hind's frontiers—as they were conceptualised by al-Bīrūnī—helped distinguish the differing historical and cultural contexts in which his expertise and scientific knowledge developed. The contexts of Khwarezm (Kāṯ and Ǧūrǧānīya), Ray and Ǧūrǧān diverged from that of Kābul and Ghazna in several ways. Kābul and Ghazna were situated on a passage between Persia and India. The Late Shahis, who used Sanskrit as an official language and worshipped Hindu deities ruled that area in pre-Islamic times. The religions of Hindu cults were thus practiced in eastern Afghanistan until at least the second half of the tenth century CE.

Surviving traditions from such religious practices may have still prevailed there at the time of al-Bīrūnī's arrival in 1017, even if Kābul and Ghazna were no longer within the territory of al-Hind by that time. In addition, families of craftsmen, slaves and probably interpreters had gathered in these two towns since at least the time of Maḥmūd's reign.

Al-Bīrūnī lived in present-day eastern Afghanistan between the years 1017 and 1030, but also travelled across several regions of al-Hind. Even if evidence for delimiting the exact territory of his travels in—and observations of—al-Hind is scant and primarily based on an analysis of his writings, thanks to the socio-historical study of the present book, I showed that his visits to this region were mostly confined to Gandhāra and Panjab. Furthermore, preliminary investigation into archaeological and literary sources related to the five specific locales visited by al-Bīrūnī that are Laghmān, Peshawar, Fort Rājagirī, Fort Lahūr and Fort Nandana, shows that this territory belonged to the kingdom of the Late Shahis.

While scholars have recently turned their attention to studying these rulers, much information is missing about their origins, society, culture and interactions with bordering regions. An in-depth understanding of the Late Shahis from this cultural perspective is in my view particularly crucial to further understanding the history of Buddhism and Brahminism in the borderlands of Central and South Asia. These kings were the successors of the Early Shahis who have been regarded as Buddhists. Archaeological sources, however, challenge this latter assumption.

Thorough research into these histories would provide new insights into the relationship into circumstances under which the areas of Ghazna, Kābul, Gandhāra and Panjab shifted from Buddhism to Brahminism, in the interface with Islam.[1]

Further, the present book also revealed that the history of Pakistani northern Panjab remains particularly unknown for the period extending from the seventh up to the eleventh century CE and corresponding to the Shahi and early Ghaznavid times.

The extent to which the *Taḥqīq mā li-l-Hind* describes the intellectual culture of the Late Shahis is also difficult to ascertain, because, as I have highlighted in the present book, al-Bīrūnī's sources are various. Determining which pieces of information are connected to the society of these kings and which others are not would constitute another way of reconstructing aspects of their society. Nevertheless, I have set forth some methodological perspectives that aimed to ease such endeavour for anyone willing to carry it on in the future. I summarize below the observations and results of this socio-historical research.

Al-Bīrūnī's scientific interests evolved over the course of his life, an evolution that parallels his personal movements in Central and South Asia, and that reflects cultural and political contexts of the society at the time. He began writing on mathematics and astronomy and later opened his field of research to history, sociology, mineralogy, pharmacology and other scientific areas. Before al-Bīrūnī visited al-Hind, his knowledge of Sanskrit and Indian sciences was mostly related to mathematics and astronomy and based on Arabic sources, either translations of Indian literature or texts dealing with India. By the time of the composition of the *Taḥqīq mā li-l-Hind*, the depth of his knowledge had grown significantly.

His understanding of Sanskrit, for instance, was rather proficient at the time he composed the *Taḥqīq mā-li-l-Hind*, the result of a lengthy process of devel-

1 The necessity to study the Shahi finds expression in the research project established in Vienna and entitled *Cultural Formation and Transformation: Shahi Buddhist Art and Architecture*. See fn. 56 of Chapter 1. Different resources, i.e., research inquiries, bibliographic references, map and objects, related to the Shahi kingdoms is available on the webpage of the project (https://shahimaterialculture.univie.ac.at/ [accessed October 2023]) including an open access database.

opment that lasted at least 30 years (1000–1030). As aforementioned, his initial studies of the Sanskrit language were mostly based on written sources of information; later on, he developed his abilities through direct collaboration with Indians. Evidence showed that at Maḥmūd's court, he encountered Indian scholars, with whom he engaged in dialogue. In order to translate several works from Sanskrit into Arabic, he must have worked with literate people well versed in Sanskrit, who may also have had some comprehension of Arabic, Persian and/or a vernacular language, used as intermediary languages.

Nevertheless, excerpts from al-Bīrūnī's writings suggested that at the Ghaznavid court, he chiefly collaborated with Brahmins, some of whom were astronomers and/or philosophers. For political reasons, the sultan encouraged the scholar to learn Sanskrit and to become acquainted with Indian culture. Al-Bīrūnī's pursuits in understanding and translating astrological and astronomical treatises were most probably due to both al-Bīrūnī's intellectual motivations and Maḥmūd's political ambitions. In the case of religious and philosophical works, however, the scholar may have been influenced by the inclinations of the Brahmins he met. His good understanding of these philosophies is in part due to his having studied with Brahmins philosophers. This observation shows that, despite the generally assumed reservation of Brahmins to disclose some of their religious, scientific and philosophical knowledge, in this case, they appear to have generously shared it with al-Bīrūnī.

The *Bhagavadgītā*, the *Viṣṇudharmottarapurāṇa*, the *Viṣṇupurāṇa*, the *Ādityapurāṇa*, the *Matsyapurāṇa* and the *Vāyupurāṇa*, from which al-Bīrūnī quoted abundantly, are certainly among the texts read by these Brahmins. On the other hand, the teachings of the Vedas were not shared with al-Bīrūnī. In the *Taḥqīq mā li-l-Hind*, the importance of the purāṇic literature as compared to the corpus of the Vedas reveals two cultural traits of the Hindu society he encountered: the sacred and confidential status of the Vedic knowledge among Indian Brahmins; and the intellectual changes occurring during the first millennium CE in South Asia, when the prominence of Vedic teachings and rituals began to wane in favour of other literary genres, such as the Purāṇas.

With regard to philosophy, Sāṅkhya and Yoga emerged as prevailing systems of thoughts among the Brahmins whom al-Bīrūnī met, as opposed to other systems of Indian philosophies. In the Indian context, historical and geographical circumstances in which the ideas of Indian philosophies were formulated, written and studied are largely unknown. Further, despite the early spread of Yoga and Sāṅkhya ideas throughout Sanskrit literature, their extant commentarial literature suggest that theses philosophies lost vitality between the early eleventh and sixteenth centuries. The latest extant classical Yoga text glossing the *Pātañjalayogaśāstra*, that is, Bhoja's *Rājamārtaṇḍa*, is dated approximately to the first half of the eleventh century, while the *Tattvakaumudī*, which is the most recent extant commentary on the *Sāṅkhyakārikā*, was composed during the second half of the tenth century.

In this context, al-Bīrūnī's translations of works related to Yoga and Sāṅkhya deserve attention, because his translations stand as evidence that these two Indian schools of thought were living traditions at a time when the production of new commentaries on their classical founding texts had ceased. Some Yoga and Sāṅkhya texts were passed on through the oral informants whom al-Bīrūnī encountered in the early eleventh century. Considering the available sources, it remains delicate to determine with confidence the geographical provenance of those Brahmins and to establish whether these Indian thinkers studied, and practiced, the two philosophies in places visited by al-Bīrūnī, that is, Gandhāra and Panjab. A thorough philological study of the Sanskrit manuscripts of the *Sāṅkhyakārikā* and some of its commentaries could help elucidate the historical and geographical developments of Sāṅkhya philosophy in these two regions.

However, one element discussed in the present book, namely the Sanskrit legend on the bilingual coins referring to the fundamental Sāṅkhya concept of *avyakta*, found in Lahore, strongly suggests that Sāṅkhya was a popular philosophy in Panjab. Thus, the use of this particular Sanskrit word in this legend—in combination with the prevalence of Sāṅkhya and Yoga rather than other Indian philosophies in al-Bīrūnī's *Taḥqīq mā li-l-Hind*—indicate that in the regions of Gandhāra and Panjab Sāṅkhya and Yoga were still popular systems of thought among Indian thinkers before approximately 1030.

In terms of religion, the questions may arise of whether the *Taḥqīq mā li-l-Hind* rather describes a śaiva or a vaiṣṇava society,[2] similarly as to what extent the Late Shahis favoured Śiva or Viṣṇu, and whether there is a connection between the content of al-Bīrūnī's book on India and the archaeological and textual data on these rulers, or not. Whereas this question lies beyond the scope of this book, investigating into it may help us understand better the cultural history of early medieval Gandhāra and Panjab. The little that has been touched upon this question in the present book tends to indicate that these categories of religious sectarian identity were more permeable than generally assumed.

The second part of this book (Chapters 3, 4, 5 and 6) took on a textual approach, examining the question of the relationship between al-Bīrūnī's Arabic translations and the literature related to Sāṅkhya and Yoga. The philo-

2 Nārāyaṇa (ناراين) and Vāsudeva (باسديو) occupy important places in al-Bīrūnī's book on India, in contrast with Mahādeva (مهاديو), who is mentioned less frequently.

logical survey in Chapter 3 constituted the first necessary step to engage with this question, as it sheds light on the conceptualisation of the two works by al-Bīrūnī and his Indian informants. The information al-Bīrūnī provides about his translations, such as the authors, titles and descriptions wholly reflect the Sanskrit textual tradition of the *Sāṅkhyakārikā* and the *Pātañjalayogaśāstra*. Al-Bīrūnī's descriptions of these two philosophies does not, however, help us understand the nature of the relations between Sāṅkhya and Yoga as philosophical schools.

Further, by the study of Arabic and Sanskrit passages, I demonstrated that both his translations, the *Kitāb Sānk* and the *Kitāb Pātanǧal*, were based on a text and a commentary and were considered one textual entity penned by one author. This conception is consistent with the earliest Sanskrit textual evidence about the authorship of the *Pātañjalayogaśāstra*. Al-Bīrūnī thus intermingled a text and a commentary in his translation. This combination of two layers of texts also reflects the viewpoint of Indian thinkers that an aphoristic text may not be dissociable from its commentary.

Chapter 4, 5, and 6 examined al-Bīrūnī's hermeneutics, which played an important part in his transmission of Sāṅkhya and Yoga. Al-Bīrūnī transformed the Sanskrit source-text in both form and substance. Both the *Kitāb Pātanǧal* and the *Kitāb Sānk* show such transformations. Based on these parallels between the two Arabic translations, I argued that al-Bīrūnī dealt with his two Sanskrit sources in a similar way.

In both cases, the many discrepancies between al-Bīrūnī's translations and their possible Sanskrit sources were due either to his choices of interpretation or to the influence of the Brahmins who assisted him. A mere literal comparison between the Arabic texts and their possible Sanskrit originals thus did not lead to significant results in terms of identifying the original source. Instead, drawing upon perspectives taken from Translation Studies made it possible to analyse al-Bīrūnī's translations in a more detailed way, while underlying causes behind discrepancies between the Sanskrit sources and their Arabic translations.

In particular, these perspectives led to several conclusions. First, al-Bīrūnī's desire to reduce the complexity of his sources accounted for the many omissions he made. Second, his idiosyncratic understanding of some concepts resulted in him substituting Indian technical notions with Islamic philosophical ones. Third, his pre-existing worldly knowledge, related to both his own and the Indian culture, enabled him to add definitions of these concepts or other elements, in his Arabic translations.

Al-Bīrūnī kept his Muslim readership in mind when composing his works on Indian philosophy, so as to facilitate their transmission to this cultural sphere.

Overall, the *Kitāb Sānk* and the *Kitāb Pātanǧal* represent original works of Sāṅkhya and Yoga, as viewed by a Perso-Muslim scholar, rather than literal translations of Sanskrit literature. In this respect, comparing al-Bīrūnī's translations of Sanskrit texts belonging to other literary genres, such as mathematical, astronomical and astrological treatises, as well as the *Bhagavadgītā* and the Purāṇas, would constitute another future research prospect, because it would supplement the results of this study on his Sanskrit sources and methods of translations.

With this approach in mind, thus, I determined several of al-Bīrūnī's transformations—formal and substantial—and potential candidates for al-Bīrūnī's original Sanskrit sources. The *Kitāb Pātanǧal* was likely based on the *Pātañjalayogaśāstra*, or a text similar to it, while the *Kitāb Sānk* shows strong connection to a commentary resembling the source of the Chinese **Suvarṇasaptati*, both of which share common features with the *Sāṅkhyavṛtti*, without however equating it.

Three facts, however, may jeopardize these conclusions. First, regarding the *Kitāb Pātanǧal*, a passage introducing the actual philosophical work and corresponding to a laudatory strophe remains unidentified. In most probability, it was a creation of al-Bīrūnī's and/or his informants, although this cannot be definitively confirmed. Second, in the case of the *Kitāb Sānk*, if a complete manuscript would be discovered, the above conclusions could be corroborated or refuted. In addition, a complete critical edition of the *Pātañjalayogaśāstra* and reconsiderations on the dating of the *Sāṅkhyakārikā*'s commentaries, may supplement this discussion.

Further, the high degree of similarity between the *Sāṅkhyavṛtti* and the **Suvarṇasaptati* on the one hand, and between the *Sāṅkhyasaptativṛtti* and the *Māṭharavṛtti* on the other hand, may help in providing a relative chronology of these commentaries. However, the question deserves a comprehensive and thorough examination of its own, including critical editions and systematic translations from the original languages.

In addition to identifying candidates for his sources, it has been possible to propose explanations for al-Bīrūnī's interpretations of some Sāṅkhya-Yoga concepts, such as Īśvara, absorption (*samādhi*), original source (*prakṛti*), conscious self (*puruṣa*), afflictions (*kleśa*), constituents (*guṇa*), mental dispositions (*cittavṛtti*) and his understanding of the *satkāryavāda* theory. Other important themes that were not dealt with in this book, such as karma, isolation (*kaivalya*) and means of knowledge (*pramāṇa*), may be the object of a further study.

Analysis of al-Bīrūnī's translations can also bring new insights into two broader discussions relevant to Indology. The first pertains to the problems and difficulties of transmission of texts and ideas from one culture to another.

Translating technical concepts of Indian philosophy into a European modern language requires the adoption of a strategy, either a direct borrowing from the source-language by transliterating a term into Roman script in supplement with a definition, or a substitution of the source-concept with an overlapping target-concept, each strategy having its own sets of benefits and shortcomings. Whatever option one chooses, cross-cultural exchanges and dialogues consequent to it are impacted by the translator's decisions, as the example of al-Bīrūnī illustrates well.

The present book did not examine questions of the reception of al-Bīrūnī's work in the Perso-Muslim intellectual sphere. Only three manuscripts containing the *Taḥqīq mā li-l-Hind* are extant, while only one of the *Kitāb Pātanǧal* is available, and none of the *Kitāb Sānk*. Al-Bīrūnī's works were never translated into Latin, whereas those of al-Ḵwarizmī and Ibn Sīnā were. His Latinized name is Alberonius, while the western name Aliboron is equally connected to al-Bīrūnī. Aliboron designates a "*stupid and/or pretentious person*."[3]

Two hypotheses may be worth exploring regarding al-Bīrūnī's legacy, or lack of it. First, the scholar addressed unconventional topics, by describing beliefs, sciences and literature of Indians considered to be heathen so that his audience would have regarded his work as controversial and heretical. Second, his descriptions and analyses of Indian philosophy, religion, mathematics and astronomy are complex and elaborated, while his Arabic is at times unusual or convulsed, perhaps because it was not his native language. These elements may have discouraged his successors from continuing or developing his work on these subjects. Despite his efforts to transmit Indian culture to his peers, his endeavour does not appear to have been successful.

Nevertheless, my objective with the present book was to fill some gaps in understanding al-Bīrūnī's life and interpretations of Sāṅkhya and Yoga. Thanks to the piecing together of pieces of evidence from various fields and by presenting relatively unknown material, I explored the circumstances of al-Bīrūnī's encounter with early medieval India and its philosophy. I also connected his personal and intellectual journey to historical, social and political events of his time. Lastly, the analyses presented in this book, I hope, have also highlighted the need to transcend boundaries between academic disciplines and to explore different fields of study in order to contextualize primary sources, to have them speak about history, and thus to reach new results.

---

3 See the article on the CNRTL's website, http://www.cnrtl.fr/definition/aliboron [accessed October 2023].

APPENDIX

Passages Related to the Kitāb Sānk Found in the Taḥqīq mā li-l-Hind

Introductory Remarks

The excerpts below are English translations of passages which Junjiro Takakusu (1904a: 27–35) connected to the *Kitāb Sānk*. The last excerpt (passage no. XXI) of this appendix has been added to Takakusu's original list.

1 Six Views on Action and Agent

In the *Kitāb Sānk*, the ascetic man (ناسك) says: "Is there a difference [of opinion] about action (فعل) and agent (فاعل), or not?"

The sage (حكيم) says: "[Some] people [say] that the soul (نفس) is not agent and the matter (مادّة) not alive; that Allah, being self-sufficient, is the one who unites and separates the two of them, and therefore He is the agent. Action emerges from Him in order to set both [soul and matter] in motion, just like what is alive and powerful sets in motion what is dead and impotent.

Others say that the union of the two is [produced] by nature, and such is a common phenomenon in everything which grows and decays.

Others say that the agent is the soul, because in the Vedas (بيذ) [it is said] that every existent comes from *puruṣa* (پورش).

Some others say that the agent is time, since the world is tied to it (i.e., time), [in the same way] as the sheep's tie [is] with a tight rope, since its movement depends upon the pulling or the loosening of the [rope].

Others say that action is nothing but the recompense for a preceding deed.

All these views are deviating from what is correct. The truth about it is only that action entirely belongs to matter because it is what ties [the soul], causes [its] return in the [bodily] shape and releases [it]. It is the agent, and the rest [of the existents], which is below it, are its assistants in the completion of the action. Because the soul is devoid of the different forces (قوى), it is not an agent."[1]

1 Here, the Arabic term قوى (*quwan*) refers to the three constituents (*guṇa*). On al-Bīrūnī's dif-

11 Enumeration of the Twenty-Five Principles (*tattva*)

As for those [among the Indians] who deviate from allusions [but direct themselves] to investigation, they call the soul (نفس) *puruṣa* (پورش). It means the "man" because it is alive in what exists. They do not consider it as anything other than life. They ascribe to it the succession of knowledge and ignorance. Indeed, ignorant in actuality and intelligent in potentiality (جاهلة بالفعل وعاقلة بالقوّة), it receives knowledge by acquisition. Its ignorance causes the occurrence of the action, and its knowledge causes its removal.

The absolute matter (المادّة المطلقة), namely the pure matter (الهيولى المخرّدة), follows it (i.e., *puruṣa*). They call it *avyakta* (أبيكت), that is, [something] without shape (بلا صورة). It is dead (موات) [but] it owns the three forces (القوى الثلاث) in potentiality without actuality (بالقوّة دون الفعل). Their names are *sattva* (ست), *rajas* (رج) and *tamas* (تم).

I have heard the expression of Śuddhodana (بدّهودن) about these [forces, speaking] to his adherent, the Šamaniyya (شمنيّة) [that the forces] are *buddha*, *dharma* and *saṅgha* (بدّ دهرم سنگ), as if they were [respectively] intellect (عقل), religion (دين) and ignorance (جهل). The first of the [forces] is quietude and goodness, from which existence and growth [arise]. The second is exertion and labour, from which constancy and continuation [arise]. The third is languor and indecisiveness, from which decay and annihilation [arise]. Therefore, the first [force] is attributed to angels, the second to men and the third to animals. The ideas of "before," "after" and "thereupon" lie in these [three forces], from the perspective of sequence and restriction of language, not from the perspective of time (هذه أشياء تقع فيها قبل و بعد و ثمّ من جهة الرتبة و تضايق العبارة لا من جهة الزمان).[2]

As for the matter which goes out to actuality with shape and with the three primordial forces, they call it *vyakta* (بيكت), i.e. the shaped [one] (المتصوّرة), and they call *prakṛti* (پركرت) the whole of pure primary (الهيولى المخرّدة) and shaped matter (المادّة المتصوّرة).[3] There is, however, no use of this word (i.e., *prakṛti*),

---

ferent uses of this word, see above pp. 148–150. *Taḥqīq* (1887), pp. 14.20–15.7; *Taḥqīq* (1958), pp. 22.12–23.5; Sachau 1910: 1/30–31. Some portions of the whole quotation have been translated by Kozah (2016: 67–70).

2 The exact meaning of this sentence is difficult to grasp. Sachau interprets as follows: "The ideas *before*, *afterwards*, and *thereupon* may be predicated of all these things only in the sense of a certain sequence and on account of the inadequacy of language, but not so as to indicate any ordinary notions of time." (Sachau 1910: 1/41).

3 Sachau's translation of this section is the following: "the union of the *abstract* ὕλη and of the *shaped matter* is called *prakṛiti*." (Sachau 1910: 1/41). The Arabic term which he interprets as union (مجموع) is in fact a passive participle of the verb to gather and literally means "what is gathered."

because we do not need to mention the absolute [one] (مطلقة)[4] and because we [consider] sufficient to express the [term] matter, since one does not exist without the other.

The [innate] temperament (طبيعة)[5] follows the [shaped matter]. They call it *ahaṃkāra* (آهنكار). Its etymology is superiority (غلبة), increase (ازدياد) and self-conceit (صلف), because the matter, when taking on shape, starts the growth of existents it. The growth (نموّ) only consists of the transformation of another [principle] and of its assimilation with what is growing. Thus, it is as if the temperament tries to defeat [other principles] in this transformation and subdues what is transformed.

It is clear that each composite [principle] has simple [ones] from which the composition comes and to which the dissolution returns. The totality of existents (الموجودات الكلّيّة) in the world consists of the five elements (عناصر). According to the [Indians], they are the sky, the wind, the fire, the water and the earth. They are called *mahābhūta* (مهابوت), meaning "the great natures" (كبار الطبائع). [...] These elements are composite. Thus, they have simple [ones], which precede them and are called *pañca* [*tan*]*mātra* (پنج ماتر), meaning five mothers (أمّهات خمسة).[6] They (i.e., the Indians) attribute to them the five senses: the simple [element] of sky is *śabda* (شبد), what is heard; the simple [element] of wind is *sparśa* (سپرس), what is touched; the simple [element] of fire is *rūpa* (روپ), what is seen; the simple [element] of water is *rasa* (رس), what is tasted; and the simple [element] of earth is *gandha* (گند), what is smelled.

Each one of these simple [elements] possesses what is connected to it as well as the totality of what was connected to the [element] below it. In this way, earth has the five qualities, then water falls short of the sense of smell as compared to [earth], then fire falls short of the sense of taste as compared to [water], then wind [falls short] of the two (i.e., senses of smell and taste) and of the colour (i.e., the sense of seeing), and the sky of the [three] and of the sense of touch [...].

The five senses are called *indriyāni* (اندريان). They are hearing by the ear, sight by the eye, smelling by the nose, tasting by the tongue and touching by the skin. Then, the will (إرادة) directs them to [their] various locations (i.e., of action).

---

4 The reading is clear from Sachau's edition (*Taḥqīq* [1887], p. 20.10).

5 The Arabic term طبيعة literally means "nature." In this context, i.e., as the rendering of the Sanskrit *ahaṃkāra* ("I" consciousness) however, it probably refers to the natural or innate character, i.e., the temperament, of a person.

6 Sachau translates the sentence in the following way: "As these elements are compound, they presuppose simple ones which are called *pañca mâtâras*, i.e. five mothers" (Sachau 1910: I/42). See also above fn. 113 in Chapter 2 of this book.

Its residence in a [person] (منه) is the heart (قلب). They call it *manas* (من). The animal nature (حيوانيّة) is rendered perfect by five necessary actions (بأفاعيل خمسة ضروريّة) belonging to a [person] (له),[7] which they call *karmendriyāni* (كرم اندريان), meaning "the senses by action" (الحواسّ بالفعل), because the outcome of the former [senses] is learning and knowledge, while that of the latter [senses] is action and work. We shall call them the necessities (ضروريّات). They are: production of a sound for [different] kinds of needs and wishes; strength by the hands for fetching and putting away; walking with the feet so as to seek [something] or to flee [from it]; and shaking off the excess of food (فضول الأغذية) through each of the two holes destined for it.

These [principles] are twenty-five: the universal soul (النفس الكلّيّة), the pure primary (الهيولى المجرّدة) and shaped (المتصوّرة) matter, the superior temperament (الطبيعة الغالبة), the simple mothers (الأمّهات البسيطة), the primary elements (العناصر الرئيسيّة), the intelligent senses (الحواسّ المدركة), the directing will (الارادة المصرّفة), and the instrumental necessities (الضروريّت الآليّة). The name of the whole [of these principles] is *tattva* (تتو). [All] knowledge is confined to them.

Therefore, Vyāsa, the son of Parāśara (بياس بن پراشر) said: "learn the twenty-five [principles] in detail, with [their exact] definitions and with [their] divisions, by a knowledge [based on] evidence and ascertainment, not by oral instruction. Then, adhere to whatever religion you want, your end will be deliverance (النجاة)."[8]

III The Five Vital Breaths

When the mixed and varied bodies come into being, made up of male and female [elements]—male [elements] are for instance bones, veins and sperm and female [elements] flesh, blood and hair—[and when] they are ready to receive life, then these vital breaths (ارواح) are combined to the [bodies]. The [bodies] are [prepared] for the [vital breaths] in the same way castles are prepared for the various affairs of the kings. The five vital breaths enter the [bodies]; among [these five breaths], by two of them the inhaling and exhaling of breath (جذب النفس وإرساله) [occurs], by the third [one] the mixing of food in the stomach; by the fourth [one] the motion of the body from one place to another; and by the fifth [one] the transfer of the senses from an extremity of

7 The referent to the pronouns in *min-hu* (منه) and *la-hu* (له) is understood in the present translation as referring to a person or human being. Sachau (1910: 1/44) does not translate these two elements.

8 *Taḥqīq* (1887), pp. 19.19–22.8; *Taḥqīq* (1958), pp. 30.10–34.4; Sachau 1910: 1/40–44.

the body to another. According to them (i.e., the Indians), the vital breaths, not being different [from each other] in essence, are same by their very nature. Only their dispositions and traces (أخلاقها وآثارها) vary with regard to the differences of bodies connected to them, because of the three forces which are struggling [against each other] inside the [bodies] and which become at variance due to envy and anger. This is the highest cause for the arising of action (فهذا هو السبب الأعلى فى الانبعاث للفعل).[9]

IV The Soul as a Female Dancer

As for the lower cause (السبب الأسفل) with regard to matter, its quest for perfection and its preference for excellence is the emergence from potentiality to actuality (الخروج من القوّة إلى الفعل). Because pride and attachment to superiority are in the foundation of the temperament (طبعية; i.e., *ahaṃkāra*), all possible [things] occur in the [matter], so that the [matter] teaches and causes the soul to return in [various] kinds of plants and animals.[10] The [Indians] compare it (i.e., the matter)[11] to a female dancer, skilled in her art, knowledgeable of the effect of each union and separation [in her movements?] which she has (بأثر كلّ وصل وفصل فيها), [and] who he is in presence of [someone] living in luxury with a strong desire to see what is with her (i.e., her dance). Then, she begins with various [movements] of her art [and] shows them one after the other. The host of the gathering looks at her, until what is with her (i.e., her dance) comes to an end, and [until] the cessation of the spectator's desire. Then, she becomes cut off, perplexed, as if what is with her (i.e., her dance) would be nothing but a repetition; and a repetition would be undesirable. He then leaves her, and the action disappears.[12]

---

9 *Taḥqīq* (1887), pp. 22.20–23.6; *Taḥqīq* (1958), p. 35.2–12; Sachau 1910: I/46. Passages Numbers III to V are immediately succeeding each other.

10 The exact meaning of this sentence is unclear to me (بما فى سنخ الطبيعة من المباهاة ومحبّة الغلبة تعرض ما فيها من أصناف الممكن على من تعلّم و تردّد النفس فى ضروب النبات و أنواع الحيوان). See Sachau: "In consequence of the vainglory and ambition which are its pith and marrow, matter produces and shows all kinds of possibilities which it contains to its pupil, the soul, and carries it round through all classes of vegetable and animal beings." (1910: I/47).

11 According to Sachau's and Hyderabad's editions, the original manuscript reads "they compare the two" (شبّهوهما), i.e., matter and soul, which both printed editions emended for "they compare to it" (شبّهوها) (*Taḥqīq* [1887], p. 23.16; *Taḥqīq* [1958], p. 35.9).

12 *Taḥqīq* (1887), p. 23.6–13; *Taḥqīq* (1958), pp. 35.12–36.3; Sachau 1910: I/47.

V The Blind and the Lame

The action disappears such as in the example of a caravan (رفقة)[13] in a desert, which was intercepted [by robbers]. Its people fled, except a blind [person] who was in the [caravan] and a lame [person], both remaining in the open country despairing of [their] escape. After they have met and got acquainted with one another, the chronically ill [one] said to the blind [one]: "I am unable of movement, but able to lead [us]. Your case in these two [matters] is the contrary to my case. Thus, put me on your shoulders and carry me, so that I [can] show you the way and [so that] we [can] get out together from the danger." He (i.e., the blind one) did [so] and [their] wish was fulfilled, through their mutual help. They separated when going out of the desert.[14]

VI The Soul as a Passer-by

c) As for the *Kitāb Sānk*, it ascribes action to matter because the shape which becomes visible is much different from the [matter] due to the three primary forces and to their dominance of one or two [over the other(s)], that are angelic, human and animal. These are the forces which belong to the [matter] not to the soul.

d) The soul shall know the actions of [matter], with the position of a spectator, in the manner of a passer-by who sits down in a village to rest. Everyone among the villagers is busy with his own particular work. The [passer-by] looks at them and considers their conditions, disliking some, liking others, and learning from them. Thus, he is busy without having himself any share in their [business] and without being the cause for the consequence of their [business].[15]

VII The Innocent Man among Thieves

It (i.e., the *Kitāb Sānk*) ascribes action to soul, even if the [soul] is exempt from it (تبرّئها),[16] in the manner of a man who happens [to get in] a company of a group

13 See *Taḥqīq* (1887), p. 23.13.

14 *Taḥqīq* (1887), pp. 23.13–16; *Taḥqīq* (1958), p. 36.3–8; Sachau 1910: I/47.

15 *Taḥqīq* (1887), pp. 24.1–6; *Taḥqīq* (1958), pp. 36.16–37.4; Sachau 1910: I/48. Passages Numbers VI to X follow upon each other.

16 This reading is that of the two Arabic editions used in the present book, while the manuscript has تبروه.

[of people] whom he does not know. They are thieves returning from a village which they have attacked and destroyed. He only marched with them a short distance, until the pursuers caught them. He is locked up among the group. The innocent [man] (البرىء) is taken in their group and, as [if he were] in their situation, [pursuers] hit him, as they hit the [group], without him having taken part in their actions.[17]

VIII The Rainwater Which Varies in Taste

They (i.e., the Indians or the adherents of the *Kitāb Sānk*) narrate the example of the soul, which resembles the rainwater falling down from the sky, being as it is and of one nature. When it is collected in receptacles placed for it in different materials, such as gold, silver, glass, ceramic, clay and saline earth (سبخة), it differs in appearance, taste and smell due to these. In this way, the soul does not affect the matter (i.e., *prakṛti*) except for life [which it gives to matter] because of the proximity [between the two].[18]

IX The Production of Light from Oil, Wick and Fire

When matter begins action, what arises from it is different [from it] due to the force which dominates among the three forces (القوى الثلاث). The assistance which the other two provide to it (i.e., the dominant force) exists in various ways; similarly, the damp oil, the dry wick and the smoking fire help each other to [produce] light.[19]

X The Driver of a Chariot

The soul is in the matter, like the rider of a vehicle, which the senses serve in driving it according to his will. But the intellect (عقل) guides it, inundating them [with what] consists in Allah, glorified be He—the [Indians] described [intellect] as that by which [one] perceives realities, and [that] which leads to the knowledge of Allah, the Sublime—and consists in the actions [striving] for whatever is loved [and] for the totality of what is praised among all.[20]

---

17 *Taḥqīq* (1887), p. 24.6–10; *Taḥqīq* (1958), p. 37.5–9; Sachau 1910: I/48–49.

18 *Taḥqīq* (1887), p. 24.10–13; *Taḥqīq* (1958), p. 37.9–13; Sachau 1910: I/49.

19 *Taḥqīq* (1887), p. 24.13–15; *Taḥqīq* (1958), p. 37.13–16; Sachau 1910: I/49.

20 *Taḥqīq* (1887), p. 24.15–18; *Taḥqīq* (1958), p. 37.16–17; Sachau 1910: I/49. Sachau has a diff-

XI Rewards of Attaining Heaven Being No Special Gain

This domain is [where] they (i.e., the Indians) turn away from theoretical investigation toward religious account, that is, of the two places of reward and punishment, the existence in these two [places] for [someone] who does not assume a bodily form and the return to embodiment and incarnation [as a human] after completion of the last action so as to be ready to what one will have [there] (ليستعدّ لما هو له).[21] Therefore, the author of the *Kitāb Sānk* does not consider the reward of heaven as good, because it has an end and it is not eternal and because it resembles the condition of life of [our] world, consisting of mutual competition and envy, due to the rivalry for superiority of ranks and classes, since thirst and desire do not stop unless there is balance.[22]

XII Births Depending on Virtue and Vice

In the *Kitāb Sānk* [it is stated]: "the one who deserves ascension and reward becomes just like one angel intermingling with the communities of spiritual [beings], not prevented from behaving freely in heavens nor from living with their inhabitants, or just like one of the eight categories of spiritual [beings]. But the one who deserves to descend because of [his] sins and crimes becomes an animal or a plant. He comes and goes until he deserves a [positive] reward and is saved from adversity (i.e., punishment) or until he himself becomes conscious,[23] and he gives up his vehicle (i.e., the body) and he is liberated."[24]

---

erent interpretation of the second sentence of this passage: "But the soul for its part is guided by the *intelligence* with which it is inspired by God. This intelligence they describe as that by which the reality of things is apprehended, which shows the way to the knowledge of God, and to such actions as are liked and praised by everybody." (1910: 1/49).

21 Sachau's interpretation slightly differs: "as regards the two places where reward or punishment is given, e.g. that man exists there as an incorporeal being, and that after having received the reward of his actions he again returns to a bodily appearance and human shape, in order to be prepared for his further destiny." (1910: 1/62).

22 *Taḥqīq* (1887), p. 31. 5–9; *Taḥqīq* (1958), p. 47.10–16; Sachau 1910: 1/62.

23 Sachau translates this expression as "or until he offers himself as expiation", which diverges from the Arabic (أو يعقل ذاته). A literal translation, however, concurs with the content of Sāṅkhya-Yoga as well as with al-Bīrūnī's understanding of it, which notably focuses on their cognitive aspects.

24 *Taḥqīq* (1887), p. 32.3–6; *Taḥqīq* (1958), pp. 48.16–49.2; Sachau 1910: 1/64.

XIII The Eight Powers

The author of the *Kitāb Pātanǧal* says: "The devotion of [one's] reflection to the oneness of Allah engages someone in being aware of a thing different from what they are [generally] occupied with. Whoever wishes Allah wishes the good for the whole creation without any exception [and] for any reason. Whoever occupies oneself with one's soul, being away from whatever [exists] except the [soul], he does not do for it (i.e., soul) any inhaling or exhaling breath. Whoever reaches this goal, their spiritual faculty (قوّة) overcomes their bodily faculty (قوّة), and then they are gifted with the ability (اقتدار) of [doing] eight things, through the outcome of which occurs renunciation. It is impossible that someone be able to dispense with what he lacks (محال أن يستغنى أحد عمّا يعجزه).

The first of these eight is the power (تمكّن) to render [one's] body subtle, in such a way that it disappears from the eyes' [perception]. The second power is to render it light, in such a way that a soil of thorn, mud or dust is all the same to him. The third power is to render it large, in such a way that he appears in a dreadful astonishing shape. The fourth power is to [fulfil one's] desires. The fifth power is to know whatever one desires. The sixth power is to preside over any desired group [of men] (فرقة). The seventh [power] is the submission of the subordinates and their obedience. The eighth [power] is the disappearance of distances between him and the remote destinations."[25]

XIV The Three Types of Knower

In their (i.e., Indians) view, knowledge occurs to the knowing [one] in one of three ways. The first of them is by divine inspiration, not over time but by birth and in the cradle, like the sage Kapila (كپل), as he was born with knowledge and wisdom. The second [way] is by divine inspiration after some time, like the children of Brāhma (براهم), as they were inspired when they attained their full maturity. The third [way] is by learning and after some time, like the rest of the men who learn when they reach maturity.[26]

---

25 *Taḥqīq* (1887), p. 34.5–12; *Taḥqīq* (1958), p. 52.5–17; Sachau 1910: 1/68–69. This passage does not relate to *the Kitāb Sānk*. I list it here as it counts among the excerpts dealt with by Takakusu.

26 *Taḥqīq* (1887), p. 35.19–20; *Taḥqīq* (1958), pp. 54.17–55.2; Sachau 1910: 1/72.

XV Nine Rules How to Conduct One's Life

An excellent conduct [of life] is that which religion (الدين) determines. Principles of [religion], aside from the numerous [practical] applications [found] among the [Indians], come down to a group of habits, which are: [1] not to kill, [2] not to lie, [3] not to steal, [4] not to commit adultery, [5] not to accumulate [goods], [6] to keep practicing holiness and purity, [7] to keep fasting and practicing asceticism, [8] to resort to the devotion to Allah with glorification and praise, [9] to keep in mind the [syllable] *aum* (اوم), which is the word of origination and creation, with one's heart [but] without uttering it.[27]

XVI The Limitations of Humankind

In the *Kitāb Sānk* [it is stated] that everything which men [can] imagine is their limit, because they [can] not transcend it.[28]

XVII The Movement of a Weaver's Wheel

The hermit (ناسك) asks in the *Kitāb Sānk*: "Why does not death take place at the cessation of action?"

The sage (حكيم) answers: "Because the cause of the separation is a condition related to the soul (نفسانيّة) and the vital breath is still in the body. There is no separation between the two [i.e., the soul and the body] except by a natural condition which separates their union. Sometimes, the impression remains for a certain time, after the end of what caused the impression; it subsides in [time] and diminishes until it disappears, just like the silk weaver who turns his wheel with a piece of wood until its rotation becomes fast. Then, he leaves it, but it is not still, although the piece of wood causing the rotation has been removed from it. Its motion only decreases little by little until it ceases. In the same way is the body after the cessation of action, the impression remains in it until it turns, in intensity and calm, toward the cessation of the natural force. Then, the past impression disappears and thus the perfect accomplishment of liberation occurs when the body falls down."[29]

---

27 *Taḥqīq* (1887), pp. 36.20–37.2; *Taḥqīq* (1958), p. 56.12–16; Sachau 1910: I/74–75.

28 *Taḥqīq* (1887), p. 37.8–9; *Taḥqīq* (1958), p. 57.5–6; Sachau 1910: I/75.

29 *Taḥqīq* (1887), p. 40.5–11; *Taḥqīq* (1958), p. 62.1–10; Sachau 1910: I/81–82.

XVIII States Other Than Liberation

A statement in the [*Kitāb*] *Sānk* [is as follows]: "he who enters upon the world with an excellent conduct [of life], who is generous with what he possesses among [any goods, his] reward in the world is the obtention of [his] wishes and desires, and [his] return in the [world] with happiness, happy in body, soul and the condition [of life], because, in reality, [good] fortune is a reward of preceding deeds [done] in this [bodily] form, or in another one. The ascetic in the world [living] without knowledge obtains elevation and reward, but he is not liberated because [the other] means [to reach liberation] are lacking. The content one, self-sufficient, when he is a master of the eight aforementioned conditions (i.e., powers), he is fooled by them, he carries out [his activities] easily and successfully and he considers them (i.e., the eight powers) as liberation, then he remains in these [powers]."[30]

XIX The Four Levels of Knowledge

And to give an example for those who differ in [their] degrees of intellection [i.e., *buddhi*]: a man journeyed at night with his disciples for some business. Then, an upright figure, whose exact knowledge was prevented by the darkness of the night, befell them in the way. The man turned toward his disciples and asked them about it, one after the other.

The first [disciple] said: "I don't know what it is."

The second said: "I don't know it and I am unable to know it."

The third said: "It is useless to learn about it [now], as the rising of the day will reveal it. If it is frightful, it [will] go away by the morning and if it is different, its case will become clear to us."

All three [disciples] failed to [obtain] the knowledge [about the figure]: the first of them because of ignorance, the second because of disability and damage of organ, the third because of indolence and of satisfaction in ignorance.

As for the fourth [disciple], he did not find any answer before ascertaining [the identity of the figure]. He went to see it. While approaching it, he saw a squash (يقطين) intertwined over it. He knew that a living man would not remain still in this standing position of [his own] free will, until such a thing gets entangled around him, and was sure that a lifeless [object] was erected. Then, he could not trust whether it was a hiding place for garbage [or not]. Therefore,

30 *Taḥqīq* (1887), p. 41.2–7; *Taḥqīq* (1958), p. 63.7–13; Sachau 1910: 1/83–84.

he came close to it, kicked it with his foot until it fell to the ground. The doubt [then] disappeared on the case of the [figure] and he returned to his master with absolute certainty. Thus, the [student] obtained knowledge by his effort.[31]

XX Categories of Beings

First, we tell what is in the *Kitāb Sānk* about this [topic].

The hermit said: "How many categories of living bodies are there?"

The sage replied: "There are three categories of them. The spiritual [ones] in the higher [world], man in the middle [world] and the animals in the lower [one]. As for their species, they are fourteen, eight of which belong to the spiritual [beings]: Brāhma, Indra, Prajāpati, Saumya, Gāndharva, Yakṣa, Rākṣasa and Piśāca. Among the [species], five belong to the animals: cattle, wild beasts, birds, reptiles and plants, I mean the trees. The human species is [only] one."

The author of this book enumerates them in another place of the [*Kitāb Sānk*] with other names: Brāhma, Indra, Prajāpati, Gāndharva, Yakṣa, Rākṣasa, Pitra and Piśāca.[32]

XXI Criticism of a List of Spiritual Beings

Because we learn from what we reported according to the [*Kitāb*] *Sānk* that this (i.e., a general view on spiritual beings) cannot be reached. Brāhma, Indra and Prajāpati are not names of species. Brāhma and Prajāpati are only similar in meaning, but their names differ due to some characteristics. Indra is the ruler of the worlds.[33]

---

31 *Taḥqīq* (1887), p. 41.7–15; *Taḥqīq* (1958), pp. 63.14–64.8; Sachau 1910: 1/84–85.

32 *Taḥqīq* (1887), p. 43.14–20; *Taḥqīq* (1958), p. 67.11–19; Sachau 1910: 1/89–90.

33 *Taḥqīq* (1887), p. 45.1–2; *Taḥqīq* (1958), p. 69.15–18; Sachau 1910: 1/92. This passage, which is a mere reference, was not mentioned by Takakusu. My translation of the first portion of the excerpt differs from that of Sachau: "However, we can learn from the extract from Sâṁkhya that this view is not correct. For Brahman, Indra, and Prajâpati are not names of species, but of individuals. Brahman and Prajâpati very nearly mean the same, but they bear different names on account of some quality or other. Indra is the ruler of the worlds" (Sachau 1910: 1/92). The context indicates that al-Bīrūnī merely presents discrepancies among the positions on spiritual beings within Sanskrit literature and in the Indian perception in general and does not intend to point out the correct view among them, as Sachau's interpretation suggests.

Glossary of Sanskrit Terms as al-Bīrūnī Interpreted Them into Arabic

This glossary covers terminology related to philosophy, theology and religion, since these domains are the focus of the textual study done in the present book. As demonstrated in Chapters 4, 5 and 6 of this book, al-Bīrūnī did generally not literally translate Sanskrit terms and concepts into Arabic, but often made use of translational strategies and paraphrases in order to render his sources into his language. The glossary and the English translations of the Arabic and Sanskrit terms below are based on analyses which I have detailed in this study. Readers can find each glossary term in the index of the present book and examine them in the text of the book.

Sanskrit		**Arabic**	
adharma	unrighteousness	الأوزار والآثام	sins and crimes
aṅga	component, limb	خصلة	quality, disposition
anumāna	inference	استدلال	reasoning
apavarga	emancipation	خلاص	liberation
abhiniveśa	clinging to life	علايق	attachment, devotion
abhyāsa	repeated practice	تعويد	habituation
artha	wealth, property	المال والنعمة	possessions and comfort
avidyā	ignorance	جهل	ignorance
avyakta	unmanifest	بلا صورة	without shape
aśakti	inability	عجز	disability
asaṃprajñāta	non-cognitive	المعقول المجرد عن المادة	of the intelligible free from matter
ahaṃkāra	'I' consciousness	طبيعة	[innate] temperament
āgama	authoritative tradition	سماع	oral tradition
āsana	posture	سكون	state of rest
īśvara	Īśvara	الله	Allah
īśvarapraṇidhāna	profound contemplation on Īśvara	عبادة	devotion
aiśvarya	mastery	القانع المستغني	content one, having no need
karma	action	إعمال	actions, deeds
karmendriya	organs of action	أفاعيل خمسة ضرورية	five necessary actions
kāma	pleasure	العيش و اللذّة	[enjoyable] way of living and pleasure
kāla	time	زمان	time
kevala/kevalin	isolated	متخلص	liberated one
kaivalya	isolation	خلاص	liberation
kleśa	afflictions	اثقال	burdens
guṇa	constituents	القوى الثلاث	three forces

(*cont.*)

grahaṇa	act of perceiving	علم	act of knowing,
		عقل	act of intellecting,
grahītṛ	perceiver	عالم	knower
		عاقل	one who intellects
grāhya	perceptible	معلوم	known
		معقول	intellected
citta (Yoga); *manas* (Sāṅkhya)	mind	نفس, قلب	soul, heart
cittavṛtti	mental dispositions	قوى النفس	faculties of the soul
jñāna	knowledge	معرفة	knowledge
tattva	principle	المبادئ	fundamental elements
tuṣṭi	satisfaction	تراجٍ	indolence
deva	deity	ملك	angel
dveṣa	aversion	عداوات	enmities
dharma	righteousness	حسن السيرة	an excellent conduct of life
dharma	moral values	الدين والسيرة	religion and conduct of life
dhāraṇā	fixation	السكينة والطمأنينة	quietude and tranquillity
dhyāna	meditation	ادامة الفكرة في	maintenance of thought upon [something]
nidrā	deep sleep	رؤيا	dream
niyama	observances	القدس ظاهرا و باطنا	holiness, outward and inward
nirodha	cessation, holding	قبض	compression /holding
puruṣa	conscious self	نفس, قلب	soul, heart
puruṣārtha	human goals	المطّالب الاربعة	human objectives
prakṛti	original cause	المادة المطلقة	absolute matter
		الهيولى المجردة	pure primary matter
pratyakṣa	direct perception	بالحواس الخمس	by the five senses
pratyāhāra	withdrawal [from the senses]	قبض الحواسّ	compression of the senses
pradhāna	primary matter	cf. *prakṛti*	
pramāṇa	means of knowledge	ادراك	apprehension
prāṇāyāma	breath control	تكسين التنفّس	quieting the breath
buddhi	intellect	عقل	intellect
		معرفة	intellection
buddhīndriya	senses of perception (related to the intellect)	الحواسّ الخمس	five senses
mahābhūta	gross elements	كبار الطبائع	great natures
mūlaprakṛti	primordial original cause	cf. *prakṛti*	
mokṣa	liberation	الخلاص والديمومة	liberation and permanence
		خلاص	liberation
yogi	yogi	زاهد	ascetic
yama	ethical rules	الكفّ عن الشرّ	refraining from evil
rāga	passion	رغبة	desire
vairāgya (also *virāga*)	dispassion	زهد	asceticism

(*cont.*)

vikalpa	conceptual thinking	ظنّ	[false] assumption
viparyaya	error	تخيل	imagination
		جهل	ignorance (as a *bhāva*)
vyakta	manifest	متصورة	shaped [matter]
saṃprajñāta	cognitive	المحسوس ذى المادة	of the perceptible with matter
saṃyama	mental concentration	فكرة	thought
		تفكر	reflection
samādhi	absorption	تصور	contemplation
		اخلاص	purity (*aṅga*)
samāpatti	contemplative state	الرتبة قبل الخلاص	Stage before emancipation
siddhi	supernatural powers	زهادة الزاهد	*siddha-hood*; lit. The asceticism of the ascetic
siddhi	accomplishment		knowledge (*bhāva*)
smṛti	memory	ذكر	memory
svabhāva	nature	طباع	nature

Bibliography

The bibliography is ordered alphabetically without consideration of the Arabic definite article *al-* used with the proper names of Arab authors.

Abbreviations of Journals

AS	Asiatische Studien / Études Asiatiques
BSOAS	Bulletin of the School of Oriental and African Studies
EI	*Encyclopaedia of Islam*, Second and Third Editions, BrillOnline. http://referenceworks.brillonline.com/entries/encyclopaedia-of-islam-2; http://referenceworks.brillonline.com/entries/encyclopaedia-of-islam-3.
EIr	*Encyclopaedia Iranica*, Online Version. http://www.iranicaonline.org.
JAOS	Journal of the American Oriental Society
JIPh	Journal of Indian Philosophy
JNSI	Journal of the Numismatic Society of India
WZKS	Wiener Zeitschrift für die Kunde Südasiens

Bibliographic References

Diwakar Nath Acharya, 2006 *Vācaspatimiśra's Tattvasamīkṣa: the earliest commentary on Maṇḍanamiśra's Brahmasiddhi / critically ed. with an introduction and critical notes Diwakar Acharya*, Franz Steiner Verlag 2006.

Andrea Acri, 2011 *Dharma Pātañjala. A Śaiva Scripture from Ancient Java Studied in the Light of Related Old Javanese and Sanskrit texts*, Groningen 2011.

Andrea Acri, 2012 "Yogasūtra 1.10, 1.21–23, and 2.9 in the Light of the Indo-Javanese Dharma Pātañjala," JIPh 40 (2021), pp. 259–276.

K.Ś. Āgāśe (ed.), 1904 *Vācaspatimiśraviracitaṭīkāsaṃvalita Vyāsabhāṣyasametāni Pātañjalayogasūtrāṇi [part 1], tathā Bhojadevaviracita Rājamārtaṇḍābhidhavṛttisametāni Pātañjalayogasūtrāṇi [part 2]*, Puṇyākhyapattana [Pune] 1904. (On-line version: http://gretil.sub.uni-goettingen.de/gretil/1_sanskr/6_sastra/3_phil/yoga/patyogbu.htm) [accessed October 2023].

V.S. Agrawala, 1943 "The Sanskrit Legend on the Bilingual Ṭankas of Mahmūd Ghaznī," JNSI 5 (1943), pp. 155–161.

Imtiaz Ahmad, 2005 "Concept of India: Expanding Horizons," in: Irfan Habib (ed.), *Early Medieval Arabic and Persian Writing, in India. Studies in the History of an Idea*, Delhi 2005, pp. 98–109.

Akbar S. Ahmed, 1984 "Al-Beruni: The First Anthropologist," *Royal Anthropological Institute Newsletter* 60 (1984), pp. 9–10.

Anna Akasoy, 2011 "Tibet in Islamic Geography and Cartography: A Survey of Arabic and Persian Sources," in: Anna Akasoy, Charles Burnett and Ronit Yoeli-Tlalim (eds.), *Islam and Tibet—Interactions along the Musk Routes*, Farnham 2011 [reprint 2017], pp. 17–41.

Jamil Ali, 1967 *The Determination of Coordinates of Positions for the Correction of Distances between Cities. A translation from the Arabic of al-Bīrūnī's Kitāb taḥdīd nihāyāt al-amākin li-taṣḥīḥ masāfāt al-masākin*, Beirut 1967.

Michael Alram (et al.), 2016 *Das Antlitz des Fremden. Die Münzprägung der Hunnen und Westtürken in Zentralasien und Indien*. Vienna 2016.

Michael Alram and Matthias Pfisterer, 2010 "Alkhan and Hephtalite Coinage," in: Alram et al. (2010), pp. 13–38.

Michael Alram, 2010) Deborah Klimburg-Salter, Minoru Inaba and Matthias Pfisterer (eds.), *Coins, Art and Chronology II. The First Millennium CE in the Indo-Iranian Borderlands*, Vienna 2010.

Arturo Annucci, 2023 "Hindu Śāhi Coinage: Towards a New Chronology," *Sourcebook for the Shahi Kingdoms*, Vienna 2023 (https://shahi.openatlas.eu/api/display/6711) [accessed October 2023].

Vāman Shivarām Apte, 1957 *The Practical Sanskrit-English Dictionary*, Pune 1957 [revised and enlarged edition; reprint 1992].

Kemal Ataman, 2005 "Re-Reading al-Bīrūnī's India: a Case for Intercultural Understanding," *Islam and Christian–Muslim Relations* 16.2 (2005), pp. 141–154.

Ali Athar, 1992 "Translations of Sanskrit Works at Akbar's Court," *Social Scientist* 20.9–10 (1992), pp. 38–45.

Azad Arezou, 2017 "The Beginnings of Islam in Afghanistan: Conquest, Acculturation, and Islamization," in: Nile Green (ed.), *Afghanistan's Islam: From Conversion to the Taliban*, Oakland 2017, pp. 41–55.

Blain Auer, 2012 Symbols of Authority in medieval Islam: History, Religion and Muslim Legitimacy in the Delhi Sultanate, London/New York 2012.

Parviz Azkaei, 2001 *Al-āṯār al-bāqiya ʿan il-qurūn al-ḵāliya. Edited and Annotated by Parviz Azkaei*, Tehran 2001.

James Robert Ballantyne, 1885 *The Sànkhya Aphorisms of Kapila with Illustrative Extracts from the Commentaries*, 1885.

N.A. Baloch, 1973 *Ghurrat al-Zijat or Karana Tilaka. A Handbook of Astronomy by Bijayanand of Benares. Translated from the Original into Arabic with an Exposition of the underlying Principles by Abu Raihān Muhammad b. Ahman al-Beruni*, Sind 1973.

N.A. Baloch, 1987 "Beruni and His Experiment at Nandana," *Erdem* 9.3 (1987), pp. 673–729 (https://dergipark.org.tr/en/download/article-file/693584) [accessed October 2023].

Marie-Geneviève Balty-Guesdon, 1992 "Le Bayt al-ḥikma de Baghdad," *Arabica* 39.2 (1992), pp. 131–150.

L. Bazin, 2012 P.B. Golden, P.B. Golden, E.J. Zürcher, P.A. Andrews, P. Zieme, A.T. Karamustafa, Barbara Flemming, Edith G. Ambros, Çiğdem Balim et al., "Turks," EI 2nd ed., First published online 2012 [accessed November 2023].

Anna Livia Beelaert, 2008 "Ṭūṭī-nāma (Jawāher al-asmār). Ed. Šams al-Dīn Āl-e Aḥmad, Tehrān, Entešārāt-e Ferdows, 1385/2006, lxiv, 720 p," *Abstracta Iranica* 31, (on-line release on 11 October 2012, https://journals.openedition.org/abstractairanica/39799) [accessed October 2023].

Shripad Krishna Belvalkar, 1917 "Māṭharavṛtti and the Date of Īśvarakṛṣṇa," in: *Commemorative Essays presented to Sir Ramkrishna Gopal Bhandarkar*, Pune 1917, pp. 171–184.

C.H. Becker and F. Rosenthal, 2012 "al-Balādhurī," EI 2nd ed., First published online 2012 [accessed November 2023].

Asoke Kumar Bhattacharyya, 1954 "Hindu Elements in Early Muslim Coinage in India," JNSI 16.1 (1954), pp. 112–121.

Asoke Kumar Bhattacharyya, 1964 "Bilingual Coins of Mahmud of Ghazni," JNSI 26.1 (1964), pp. 53–56.

Narendra Nath Bhattacharyya, 1991 *The Geographical Dictionary of Ancient and Early Medieval India*, Delhi 1991 [reprint 1999].

A.D.H. Bivar, 1979 "The Stations of al-Bīrūnī on the Journey from Ghazna to Peshawar," in: Said 1979, pp. 160–176.

A.D.H. Bivar, 2010 "Weights and Measures i. Pre-Islamic Period," EIr, Last update 2010 [accessed January 2021].

T.J. de Boer and J. Arnaldez, 2012 "Ḳuwwa," EI 2nd ed., First published online 2012 [accessed November 2023].

Jacques-Dominique Boilot, 1955 "L'œuvré d'al-Beruni : essai bibliographique," *Mélanges de l'Institut dominicain d'études orientales du Caire* 2 (1955), pp. 161–256.

O.G. Bolshakov, 2012 "Eṣṭaḵrī, Abū Esḥāq Ebrāhīm," EIr, Last update 2012 [accessed November 2023].

Clifford Edmund Bosworth, 1963 *The Ghaznavids. Their Empire in Afghanistan and Eastern Iran, 994:1040*, Edinburgh 1963.

Clifford Edmund Bosworth (ed.), 1970 *Ḥudūd al-ʿālam. 'The Regions of the World. A Persian Geography. 372 A.H.–982 A.D. Translated and explained by V. Minorsky*, London 1970 [first edition 1937].

Clifford Edmund Bosworth, 1976 "The Khwarazmian Historical Background to Biruni's Life," in: *The Commemoration Volume of Biruni International Congress in Tehran. Center for Research and Cultural Co-ordination*, 38, Tehran 1976, pp. 11–27.

Clifford Edmund Bosworth, 1977 *The Later Ghaznavids: Splendour and Decay. The dynasty in Afghanistan and Northern India. 1040–1186*, New York 1977.

Clifford Edmund Bosworth, 1988 "Balāḏorī, Abu' l-Ḥasan or Abu Bakr Aḥmad b. Yaḥyā b. Jāber," EIr, Last update 1988 [accessed November 2023].

Clifford Edmund Bosworth, 1991 "Farrukhī's Elegy of Maḥmūd of Ghazna," *Iran. Journal of the British Institute of Persian Studies* 29 (1991), pp. 43–50.

Clifford Edmund Bosworth, 2004 "An Oriental Samuel Pepys? Abu'l-Faḍl Bayhaqī's Memoirs of Court Life in Eastern Iran and Afghanistan, 1030–1041," *Journal of the Royal Asiatic Society* 14.1 (2004), pp. 13–25.

Clifford Edmund Bosworth, 2007 *Historic Cities of the Islamic World*, Leiden/Boston 2007.

Clifford Edmund Bosworth, 2011a "Chorasmia ii. In Islamic times," EIr, Last update 2011 [accessed November 2023].

Clifford Edmund Bosworth, 2011b "Āl-e Afrīḡ," EIr, Last update 2011 [accessed November 2023].

Clifford Edmund Bosworth, 2011c "Ālī b. Ma'mūn," EIr, Last update 2011 [accessed November 2023].

Clifford Edmund Bosworth, 2011d "Ebn Ḵordāḏbeh, Abu ' l-Qāsem 'Obayd-Allāh," EIr, Last update 2011 [accessed November 2023].

Clifford Edmund Bosworth (ed.), 2011e *The History of Beyhaqi (The History of Sultan Mas'ud of Ghazna, 1030–1041) by Abu'l-Fażl Beyhaqi. Translated with a historical, geographical, linguistic and cultural commentary and notes by C.E. Bosworth. Fully revised and with further commentary by Mohsen Ashtiany. Vol. II, Translation of Years 424–432 A.H. = 1032–1041 A.D and the history of Khwarazm*, Boston 2011.

Clifford Edmund Bosworth, 2012a "Khwārazm," EI 2nd ed., First published online 2012 [accessed November 2023].

Clifford Edmund Bosworth, 2012b "Kāth," EI 2nd ed., First published online 2012 [accessed November 2023].

Clifford Edmund Bosworth, 2012c "Gorgān vi. History from the Rise of Islam to the Beginning of the Safavid Period," EIr, Last update 2012 [accessed November 2023].

Clifford Edmund Bosworth, 2012d "Ḳābūs b. Wushmagīr b. Ziyār," EI 2nd ed., First published online 2012 [accessed November 2023].

Clifford Edmund Bosworth, 2012e "Ghazna," EI 2nd ed., First published online 2012 [accessed November 2023].

Clifford Edmund Bosworth and Yolande Crowe, 2012 "Sāmānids," EI 2nd ed., First published online 2012 [accessed November 2023].

Gerhard Böwering, 2012 "Hojviri, Abu'l-Ḥasan 'Ali b. 'Oṯmnān b. 'Ali al-Ḡaznavi al-Jollābi," EIr, Last update 2012 [accessed November 2023].

Ralph W. Brauer, 1995 *Boundaries and Frontiers in Medieval Muslim Geography*, Philadelphia 1995.

Rémi Brague, 2003 *Muḥammad Ibn Zakkariyā Rāzī (Rhazès). La médecine spirituelle. Traduction de l'arabe, introduction, notes et bibliographie par Rémi Brague*, Paris 2003.

Luc Brisson, 2004 *Platon. Phèdre. Traduction inédite, introduction et notes par Luc Brisson. Suivi de la Pharmacie de Platon de Jacques Derrida. Édition augmentée de suppléments bibliographiques (1992–1996 et 1996–2004)*, Paris 2004.

Carl Brockelmann, 2012 "Kalīla Wa-Dimna," EI 2nd ed., First published online 2012 [accessed November 2023].

John Brockington, 1999 "Epic Sāṅkhya: Texts, Teachers, Terminology," AS 53.3 (1999), pp. 473–490.

Johannes Bronkhorst, 1981 "Yoga and seśvara Sāṅkhya," JIPh 9 (1981), pp. 309–320.

Johannes Bronkhorst, 1983 "God in Sāṅkhya," WZKS 27 (1983), pp. 149–164.

Johannes Bronkhorst, 1985 "Patañjali and the Yoga Sūtras," *Studien zur Indologie und Iranistik* 10 (1985), pp. 191–212.

Johannes Bronkhorst, 1993 *The Two Traditions of Meditation in Ancient India*, Delhi 1993 [reprint 2000].

Johannes Bronkhorst, 1994 "The Qualities of Sāṅkhya," WZKS 38 (1994), pp. 309–322.

Johannes Bronkhorst, 2003 "Review of Yuktidīpikā. The Most Significant Commentary on the Sāṅkhyakārikā. Critically edited by Albrecht Wezler and Shujun Motegi, Vol. 1. Stuttgart: Franz Steiner Verlag. 1998," *Zeitschrift der Deutschen Morgenländischen Gesellschaft* 153.1 (2003), pp. 242–247.

Johannes Bronkhorst, 2008 *Aux origines de la philosophie indienne*, Gollion 2008.

Johannes Bronkhorst, 2011a *Buddhism in the shadow of Brahmanism*, Leiden 2011.

Johannes Bronkhorst, 2011b *Language and Reality. On an Episode in Indian Thought*, Leiden 2011.

Johannes Bronkhorst, 2007 *Greater Magadha. Studies in the Culture of Early India*, Delhi 2007.

Sara Cappelletti, 2015 *Le Monete Bilingui Di Maḥmūd di Ghazna*, Pisa 2015.

Pulinbehari Chakravarti, 1938 *Yuktidīpikā. Critically edited for the first time from Original Manuscripts by Pulinbehari Chakravarti*, Calcutta 1938.

Pulinbihari Chakravarti, 1951 *Origin and development of the Sāṃkhya system of thought*, Kolkata 1951.

Suniti Kumar Chatterji, 1951 "Al-Bīrūnī and Sanskrit," in: Comm. Volume 1951, pp. 83–100.

Brajadulal Chattopadhyaya (ed.), 2006 *Studying Early India. Archaeology, Texts and Historical Issues*, London 2006.

J.W. Clinton, 2012 "Masʿūd-i Saʿd-i Salmān," EI 2nd ed., First published online 2012 [accessed in November 2023].

Henry Thomas Colebrooke, 1824 "On the Philosophy of the Hindus, Part 1: Sánkhya," *Transactions of The Royal Asiatic Society* 1 (1824), pp. 19–43.

Collected Volume, 1951 *Al-Bīrūnī Commemoration Volume*. A.H. 362–A.H. 1362, Kolkata 1951

V. Courtois, 1952 *Al-Biruni: a Life Sketch*, Kolkata 1952.

Alexander Cunningham, 1871 *The Ancient Geography of India. I. Buddhist Period including the Campaigns of Alexander and the Travels of Hwen-Thsang*, London 1871.

Alexander Cunningham, 1875 *Archaeological Survey of India. Report for the Year 1872–1873*, Kolkata 1875.

Bruno Dagens, 1964 Marc Le Berre and Daniel Schlumberger, *Monuments préislamiques d'Afghanistan*, Paris 1964.

H. Daiber, 2012 "Masāʾil Wa-Adjwiba," EI 2nd ed., First published online 2012 [accessed November 2023].

Ahmad Hasan Dani, 1969 *Peshawar. Historic City of the Frontier*, Peshawar 1969.

Ahmad Hasan Dani, 1979 "Al-Bīrūnī's Indica a Re-evaluation," in: Said 1979 (ed.), pp. 182–189.

Ahmad Hasan Dani, 1964 Helmut Humbach and Robert Gobl, "Tochi Valley Inscriptions in the Peshawar Museum," *Ancient Pakistan* 1 (1964), pp. 125–135.

Alain Daniélou, 1983 *Histoire de l'Inde*, Paris 1983.

Saifur Rahman Dar, 1994 "The Earliest Sculpture from Lahore now in the Lahore Museum," *Lahore Museum Heritage* 3.1 (1994), pp. 29–42.

Saifur Rahman Dar, 2001 "The Primordial Site of the Ghaznavid Lahore. Its Location and Limits," *Journal of Asian Civilizations* 24.2 (2001), pp. 53–86.

Surendra Nath Dasgupta, 1920 *A Study of Patañjali*, Kolkata 1920.

Surendra Nath Dasgupta, 1922 *History of Indian Philosophy*, Vol. 1, Cambridge 1922.

Surendra Nath Dasgupta, 1924 *Yoga as philosophy and religion*, London 1924.

Surendra Nath Dasgupta, 1941 *Philosophical Essays*, Kolkata 1941.

Surendra Nath Dasgupta, 1930 *Yoga Philosophy in Relation to Other Systems of Indian Thought*, Delhi 1930 [reprint 1979].

Etienne De La Vaissière, 2003 "Is There a 'Nationality of the Hephtalites?'," *Bulletin of the Asia Institute* 17 (2003), pp. 119–132.

Etienne De la Vaissière, 2008 *Islamisation de l'Asie centrale. Processus locaux d'acculturation du VIIe au XIe siècle*, Leuven 2008.

Jean Deloche, 1968 *Recherches sur les routes de l'Inde au temps des mogols* (*Étude critique des sources*), Paris 1968.

Marie-Thérèse Debarnot (ed. & trad.), 1985 *Kitāb maqālīd ʿilm al-hayʾa. La Trigonométrie sphérique chez les Arabes de l'Est à la fin du Xe siècle*, Damas 1985.

Marcel L. Devic, 1878 *Les Merveilles de l'Inde. Ouvrage inédit du Xe siècle. Traduit pour la première fois d'après un manuscrit de la mosquée de Sainte-Sophie, à Constantinople. Avec introduction, notes, index analytique et géographique par L. Marcel Devic*, Paris 1878.

Nundo Lal Dey, 1927 *The Geographical Dictionary of Ancient and Mediaeval India*, London 1927.

A. Dietrich, 2012 "al-Masīḥī," EI 2nd ed., First published online 2012 [accessed November 2023].

Siglinde Dietz, 2007 "Buddhism in Gandhara," in: Ann Heirman and Stephan Peter Bumbacher (eds.), *The Spread of Buddhism*, Leiden/Boston, pp. 49–74.

Akhilesh K. Dubey, 2005 *Sanskrit Inscriptions of Northern India (Economic & Social Studies of 10th to 12th Century)*, Delhi 2005.

Marie-Paule Duminil and Annick Jaulin, 2008 *Aristote. Métaphysique. Introduction, traduction, notes, bibliographie et index par Marie-Paule Duminil et Annick Jaulin*, Paris 2008.

S. Dvivedin (ed.), 1901 *Brāhmasphuṭasiddhānto Dhyānagrahopadeśādhyāyaś ca, Gaṇakacakracūḍāmaṇi Srībrahmaguptaviracitaḥ = Brāhmasphuṭasiddhānta and Dhyānagrahopadeśādhyāya by Brahmagupta*, Benares (1901–1902).

Richard M. Eaton (ed.), 2003 *India's Islamic Traditions, 711–1750*, New Delhi 2003 [reprint 2006].

Richard M. Eaton, 2020 *India in the Persianate Age, 1000–1765*, UK 2020.

Umberto Eco, 2010 *Dire presque la même chose : expériences de traductions*, Paris 2010 [reprint 2011].

Franklin Edgerton, 1924 "The Meaning of Sankhya and Yoga," *The American Journal of Philology* 45 (1924), pp. 1–46.

Sir Henry Miers Elliot and John Dowson, 1867 *History of India as Told by its own Historians. The Muhammadan Period*, Vol. 1, London 1867 [reprint 2013].

Vincent Eltschinger, 2010 "Dharmakīrti," *Revue Internationale de philosophie* 253.3 (2010), pp. 397–440.

Johan Elverskog, 2010 *Buddhism and Islam on the Silk Road*, Philadelphia/Oxford 2010.

Carl W. Ernst, 2003 "Muslim Studies of Hinduism? A Reconsideration of Arabic and Persian Translations from Indian Languages," *Iranian Studies* 36.2 (2003), pp. 173–195.

Anne-Marie Esnoul, 1964 *Les Strophes de Sāṃkhya (Sāṃkhya-Kārikā): avec le commentaire de Gauḍapāda*, Paris 1964.

Georg Feuerstein, 1979 *The Yoga-Sūtra of Patañjali. An Exercise in the Methodology of Textual Analysis*, Delhi 1979.

Georg Feuerstein, 1987 "The Concept of God (Īśvara) in Classical Yoga," JIPh 15 (1987), pp. 385–397.

Anna Filigenzi, 2015 *Art and Landscape. Buddhist Rock Sculptures of Late Antique Swat/Uḍḍiyāna. With Contributions by Luca Maria Olivieri and a Note by Peter Rockwell*, Vienna 2015.

Pierre-Sylvain Filliozat, 2005 *Le Yogabhāṣya de Vyāsa sur le Yogasūtra de Patañjali*, Palaiseau 2005.

Ghassan Finianos, 1975 *Les grandes divisions de l'être "mawjūd" selon Ibn Sīnā*, Fribourg 1975.

Klaus Fischer, 1964 "Une tête śivaïte en marbre de l'Afghanistan oriental," *Arts asiatiques* 10.1 (1964), pp. 35–42.

Finbarr Barry Flood, 2009 *Objects of Translation. Material Culture and Medieval "Hindu-Muslim" Encounter*, Princeton 2009.

Miquel Forcada, 2014 "Ibn Bājja on Taṣawwur and Taṣdīq: Science and Psychology," *Arabic Sciences and Philosophy* 24 (2014), pp. 103–126.

Alfred Foucher, 1947 *La vieille Route de l'Inde de Bactres à Taxila*, Vol. 2, Paris 1947.

Eli Franco, 2013 "On the Periodization and Historiography of Indian Philosophy," in: Franco 2013 (ed.), pp. 53–90.

Eli Franco (ed.), 2013 *Historiography and Periodization of Indian Philosophy*, Vienna 2013.

Eli Franco and Karin Preisendanz, 2010 "Vorwort," in: Erich Frauwallner, *Die Philosophie des Buddhismus*, Berlin 2010, pp. xi–xxx.

Erich Frauwallner, 1973 *History of Indian Philosophy. The Philosophy of the Veda and of the Epic. The Buddha and the Jina. The Sāṅkhya and the Classical Yoga-System (Transl. from Original German into English by V.M. Bedekar), Vol. 1*, Delhi/Patna/Varanasi 1973

Yohanan Friedmann, 2012 "Čač-Nāma," EI 2nd ed., First published online 2012 [accessed November 2023].

Johann W. Fück, 2012 "Buzurg b. Shahriyār," EI 2nd ed., First published online 2012 [accessed November 2023].

Toru Funayama, 1995 "Acaṭa, Śāntarakṣita, Jinendrabuddhi, and Kamalaśīla on the Aim of a Treatise (prayojana)," WZKS 39 (1995), pp. 181–201.

Toru Funayama, 2010 "The Work of Paramārtha. An example of Sino-Indian crosscultural exchange," *Journal of the International Association of Buddhist Studies* 31.1–2 (2010), pp. 141–183.

Francesco Gabrieli, 2012 "Ibn al-Muḳaffaʿ," EI 2nd ed., First published online 2012 [accessed November 2023].

Hans-Georg Gadamer, 1976 *Vérité et méthode : les grandes lignes d'une herméneutique philosophique*, Paris 1976 [reprint 1996].

Richard Garbe, 1894 *Die Sâṃkhya-Philosophie. Eine Darstellung des Indischen Rationalismus*, Leipzig 1894.

Richard Garbe, 1896 *Sāṅkhya und Yoga*, Strasbourg 1896.

Richard Garbe *Die Sâṃkhya-Philosophie. Eine Darstellung des indischen Rationalismus*, 1917 Leipzig 1917 [first edition 1894].

Tuvia Gelblum, 1992 "Notes on an English Translation of the 'Yogasūtrabhāṣyavivaraṇa,'" BSOAS 55 (1992), pp. 76–89.

Charles Genequand, 1996 "La mémoire de l'âme : Porphyre et la Théologie d'Aristote," *Bulletin d'études orientales* 68 (1996), pp. 103–113.

Finnian Gerety, 2015 *This Whole World Is OM: Song, Soteriology, and the Emergence of the Sacred Syllable*, PhD thesis 2015, Department of South Asian Studies, Harvard University, Unpublished.

H.A.R. Gibb, 2012 "Abd al-Raḥmān b. Samura," EI 2nd ed., First published online 2012 [accessed November 2023].

André Godard, 1925 "Ghazni," *Syria* 6.1 (1925), pp. 58–60.

Amélie-Marie Goichon, 1933 *Introduction à Avicenne : son Epître des définitions / traduction avec notes [par] A.-M. Goichon*, Paris 1933.

Amélie-Marie Goichon, 1938 *Lexique de la langue philosophique d'Ibn Sīnā (Avicenne)*, Paris 1938.

Bernard R. Goldstein, 2012 "Ibn ʿIrāḳ," EI 2nd ed., First published online 2012 [accessed November 2023].

Lallanji Gopal, 1989 *The Economic Life of Northern India, c. A.D. 700–1200*, Delhi 1989.

Jan Gonda, 1951 "Remarks on al-Biruni's Quotations from Sanskrit Texts," in: Comm. Volume 1951, pp. 110–118.

Frantz Grenet, 2002 "Regional Interaction in Central Asia and Northwest India in the Kidarite and Hephtalite periods," in: Nicholas Sims-Williams (ed.), *Indo-Iranian Languages and Peoples*, Oxford 2002, pp. 203–224.

Amrita Grover, 2006 *Medieval Punjab: Perspectives on Historiography and Polity*, Delhi 2006.

Giorgio Gullini and Dominico Facenna, 1962 *Reports on the Campaigns 1956–1958 in Swat (Pakistan)*, Roma 1962.

Dimitri Gutas, 2011 "Avicenna ii. Biography," EIr, Last update 2011 [accessed November 2023].

Paul Hacker, 1968 "Śaṅkara der Yogin und Śaṅkara der advaitin. Einige Beobachtungen," WZKS 12–13 (1968–1969), pp. 119–148.

M. Hadj-Sadok, 2012 "Ibn K͟hurradād͟hbih," EI 2nd ed., First published online 2012 [accessed November 2023].

Wilhelm Halbfass, 1976 "Review: Sāṅkhya-Saptati-Vṛtti (V_1) by Esther A. Solomon: Sāṅkhya-Vṛtti (V_2) by Esther A. Solomon," JAOS 96.1 (1976), pp. 144–145.

Wilhelm Halbfass, 1977 "Review: The Commentaries of the Sāṅkhya Kārikā-A Study by Esther A. Solomon," JAOS 97.3, pp. 371–372.

Wilhelm Halbfass, 1983 *Studies in Kumārila and Śaṅkara*, Reinbeck 1983.

Wilhelm Halbfass, 1988 *India and Europe. An Essay in Understanding*, Albany 1988.

Wilhelm Halbfass, 1991 *Tradition and Reflection. Explorations in Indian Thought*, Albany 1991.

Madeleine Hallade, 1968 *Inde : un millénaire d'art bouddhique : rencontre de l'Orient et de l'Occident. Photos par Hans Hinz*, Paris 1968.

Kengo Harimoto, 1999 *A Critical Edition of the Pātañjalayogaśāstravivaraṇa. First Pāda Samādhipāda*. With an Introduction, Ann Arbor 1999.

Kengo Harimoto, 2004 "Review: Yogasūtrabhāṣyavivaraṇa of Śaṅkara: Vivaraṇa Text with English Translation and Critical Notes along with Text and English Translation of Patañjali's Yogasūtras and Vyāsabhāṣya by T.S. Rukmani," JAOS 124 (2004), pp. 176–180.

Kengo Harimoto, 2014 *God, Reason and Yoga: a Critical Edition and Translation of the Commentary Ascribed to Śaṅkara on Pātañjalayogaśāstra 1.23–28*, Hamburg 2014.

R. Hartmann and John Andrew Boyle, 2012 "Gurgān," EI 2nd ed., First published online 2012 [accessed November 2023].

Edeltraud Harzer, 2006 *The Yuktidīpikā: A Reconstruction of Sāṅkhya Methods of Knowing*, Aachen, 2006.

Dag Hasse, 2010 "The Soul's Faculties," in: Robert Pasnau (ed.), *The Cambridge History of Medieval Philosophy*, Vol. 1, Cambridge 2010, pp. 305–319.

Masaaki Hattori, 1999 "On Seśvara Sāṅkhya," AS 53.3 (1999), pp. 609–618.

Jakob Wilhelm Hauer, 1930 "Das neugefundene arabische Manuskript von al-Bīrūnīs Übersetzung des Pātañjala. (Ein vorläufiger Bericht)," *Orientalische Literaturzeitung* 33.4 (1930), pp. 274–282.

S.J. Heras, 1951 "The Advaita Doctrine in Alberuni," in: Comm. Volume 1951, pp. 119–123.

Heinrich Hermelink, 1977 "Zur Bestimmung von Biruni's Todestag," *Sudhoffs Archiv* 61.3 (1997), 298–300.

Cl. Huart and H. Massé, 2012 "Firdawsī," EI 2nd ed., First published online 2012 [accessed November 2023].

Helmut Humbach, 1986 "Hindu Shahi Inscriptions in Hybrid Sanskrit from Laghman," *Journal of the Epigraphical Society of India* 12, 63–68.

Michel Hulin, 1978 *Le principe de l'Ego dans la pensée indienne classique. La notion d'ahaṃkāra*, Paris 1978.

Minoru Inaba, 2004 "Appendix: Nāy Qalʿa, Wujīristan and the Khalaj," in: Giovanni Verardi and Elio Paparatti (eds.), *Buddhist Caves of Jāghūrī and Qarabāgh-e Ghaznī, Afghanistan*, Rome 2004, pp. 105–111.

Minoru Inaba, 2005 "The Identity of the Turkish Rulers to the South of Hindukush from the 7th to the 9th Centuries A.D," *Zinbun* 39 (2005), pp. 1–19.

Minoru Inaba, 2010 "From Kesar the Kābulšāh and Central Asia," in: Alram et al. (2010), pp. 443–455.

Minoru Inaba, 2013 "Sedentary Rulers on the Move: the Travels of the Early Ghaznavid Sultans," in: David Durand-Guédy (ed.), *Turko-Mongol Rulers, Cities and City Life*, Leiden/Boston 2013, pp. 75–98.

Minoru Inaba, 2015a "A Venture on the Frontier: Alptegin's conquest of Ghazna and its Sequel," in: Andrew Charles Spencer Peacock and Deborah G. Tor (eds.), *Medieval Central Asia and the Persianate World. Iranian Tradition and Islamic Civilisation*, London/New York 2015, pp. 108–128.

Minoru Inaba, 2015b "From Caojuzha to Ghazna/Ghaznīn: Early Medieval Chinese and Muslim Descriptions of Eastern Afghanistan," *Journal of Asian History* 49.1–2 (2015), pp. 97–117.

Minoru Inaba, 2016 "Ghaznavids," EI 3rd ed., First published online 2016 [accessed November 2023].

Vladimir Ivir, 1987 "Procedures and strategies for the translation of culture," *Indian Journal of Applied Linguistics* 13.2 (1987), pp. 35–46.

P. Jackson and P.A. Andrews, 2012 "Lāhawr," EI 2nd ed., First published online 2012 [accessed November 2023].

Colonel G.A. Jacob, 2004 *Laukikanyāyāñjaliḥ. A Handful popular Maxims current in Sanskrit Literature*, Delhi 2004.

Hermann Jacobi, 1970 "Über das Ursprüngliche Yogasystem," *Kleine Schriften*, Teil 2, 682–725 (1970) [first print *Sitzungberichte der Preussischen Akademie der Wissenschaften, phil.-hist. Kl.* 1929].

Knut A. Jacobsen, 2008 *Kapila, founder of Sāṃkhya and avatāra of Viṣṇu: with a Translation of Kapilāsurisaṃvāda*, New Delhi 2008.

G. Kaur, 1982 "Al-Biruni: An Early Student of Comparative Religions," *Islamic Culture* 56.2 (1982), pp. 149–163.

A. Berriadele Keith, 1924 "The Mathara-Vrtti," *Bulletin of the School of Oriental Studies* 3 (1924), pp. 551–554.

S.E. Kennedy, 1970 "Al-Bīrūnī," in: Charles Coulston Gillispie (ed.), *Dictionary of Scientific Biography*, Vol. 2, New York 1970, pp. 147–158.

Anas B. Khalidov, 2011 "Ebn Ḥawqal, Abu' l-Qāsem Moḥammad," EIr, Last update 2011 [accessed November 2023].

M.S. Khan, 1976 "Al-Bīrūnī and the Political History of India," *Oriens* 25–26 (1976), pp. 86–115.

Waliullah Muhammad Khan, 1979 "Lahore and al-Bīrūnī," in: Said 1979, pp. 221–226.

M.A. Saleem Khan, 2001 *Al-Biruni's Discovery of India: an Interpretative Study*, Delhi 2001.

Ijaz Khan, 2017 *Exploring the Settlement Archaeology of the Hindu Shahi Dynasty* (*c. 822 CE to c. 1026 CE*) *in North-Western Pakistan*, PhD Thesis 2017, School of Archaeology and Ancient History, University of Leicester, Unpublished.

Natasha Kimmet, 2020 "What Archaeological Evidence is there for the Śāhi Political Center at Hund (Udabhāṇḍapura)?," *Sourcebook for the Shahi Kingdoms*, Vienna 2020 (https://shahi.openatlas.eu/api/display/2272) [accessed October 2023].

Willibald Kirfel, 1920 *Die Kosmographie der Inder. Nach den Quellen dargestellt*, Bonn/Leipzig 1920.

Mario Kozah, 2016 *The Birth of Indology as an Islamic Science: al-Bīrūnī's Treatise on Yoga Psychology*, Leiden/Boston 2016.

Mario Kozah, 2020 *The Yoga Sutra of Patañjali. Abū Rayḥān al-Bīrūnī. Edited and Translated by Mario Kozah*, New York 2020.

Hermann Kulke and Dietmar Rothermund, 1986 *A History of India*, Kent 1986.

Shiv Kumar and D.N. Bhargava, 1990 *Yuktidīpikā, Vol. I*, Delhi 1990

Shiv Kumar and D.N. Bhargava, 1992 *Yuktidīpikā, Vol. II*, Delhi 1992

Shoshin Kuwayama, 2002 *Across the Hindukush of the First Millennium: a collection of the papers*, Kyoto 2002 (http://hdl.handle.net/2433/120966) [accessed October 2023].

Louis de La Vallée Poussin, 1935 *Dynasties et Histoires de l'Inde depuis Kanishka jusqu'aux invasions musulmans*, Paris 1935.

Gerald James Larson, 1979 *Sāṅkhya. An Interpretation of its History and Meaning*, Delhi 1979 [second revised edition; first edition 1969].

Gerald James Larson, 1999 "Classical Yoga as Neo-Sāṅkhya: a Chapter in the History of Indian Philosophy," AS 53.3 (1999), pp. 723–732.

Gerald James Larson and Ram Shankar Bhattacharya, 1987 *Encyclopedia of Indian Philosophies. Vol. IV. Sāmkhya. A Dualist Tradition in Indian Philosophy*, Delhi 1987.

Gerald James Larson and Ram Shankar Bhattacharya, 2008 *Encyclopedia of Indian Philosophies. Vol. XII. Yoga. India's Philosophy of Meditation*, Delhi 2008.

Bruce B. Lawrence, 1978 "Al-Biruni's Approach to the Comparative Study of Indian Culture," *Studies in Islam* 15.1 (1978), pp. 1–13.

Guy Le Strange, 1930 *The Lands of the Eastern Caliphate. Mesopotamia, Persia, and Central Asia from the Moslem conquest to the time of Timur*, Cambridge 1930.

Maurice Lombard, 1971 *L'islam dans sa première grandeur* (VIIIe–XIe *siècle*), Paris 1971.

Bimala Churn Law, 1955 "Al-Bīrūnī's knowledge of Indian Geography," *Indo-Iranica* 7.4 (1955), pp. 1–26.

Olena Łucyszyna, 2016 "Classical Sāṁkhya on the Relationship between a Word and Its Meaning," JIPh 44 (2016), pp. 303–323.

Philipp André Maas, 2006 *Samādhipāda. Das erste Kapitel des Pātañjalayogaśāstra zum ersten Mal kritisch ediert, = The First Chapter of the Pātañjalayogaśāstra for the First Time Critically Edited*, Aachen 2006.

T.G. Mainkar, 1972 *Sāṃkhyakārikā of Īśvarakṛṣṇa with the Commentary of Gauḍapāda Translated into English with Notes*, Pune 1972 [second revised and enlarged edition; first edition 1964].

Philipp André Maas, 2009 "The So-called Yoga of Suppression in the Pātañjala Yogaśāstra," in: Eli Franco (ed.) in collaboration with D. Eigner, *Yogic Perception, Meditation, and Altered States of Consciousness* Vienna 2009, pp. 263–282.

Philipp André Maas, 2013 "A Concise Historiography of Classical Yoga Philosophy," in: Franco 2013 (ed.), pp. 53–90.

Philipp André Maas and Noémie Verdon, 2018 "On al-Bīrūnī's *Kitāb Pātanǧal* and the *Pātañjalayogaśāstra*," in: Karl Baier, Philipp André Maas and Karin Preisendanz (eds.), *Yoga in Transformation: Historical and Contemporary Perspectives*, Vienna 2018, pp. 283–334.

David W. MacDowall, 1968 "The Shahis of Kabul and Gandhara," *The Numismatic Chronicle (1966–)* 8 (1968), pp. 189–224.

Derryl N. Maclean, 1989 *Religion and Society in Arab Sind*, Leiden 1989.

Ramesh Chandra Majumdar (ed.), 1957 *The Struggle for Empire*, Mumbai 1979 [reprint 1979].

James Mallinson and Mark Singleton, 2017 *Roots of Yoga*, London 2017.

Louis Massignon, 1954 *Essai sur les origines du lexique technique de la mystique musulmane*, Paris 1954 [enlarged edition; first edition 1922].

Michael W. Meister, 1996 "Temples Along the Indus," *Expedition*, 38.3 (1996), pp. 41–54.

Michael W. Meister, 2010 *Temples of the Indus. Studies in the Hindu Architecture of Ancient Pakistan*, Leiden 2010.

Marek Mejor, 2004 "Some observations on the date of the *Yukti-dīpikā* (Apropos of the new edition)," in: Piotr Balcerowicz and Marek Mejor (eds.), *Essays in Indian philosophy, religion and literature*, Delhi 2004, pp. 399–433.

Christopher Minkowski, 2008 "Why Should We Read the Maṅgala Verses?," in: Walter Slaje (ed.), *Śāstrārambha. Inquiries into the Preamble in Sanskrit*, Wiesbaden 2008, pp. 1–24.

Vladimir Minorsky, 1948 "Gardīzī on India," BSOAS 12.3–4 (1948), pp. 625–640.

Vladimir Minorsky, 1951 "On some of Bīrūnī's Informants," in: Comm. Volume 1951, pp. 110–118.

André Miquel, 1967 *La géographie humaine du monde musulman jusqu'au milieu du 11ᵉ siècle. Vol. I. Géographie et géographie humaine dans la littérature arabe des origines à 1015*, Paris 1967.

André Miquel, 2012 "al-Iṣṭak͟hrī," EI 2nd ed., First published online 2012 [accessed November 2023].

Yogendra Mishra, 1972 *The Hindu Sahis of Afghanistan and the Punjab, AD 865–1026: A Phase of Islamic Advance into India*, Delhi 1972.

Jai Shankar Mishra, 1985 *Al-Biruni. An Eleventh Century Historian*, Varanasi 1985.

Ram Gopal Mishra, 1983 *Indian Resistance to Early Muslim Invaders up to 1206 A.D*, Meerut City 1983.

Rājendralāl Mitra, 1883 *The Yoga Aphorisms of Patan̲jali, With the commentary of Bhoja and an English Translation*, Kolkata 1883.

Sir Monier Monier-Williams, Online *A Sanskrit-English Dictionary. Etymologically and philologically arranged. With special Reference to Cognate Indo-European Languages*, (Köln Online: https://www.sanskrit-lexicon.uni-koeln.de/monier/) [accessed October 2023].

Vincent-Mansour Monteil, 1996 *Le Livre de l'Inde* (*extraits choisis, traduits de l'arabe, présentés et annotés par l'auteur*), Arles 1996.

Shujun Motegi, 1999 "The Teachings of Pañcaśikha in the Mokṣadharma," AS 53.3 (1999), pp. 513–535.

Shujun Motegi, 2013 "The Early History of Sāṅkhya Thought," in: Eli Franco (ed.), Franco 2013 (ed.), pp. 35–52.

Yasutaka Muroya, 2011 "Bhāsaravajña's Interpretation of *bhāva eva nāśaḥ* and a related chronological problem," in: Helmut Krasser, Horst Lasic, Eli Franc and Birgit Kellner (eds.), *Religion and Logic in Buddhist Philosophical Analysis*, Vienna 2011, pp. 341–361.

Tilman Nagel, 1990 "Buyids," EIr, Last update 1990 [accessed November 2023].

Seyyed Hossein Nasr and Mehdi Mohaghegh, 2005 *Al-Bîrûnî and Ibn Sînâ, al-As'ilah wa'l-Ajwibah (Questions and Answers). Including the further answers of al-Bîrûnî and al-Maʿsûmî's defense of Ibn Sînâ. Edited with English and Persian Introduction*, Tehran 2005.

Muhammad Nazim, 1931 *The Life and Times of Sulṭān Maḥmūd of G͟hazna*, Cambridge 1931.

Andrew J. Nicholson, 2010 *Unifying Hinduism. Philosophy and Identity in Indian Intellectual History*, New York 2010.

Gerhard Oberhammer, 1960 "The Authorship of the Ṣaṣṭitantram," *Wiener Zeitschrift für die Kunde Süd- und Ostasiens* 6 (1960), pp. 71–91.

Gerhard Oberhammer, 1977 *Strukturen Yogischer Meditation*, Vienna 1970.

Karen O'Brien-Kop, 2017 "Classical Discourses of Liberation: Shared Botanical Metaphors in Sarvāstivāda Buddhism and the Yoga of Patañjali," *Religions of South Asia* 11.2–3 (2017), pp. 123–157.

Karen O'Brien-Kop, 2021 *Rethinking 'Classical Yoga' and Buddhism: Meditation, Metaphors and Materiality*, Bloomsbury 2021.

Dinabandhu Pandey, 1973 *The Shahis of Afghanistan and the Punjab*, Delhi 1973.

Ram Chandra Pandeya, 1967 *Yuktidīpikā. An Ancient Commentary on the Sāṁkhya-Kārikā of Īśvarakṛṣṇa*, Varanasi 1967.

Charles Pellat, 2012 "al-D̲jayhānī," EI 2nd ed., First published online 2012 [accessed November 2023].

Shlomo Pines and Tuvia Gelblum, 1966 "Al-Bīrūnī's Arabic Version of Patañjali's 'Yogasūtra'," BSOAS 29.2 (1966), pp. 302–325.

Shlomo Pines and Tuvia Gelblum, 1977 "Al-Bīrūnī's Arabic Version of Patañjali's 'Yogasūtra': A Translation of the Second Chapter and a Comparison with Related Texts," BSOAS 40.3 (1977), pp. 522–549.

Shlomo Pines and Tuvia Gelblum, 1983 "Al-Bīrūnī's Arabic Version of Patañjali's 'Yogasūtra': A Translation of the Third Chapter and a Comparison with Related Texts," BSOAS 46.2 (1983), pp. 258–304.

Shlomo Pines and Tuvia Gelblum, 1989 "Al-Bīrūnī's Arabic Version of Patañjali's 'Yogasūtra': A Translation of the Fourth Chapter and a Comparison with Related Texts," BSOAS 52.2 (1989), pp. 265–305.

David Erwin Pingree, 1963 "Astronomy and Astrology in India and Iran," *Isis* 54.2 (1963), pp. 229–246.

David Erwin Pingree, 1968 "The Fragments of the Works of Yaʿqūb Ibn Ṭāriq," *Journal of Near Eastern Studies* 27.2 (1968), pp. 97–125.

David Erwin Pingree, 1969 "The Later Pauliśasiddhānta," *Centaurus* 14 (1969), pp. 172–241.

David Erwin Pingree, 1981 *Census of the Exact Sciences in Sanskrit. Series A, Vol. 4*, Philadelphia 1981

David Erwin Pingree, 1983 "Brahmagupta, Balabhadra, Pṛthūdaka and Al-Bīrūnī," JAOS 103 (1983), pp. 353–360.

David Erwin Pingree, 1991 *Census of the Exact Sciences in Sanskrit, Series A, Vol. 5*, Philadelphia 1991.

David Erwin Pingree, 2012a "Sindhind," EI 2nd ed., First published online 2012 [accessed November 2023].

David Erwin Pingree, 2012b "ʿIlm al-Hayʾa," EI 2nd ed., First published online 2012 [accessed November 2023].

Venkatarama Raghavan, 1989 *Abhinavagupta and His Works*, Varanasi 1980.

Gopinatha T.A. Rao, 1997 *Elements of Hindu Iconography, Vol. 2, Parts 1 and 2*, Delhi 1997 [first edition 1914].

Abdur Rehman, 1978 "An Inscription of Jayapāla Śāhī," *The Journal of the Royal Asiatic Society of Great Britain and Ireland* 1 (1978), pp. 31–37.

Abdur Rehman, 1979 *The Last Two Dynasties of the Śāhis. An Analysis of their History, Archaeology, Coinage and Paleography*, Islamabad 1979.

Abdur Rehman, 1998 "The Zalamkot Bilingual Inscription," *East and West* 48.3–4 (1998), pp. 469–473.

Abdur Rehman, 2003 "Ethnicity of the Hindu Shāhīs," *Pakistan Historical Society* 51.3 (2003), pp. 3–10.

Louis Renou and Jean Filliozat, 1953 *L'Inde classique: manuel des études indiennes*, Vol. 2, Paris 1953.

James Reynolds, 1858 *Kitab-i-Yamini: Historical Memoirs of the Amír Sabaktagín and the Sultán Mahmúd of Ghazna. Translated from the Persian version of the contemporary Arabic chronicle of Al-Utbi, by the Rev. James Reynolds*, London 1858.

Yael Rice, 2010 "A Persian Mahabharata: The 1598–1599 Razmnama," *Manoa* 22.1 (2010), pp. 125–131.

Hellmut Ritter, 1956 "Al-Bīrūnī's Übersetzung des Yoga-Sūtra des Patañjali," *Oriens*, 9.2 (1956), pp. 165–200.

Benjamin Rowland and Mortimer Frances Rice, 1971 *Art in Afghanistan: objects from the Kabul Museum / Photographs by Frances Mortimer Rice*, London 1971.

Nasha Bin Rodziadi Khaw, 2016 *Study and Analysis of Brāhmī and Śāradā Inscription from Gandhāra: its Impact on Religious, Cultural and Historical Landscape of the Region*, PhD Thesis 2016, Department of Archaeology, University of Peshawar, Unpublished.

Nasha Bin Rodziadi Khaw, 2015 "Study and Analysis of the Proto-Śāradā and Śāradā inscriptions in the Lahore Museum (Punjab-Pakistan)," Gandhāran Studies, 9, pp. 93–121.

Trichur Subramaniam Rukmani, 1998 "The Yogasūtrabhāṣyavivaraṇa is not a Work of Śaṅkarācārya the Author of the Brāhmasūtrabhāṣya," JIPh 26 (1998), pp. 263–274.

Trichur Subramaniam Rukmani, 1999 "Sāṅkhya and Yoga: where they do not speak in one Voice," AS 53.3 (1999), pp. 733–753.

Trichur Subramaniam Rukmani, 2001 *Yogasūtrabhāṣyavivaraṇa of Śaṅkara: Vivaraṇa. Text with English and critical notes along with text and English translatison of Patanjali's Yogasūtras and Vyāsabhāṣya. Vol. 1 and 2*, Delhi 2001.

Carl Edward Sachau, 1878 *Chronologie orientalischer Völker von Albêrûni. Herausgegeben von Dr. C. Eduard Sachau*, Leipzig 1878.

Carl Edward Sachau, 1879 *The Chronology of Ancient Nations. An English Version of the Arabic text of the Athâr-ul-bâkiya of Albîrûnî or "Vestiges of the past," collected and reduced to writing by the author in 390–391, A.D. 1000*. London 1879.

Carl Edward Sachau, 1887 *Alberuni's India. An Account of the Religion, Philosophy, Literature, Chronology, Astronomy, Customs, Laws and Astrology of India about A.D. 1030, edited in the Arabic by dr. Edward Sachau*, London 1887 [reprint Frankfurt 1993].

Carl Edward Sachau, 1888 *Indo-Arabische Studien zur Aussprache und Geschichte des Indischen in der ersten Hälfte des XI. Jahrhunderts*, Berlin 1888.

Carl Edward Sachau, 1910 *Alberuni's India. An Account of the Religion, Philosophy, Literature, Geography, Chronology, Astronomy, Customs, Laws and Astrology of India about AD 1030. Vol. 1 and 2*, London 1910.

Hakim Mohammed Said (ed.), 1973 *Kitāb al-ṣaydana fī l-ṭibb; with English translation: Al-Biruni's Book on Pharmacy and Materia Medica, Vol. 1*, Karachi 1973.

Mohammed Hakim Said (ed.), 1979 *Al-Bīrūnī commemorative Volume: Proceedings of the International Congress held in Pakistan on the occasion of Millenary of Abū al-Rāiḥān Muḥammad Ibn Aḥmad al-Bayrūnī (973–ca. 1051 A.D.). November 26, 1973 thru December 12, 1973*, Karachi 1979.

Hakim Mohammed Said, 2001 *Kitāb al-ğamāhir fī maʿrifa al-ğawāhir. English Translation*, Karachi 2001.

Mohammed Hakim Said and Ansar Zahid Khan, 1981 *Al–Bīrūnī. His Times, Life and Works*, Karachi 1981.

Richard Salomon, 2018 *The Buddhist Literature of Ancient Gandhāra. An Introduction with Selected Translations*, Somerville 2018.

Hamarneh K. Sami, 1973 *Al-Biruni's Book on Pharmacy and Materia Medica. Introduction, Commentary and Evaluation*, Karachi 1973.

Floréal Sanagustin, 2003 "Abū Rayhān al-Bīrūnī : un rationaliste à l'épreuve de l'altérité," *Bulletin d'études orientales* 55 (2003), pp. 249–263.

H. Sarma, 1926 *Śrī Śaṅkarācāryavicaritā Jayamaṅgalā nāma Sāṅkhyasaptatiṭīkā. With an Introduction by Gopi Nath Kaviraj*, Calcutta 1926.

K.V. Sarma, 1985a "A Survey of Source Materials," in: S.N. Sen and K.S. Shukla (eds.), *History of Astronomy in India*, Delhi 1985, pp. 253–307.

Haradatta Sarma, 1985b "Jayamaṅgalā and the other commentaries," *The Indian Historical Quarterly* 5 (1985), pp. 417–431 [first print 1929]

Sreeramula Rajeswara Sarma, 2009 "Early Transmission of Indian Mathematics," in: G.G. Joseph (ed.), *Kerala Mathematics: History and its Possible Transmission to Europe*, Delhi 2009, pp. 205–231.

N.A. Sastri, 1944 *Suvarnasaptati Sastra. Sānkhya Kārikā Saptati of Īsvara Krsṇa with a Commentary Reconstructed into Sanskrit from the Chinese Translation of Paramārtha and Edited with English Notes, Introduction and Appendices*, Tirupati 1944.

Polakam Sri Rama Sastri and S.R. Krishnamurthi Sastri (eds.), 1952 *Pātañjala-Yogasūtra-Bhāṣya-Vivaraṇa of Śaṅkara-Bhagavatpāda. Critically Edited with Introduction*, Madras 1952.

Ḍhunḍhirāj Śāstrī, 2009 *Yogasūtram by Maharṣipatañjali. With six commentaries (1) Rājamārtaṇḍa by Bhojarāja, (2) Pradīpikā by Bhāvā-Gaṇeśa (3) Vṛtti by Nāgoji Bhaṭṭa (4) Maṇiprabhā by Rāmānandayati (5) Chandrika by Anantadeva and (6) Yogasudhākara by Sadāśivendra Sarasvatī. Edited Notes by Nyāsāchārya Kāvyatīrtha Paṇḍit Ḍhunḍhirāj Śāstrī*, Varanasi 2009.

Hartmut Scharfe, 2002 *Education in Ancient India*, Leiden 2002.

Daniel Schlumberger, 1978 Lashkari Bazar. *Une résidence royale ghaznévide et ghoride. 1A. L'architecture. Avec la collaboration de Marc Le Berre, architecte, et des contributions de Jean-Claude Gardin et Geneviève Casal*, Paris 1978.

Peter Schreiner, 1999 "What comes first (in the Mahābhārata): Sāṅkhya or Yoga?," AS 53.3 (1999), pp. 755–777.

Joseph E. Schwartzberg (et al.), 1978 *A Historical Atlas of South Asia*, Chicago 1978.

F.A. Shamsi, 1979 "Abū al-Raiḥān Muḥammad Ibn Aḥmad al-Bayrūnī. 362, 973–ca. 443, 1051," in: Said 1979, pp. 260–288.

Har Dutt Sharma, 1933 *The Sāṁkhyā—Kārikā. Is'vara Kṛṣṇa's Memorable Verses on Sāṁkhya Philosophy with the Commentary of Gauḍapādācārya. śrīmadīśvarakṛṣṇāpraṇītāḥ sagauḍapādabhāṣyāḥ sāṅkhyakārikāḥ. Critically edited with Introduction, Translation and Notes by Vidyāsudhākara Dr. Har Dutt Sharma*, Pune 1933.

Arvind Sharma, 1983 *Studies in "Alberuni's India,"* Wiesbaden 1983.

Ajay Mitra Shastri, 1975 "Sanskrit Literature Known to al-Bīrūnī," *Indian Journal of History of Science* 10.2 (1975), pp. 111–138.

Ehsan Shavarebi, 2022 "The Mosque of the Forgotten City: The Bilingual Inscription of Zalamkot Revisited," *East and West* 62.2 (2022), pp. 195–210.

Iqtidar Husain Siddiqi, 2003 "Abu Raihan Al-Beruni: His Life and Works," in: I.H. Siddiqi (ed.), *Medieval India. Essays in Intellectual Thought and Culture*. Vol. 1, Delhi 2003, pp. 17–25.

Upinder Singh, 2017 *Political Violence in Ancient India*, Cambridge 2017.

Walter Slaje, 1986 "Untersuchungen zur Chronologie einiger Nyāya-Philosophen," *Studien zur Indologie und Iranistik* 11–12 (1986), pp. 245–278.

Walter Slaje, 1993 *Śāradā. Deskriptiv-synchrone Schriftkunde zur Bearbeitung kaschmirischer Sanskrit-Manuskripte. Auf der Grundlage von Kuśalas Ghaṭakharpara -Gūḍhadīpikā und unter graphischer Mitwirkung von Eva Slaje*, Reinbeck 1993.

Esther A. Solomon, 1973a *Sāṅkhyasaptativṛtti* (*V1*), Ahmedabad 1973.

Esther A. Solomon, 1973b *Sāṅkhyavṛtti* (*V2*), Ahmedabad 1973.

Esther A. Solomon, 1974 *The Commentaries of the Sāṅkhya Kārikā—A Study*, Ahmedabad 1974.

D. Sourdel, 2012a "Bayt al-Ḥikma," EI 2nd ed., First published online 2012 [accessed November 2023].

D. Sourdel, 2012b "Dār al-Ḥikma," EI 2nd ed., First published online 2012 [accessed November 2023].

D. Sourdel, 2012 Clifford Edmund Bosworth, P. Hardy and Halil İnalcık, "Gh̲ulām," EI 2nd ed., First published online 2012 [accessed November 2023].

Janine Sourdel-Thomine, 1978 *Lashkari Bazar. Une résidence royale ghaznévide et ghoride. 1B. Le décor non figuratif et les inscriptions. Dessins de Marc Le Berre et Claire Hardy-Guilbert*, Paris 1978.

Fabrizio Speziale, 2010 "Les traités persans sur les sciences indiennes : médecine, zoologie, alchimie," in: Denis Hermann and Fabrizio Speziale (eds.), *Muslim Cultures in the Indo-Iranian World during the Early-Modern and Modern Periods*, Berlin 2010, pp. 403–447.

Srinivasa Ayya Srinivasan, 1967 *Vācaspatimiśras Tattvakaumudī. Ein Beitrag zur Textkritik bei kontaminierter Überlieferung*, Hamburg 1967.

Judith Stareček, 2003 *Die Rezeption des Yogasūtra und der Bhagavadgītā in al-Bīrūnī's al-Hind: Versuch einer Gegenüberstellung*, Master Thesis 2003, South Asian, Tibetan and Buddhist Studies, Vienna University, Unpublished.

Marc Aurel Stein, 1892 *Kalhaṇa's Rājataraṅgiṇī. A Chronicle of the Kings of Kaśmīr. Translated, with an Introduction, Commentary, & Appendices, by M.A. Stein, Vol. III. Sanskrit Text with Critical Notes*, Mumbai 1892 [reprint 1988].

Marc Aurel Stein, 1893 "Zur Geschichte der Çāhis von Kābul," in: Festgruss an Rudolf Von Roth zum Doktor-Jubilaüm 24. August 1893 von seinen Freunden and Schülern, Stuttgart 1893, pp. 195–202.

Marc Aurel Stein, 1900: I *Kalhaṇa's Rājataraṅgiṇī. A Chronicle of the Kings of Kaśmīr. Translated, with an Introduction, Commentary, & Appendices, by M.A. Stein. Vol. I. Introduction. Books I–VII*, Westminster 1900.

Marc Aurel Stein, 1900: II *Kalhaṇa's Rājataraṅgiṇī. A Chronicle of the Kings of Kaśmīr. Translated, with an Introduction, Commentary, & Appendices, by M.A. Stein. Vol. II. Book VIII. Notes, geographical Memoir, Index, Maps*, Westminster 1900.

S.M. Stern, 2012 "Abū Ḥayyān al-Tawḥīdī," EI 2nd ed., First published online 2012 [accessed November 2023].

Otto Strauss, 1925 *Indische Philosophie*, München 1925.

Gotthard Strohmaier, 1991 *In den Gärten der Wissenschaft. Ausgewählte Texte aus den Werken des muslimischen Universalgelehrten Al-Bīrūnī*, Leipzig 1991.

Mohammed Hassan Syed, 2003 *History of Muslims in India*, Vol. 1, Delhi 2003.

Junjiro Takakusu, 1904a "La Sāṅkhyakārikā, étudiée à la lumière de sa version chinoise (I)," *Bulletin de l'Ecole française de l'Extrême-Orient* 4 (1904), pp. 1–65.

Junjiro Takakusu, 1904b "La Sāṅkhyakārikā, étudiée à la lumière de sa version chinoise (II)," *Bulletin de l'Ecole française de l'Extrême-Orient* 4 (1904), pp. 978–1064.

G.E. Tetley, 2009 The Ghaznavid and the Seljuqturks: Poetry as a source for Iranian History, London 2009.

Romila Thapar, 2002 *Early India. From the Origins to AD 1300*, London 2002.

Romila Thapar, 2004 *Somanatha. The Many Voices of a History*, London 2004.

Edward Thomas, 1846 "Coins of the Hindus Kings of Kabul," *Journal of Royal Asiatic Society* (1846), pp. 177–198.

Edward Thomas, 1847 "Coins of the Kings of Ghazni," *Journal of Royal Asiatic Society* (1847), pp. 267–388.

Edward Thomas, 1859 *Coins of the Kings of Ghazni. Supplementary Contributions*, London 1859.

Houari Touati, 2000 *Islam et Voyage au Moyen Âge*, Paris 2000.

Raffaele Torella, 1999 "Sāṅkhya as *sāmānyaśāstra*," AS 53.3 (1999), pp. 553–562.

Raffaele Torella, 2011 *The philosophical traditions of India: an appraisal; translated from Italian by Kenneth Frederick Hurry*, Varanasi 2011.

Gary A. Tubb and Emery R. Boose, 2006 *Scholastic Sanskrit: a handbook for students*, New York 2006.

Giuseppe Tucci, 1957 *Storia della filosofia Indiana*, Bari 1957.

Kevin Van Bladel, 2011 "The Bactrian Background of the Barmakids," in: Anna Akasoy, Charles Burnett and Ronit Yoeli-Tlalim (eds.), *Islam and Tibet—Interactions along the Musk Routes*, Farnham 2011 [reprint 2017], pp. 43–88.

Kevin Van Bladel, 2012 "Barmakids," EI 3rd ed., First published online 2012 [accessed November 2023].

Satkāriśarmā Vaṅgīya, 1970 *Sāṅkhyakārikā. With the Māṭharavṛtti of Māṭharācārya ed. Viśṇu Prasād Śarmā. And the Jayamaṅgalā of Śrī Śaṅkara, critically ed. with an introd. by Śrī Satkāriśarmā Vaṅgīya*, Varanasi 1970.

Noémie Verdon, 2015 "Cartography and cultural encounter: conceptualisation of al-Hind of Arabic and Persian writers from the 9th to the 11th c. A.D," in: Himanshu Prabha Ray, *Negotiating Cultural Identity: Landscape in Early Medieval South Asian History*, Delhi 2015, pp. 30–59.

Noémie Verdon, 2019a "Indian Calendar and Festivals through al-Bīrūnī's Eyes," in: Tokimasa Takeda and Bill M. Mak (eds.), *East-West Encounter in the Science of Heaven and Earth*, Kyoto 2019, pp. 63–81.

Noémie Verdon, 2019b "Negotiating Sāṅkhya-Yoga concepts: al-Bīrūnī and Falsafa," *RINDAS Series of Working Papers: Traditional Indian Thoughts* 28 (2019).

Noémie Verdon, 2019c "On the Commentarial Transmission of the Suvarṇasaptati, Sāṃkhyavṛtti, and Gauḍapādabhāṣya," *The Journal of Hindu Studies*, 12 (3), pp. 292–318.

Noémie Verdon, 2021 "Wayhind (Modern Hund) in Early Medieval Historical Sources," *Sourcebook for the Shahi Kingdoms*, Vienna 2021 (https://shahi.openatlas.eu/api/display/2269) [accessed October 2023].

Noémie Verdon and Hyoung Seok Ham, 2023 "Al-Bīrūnī's Indian Philosophy: The Kitāb Sānk and the Kitāb Pātanğal (Korean)," *Korean Journal of Yoga Studies* 30 (2023), pp. 103–134.

Noémie Verdon, forthcoming "The place of Kashmir in al-Bīrūnī's writings," in: Akane Saito, Satoshi Ogura and Yuko Yokochi (eds.). *Intellectual tradition in pre-modern Kashmir*, forthcoming.

Noémie Verdon and Vladimir Lončar, 2013 *The Quill of the Invaders*, Lausanne 2013.

Noémie Verdon and Michio Yano, 2020 "Al-Bīrūnī's *India*, Chapter 14: An account of Indian Astronomical, Mathematical and other Literature," *History of Science in South Asia* (2020), pp. 57–76.

Klaus Vondrovec, 2010 "Coinage of the Nezak," in: Alram et al. (2010), pp. 169–190.

Thomas Watters, 1904 *On Yuan Chwang's travels in India. 629–645 A.D, Vol. 1 and 2*, London 1904.

Hartmut Weiss, 1986 *Quellen des Yoga. Klassische Texte der Körper- und Geisteschulung*, Bern 1986.

Albrecht Wezler, 1983 "Philological Observations on the so-called Pātañjalayogasūtrabhāṣyavivaraṇa (Studies in the Pātañjalayogaśāstravivaraṇa I)," *Indo-Iranian Journal* 25 (1983), pp. 17–40.

Albrecht Wezler and Shujun Motegi, 1998 *Yuktidīpikā. The Most Significant Commentary on the Sāṅkhyakārikā, critically edited by Albrecht Wezler and Shujun Motegi*, Stuttgart 1998.

Ian Whicher, 1999 "Classical Sāṅkhya, Yoga, and the Issue of the Final Purification," AS 53.3 (1999), pp. 779–798.

André Wink, 1990 *Al-Hind. The Making of the Indo-Islamic World. Early Medieval India and the Expansion of Islam. 7th–11th Centuries, Vol. 1*, Delhi 1990.

André Wink, 1997 *Al-Hind. The Making of the Indo-Islamic World. The slave kings and the Islamic conquest: 11th–13th centuries, Vol. 2*, Leiden/New York 1997.9

Robert Wisnovsky (et al.), 2011 "Introduction. Vehicles of Transmission, Translation, and Transformation in Medieval Textual Culture," in Robert Wisnovsky, Faith Wallis, Jamie C. Fumo and Carlos Fraenkel (eds.), *Vehicles of Transmission, Translation and Transformation in Medieval Textual Culture*, Turnhout 2011, pp. 1–22.

James Haughton Woods, 1914 *The Yoga-System of Patañjali, Or the Ancient Hindu Doctrine of Concentration of Mind, Embracing the Mnemonic Rules, Called Yoga-Sūtras, of Patañjali and the Comment, Called Yoga-Bhāshya, Attributed to Veda-Vyāsa, and the Explanation, Called Tattva-Vaiçāradī, of Vāchaspati-Miçra*, Cambridge 1914.

Sally Hovey Wriggings, 2004 *The Silk Road Journey with Xuanzang*, Oxford 2004.

R. Ramsay Wright, 1934 *The Book of Instruction in the Elements of the Art of Astrology by Abū l-Rayḥān Muḥammad bin Aḥmad al-Bīrūnī. The Translation facing the Text by R. Ramsay Wright*, London 1934.

Michio Yano, 2003 "Calendar, Astrology, and Astronomy," in: Gavin Flood (ed.), *The Blackwell Companion to Hinduism*, Oxford 2003, pp. 276–292.

Michio Yano, 2004 "Planet Worship in Ancient India," in: Jan P. Hogendijk, Kim Plofker, Michio Yano and Charles Burnett (eds.), *Studies in the History of the Exact Sciences in Honor of David Pingree*, Leiden/Boston 2004, pp. 331–348.

Michio Yano, 2005 "Medicine and Divination in India," *East Asian Science, Technology, and Medicine* 24 (2005), pp. 44–61.

Michio Yano, 2013 "al-Bīrūnī," EI 3rd ed., First published online 2013 [accessed November 2023].

Svāmī Yogindrānandaḥ, 1968 *Bhāsarvajña. Nyāyabhūṣaṇam: Śrīmadācāryabhāsarvajñapraṇitasya Nyāyasārasya svopajñaṃvyākhyānam. Svāmī Yogindrānandaḥ śrautasaugatārhatādyanekadarśanagranthasāhāyyena saṃśodhya tippaṇādibhiḥ pariṣkṛtya sampaditavān*, Vārāṇasī 1968.

Travis Zadeh, 2011 *Mapping Frontiers across Medieval Islam: Geography, Translation, and the ʿAbbāsid Empire*, London/New York 2011.

Index

The index contains words in both Sanskrit and Arabic languages. Therefore, I enumerate the terms in the alphabetic order according to English spelling, regardless of the diacritic signs of the transliterated items. The Arabic article *al-* is not taken into account for sequencing the listed words. Many technical terms are also found in the glossary of Sanskrit and Arabic concepts of the present book. Lastly, words that occur frequently in the book, such as al-Bīrūnī, Sāṅkhya and Yoga, works' titles, *Kitāb*s *Sānk* and *Pātanğal*, etc., or common place names, e.g., al-Hind, Gandhāra and Panjab, are referred to in this index in case of substantial discussion or content.

Sanskrit and Arabic Technical Terms

abhyāsa 108, 151, 160
adharma 148, 207, 219–221
āgama 81, 106, 156
ʿağz (عجز) 222–224
ahaṃkāra 86, 104, 241, 243
aiśvarya 123, 147, 219–220
ajñāna 148, 219–221
ʿālim (عالم) 186–187*n*84
ʿamal (*pl. aʿmāl*) (sg. عمل; pl. اعمال) 179, 213
anaiśvarya 148, 219–221
aṅga, cf. *aṣṭāṅga*
anumāna 106
ʿāqil (عاقل) 186–187*n*84
ʿaql (عقل) 172, 176–177, 186–187*n*84, 240, 245
artha 192
aśakti 157, 222–223
asaṃprajñāta samādhi 107, 143–145, 172
āsana 107, 175
aṣṭāṅga 107, 124, 141, 172, 175–177
aṯqāl (اثقال) 4, 117*n*102, 120, 185
aum 153–154, 158, 248
avyakta 35–37, 84, 86, 88, 104, 106, 146, 148, 216, 240
(al-)ʿayš wa l-laḏḏa (العيش و اللذّة) 192
azaliyya (ازلية) 152

(al-)barāhima (البراهمة) 89–90
bhāṣya 113–116
bhāva 123, 147, 219–220
bi-l-ḥawāssi l-Kamis (بالحواسّ الخمس) 147
bi-l-samāʿi (بالسماع) 156
bi-l-ṭibāʿ (با لطّباع) 213
brahmaloka 169
buddhi 88, 104, 123, 147, 157, 219–220, 222–224
buddhīndriya 87, 104–105
burhān (برهان) 193

citta 107, 108, 187–188
cittavṛtti 107, 140, 146, 149, 236

deva 35, 165, 192
dhāraṇā 107, 172, 176
dharma 123, 147, 192, 207, 219–221
dhyāna 107, 172, 175
ḏikr (ذكر) 140
(al-)dīn wa l-sīra (الدين والسيرة) 192
duḥkhatraya 104
dvīpa 169

fāʿil (فاعل) 239
fikra (فكرة) 169, 178–182
fiʿl (فعل) 156, 201, 207, 211, 239
Actuality, *ilā fiʿl* (إلى فعل) or *bi-l-fiʿl* (بالفعل) 86, 240, 243

ğahl (جهل) 222–224, 240
ğufūr (الجفور) 61–62
guṇa 106, 149, 208, 212, 236
grahaṇa 186–187*n*84
grahītṛ 186–187*n*84
grāhya 186–187*n*84

hayūlā (هيولى), cf. *mādda*

ʿibāda (عبادة) 151, 160
idāma l-fikra (ادامة الفكرة) 176
idrak (ادراك) 140, 156
iḵlāṣ (اخلاص) 175
ʿilm (علم) 186–187*n*84
īśvarapraṇidhāna 152, 160

jñāna 108, 123, 147, 219–220

(al-)kaff ʿani l-šarr (الكفّ عن الشرّ) 175
kaivalya 91*n*127, 105, 106, 107, 160–161, 175, 236
kāla 211, 213
kalām (كلام) 154
ḵalāṣ (خلاص) 91*n*127, 151, 160, 185, 218
(al-)ḵalāṣ wa l-daymūma (الخلاص والديمومة) 192–193
kāma 192
karma 108, 146, 153, 163, 165, 236
karman, cf. *karma*
karmendriya 87, 105, 242
ḵaṣla (خصلة) 141, 175–177
kleśa 108, 153, 163–164, 236

loka, cf. *brahmaloka*
lokāloka 169

mabādīʾ (مبادئ) 125
mādda (مادّة) 201, 213, 216, 239
Absolute (المطلقة), pure primary (الهيولى المجرّدة) 86, 148, 240, 242
mahābhūta 86, 105, 212, 241
mahat, cf. *buddhi*
(al-)māl wa l-naʿma (المال والنعمة) 192
malak (ملك) 35–36, 165, 192, 205
maʿlūm (معلوم) 186–187*n*84
manas 87, 105, 107, 242
maʿqūl (معقول) 186–187*n*84
maʿrifa (معرفة) 222
(al-)maṭālib al-arbaʿa (المطالب الاربعة) 192
mokṣa 91*n*127, 193
mufassir (مفسّر) 119–120, 167, 169, 171–173
mūlaprakṛti, cf. *prakṛti*

nafs (نفس) 85, 126, 161–162, 185, 201, 207–208, 213, 225, 240, 252
naqala (نقل) 2
naraka 169
nidrā 140
nirīśvara (-sāṅkhya) 122
niyama 107, 123–124, 165–166, 175

pātāla 169
pradhāna, cf. *prakṛti*
prakṛti 35–36, 86, 88, 104, 105–106, 148, 211, 213, 215, 236, 240
pramāṇa 106, 140, 147, 156, 193, 236
praṇava 153–154
prāṇāyāma 107, 175
pratyāhāra 107, 176
pratyakṣa 106, 147
puruṣa 85, 105–106, 124, 146, 152, 153, 158, 208, 211, 213, 215–216, 236, 239–240
puruṣārtha 192

qabḍ (قبض) 185
qabḍ al-ḥawāss (قبض الحواسّ) 175
qalb (قلب) 185, 242
(al-)quds ẓāhiran wa bāṭinan (القدس ظاهرا و باطنا) 175
quwwa (pl. *quwan*) (sg. قوة; pl. قوى) 149, 247
Potentiality, *bi-l-quwwa* (بالقوّة) or *min al-quwwa* (من القوّة) 240, 243
quwwa nafsiyya (قوة نفسية), *quwwan al-nafs* (قوى النفس) 140, 149, 186
quwan ṯalāṯ (قوى ثلاث), *al-quwan al-ṯalāṯ al-uwal* (القوى الثلث الاول) 149, 161–162, 201, 208, 239, 240, 245

rāga 148, 219–221
rajas 86, 106, 149, 211, 240
rasāyan (رساين) 160–161
rasāyana 160–161, 166
ribāṭāt (رباطات) 61–62
rūḥ (pl. *arwāḥ*) (sg. روح; pl. ارواح) 242
rūḥānī (روحانى) 192, 205–206
ruʾyā (رؤيا) 140

šahāda
Bilingual dirham 34–37
(al-)sakīna al-ṭumaʾnīna (السكينة والطمأنينة) 176
samādhi 107, 143–145, 152, 172, 175–176, 236
(al-)šamaniyya (الشمنية) 89–90, 240
samāpatti 145, 186
saṃprajñāta samādhi 107, 143–145
samudra 169
saṃyama 169, 178–182
sāṅkhya 97, 110, 112, 121, 122, 126, 127, 216, 229
śāstra 112–116, 120–122
satkāryavāda 106–107, 148
sattva 86, 106, 123, 149, 211, 240
seśvara (-sāṅkhya) 122
siddhi 157, 222

smṛti 140
sukūn (سكون) 175
sūkṣmaśarīra 141, 147, 206
sūtra 113–116
svabhāva 211, 213

ṭabīʿa (طبيعة) 86, 241, 242, 243
tafakkur (تفكّر), cf. *fikra*
tafsīr (تفسير) 119–120, 133, 167
taḵayyul (تخيّل) 140
taksīn al-tanaffus (تكسين التنفّس) 175
tamas 86, 106, 123, 149, 211, 240
tanmātra 86, 105, 241
tarāḵin (تراخى) 222–224
taṣawwur (تصور) 143–144, 172
tattva 87, 106, 146, 149, 211–212, 242
taʿwīd (تعويذ) 151, 160
tuṣṭi 157, 219, 222

vācaka 153–154
vācya 153
vairāgya 108, 123, 147, 151, 160, 219–220
vikalpa 140
viparyaya 140, 157, 222
vyakta 86, 88, 148, 216, 240

waḥdāniyya (وحدانية) 152

yama 107, 123–124, 165–166, 175
yoga 97, 107, 120–121, 122, 126, 142, 177

zāhid (زاهد) 161, 220
ẓanna (ظن) 140
zīğ (زيج) 63–64
zuhd (زهد) 151, 160

Geographical Terms

Barikot, cf. Swāt

Chota Lahore 30*n*106

Dunpūr 28, 48*n*199

Gandhāra 17–18, 20, 21, 29, 31–33, 35, 38, 45, 48, 57–58, 96, 127, 231–232, 234
  Culture and society sections 1.3 and 2.4
Gardez 71
Ghazna 15–17, 21–22*n*58, 28, 38, 41, 42, 49, 57–58, 72, 77, 80, 96, 231–232
Ghorvand River 51
Ğūrğān / Gorgan 14–15, 57–58, 66, 231
Ğūrğānīya / Kunya-Urgench 15, 57–58, 231

Himavant or Himalaya 23, 30
(al-)Hind 2, 2*n*7, 7–8, 11
  Terminology with regard to al-Bīrūnī's travels 25
  Language of al-Hind 59
Hund, cf. Wayhind

India, cf. al-Hind
Indian Ocean 22, 65–66, 69

Jhelum 29, 48*n*199

Kābul 15–18, 28, 38, 42, 48, 54, 57–58, 77, 80, 96, 231–232
Kanauj 29*n*95, 30, 32–33, 38, 41, 42, 45–46, 52, 57*n*256, 58, 81, 83–84, 156
Kangra, cf. Nagarkot
Kashmir (Valley) 23, 29, 30, 51–52, 57*n*256
  Al-Bīrūnī's knowledge of 43–45
  Intellectual centre and interactions 81–83
Kāṯ(-Kala) 13–14, 57, 231
Khwarezm 13–16, 38, 41, 66
Kindī 28, 28*n*93, 48*n*199
Kulārğak (mountain) 23
Kunya-Urgench, cf. Ğūrğānīya

Laghmān 28–30, 32, 38, 42, 58, 78, 80, 231
  Culture 48–50
Lahore 23*n*64, 26, 29, 29*n*94, 38, 49, 49*n*207, 54, 71, 78–80, 84, 96, 234
  Bilingual dirham 33–38
  Mandahūkūr 23*n*64, 46
Lahūr (Fort) 23, 23*n*64, 28, 30, 30*n*106, 31, 32, 38, 51–52, 42, 58, 78, 82, 231
  Culture 51–52
Lohara(koṭṭa), cf. Lahūr
Lohkot, cf. Lahūr

Maḥmūdpur, cf. Lahore
Mandahūkūr, cf. Lahore
Multan 29, 31, 31*n*112, 38, 45, 71, 78–80, 81, 83–84, 96

Nagarkot 18, 18*n*43, 38, 46, 78–80, 94, 96
Nandana (Fort) 29, 31, 32, 38, 54, 56, 57–58, 78–80, 96, 231
  Culture 52–53

Panjab 17–18, 20, 21–22, 29, 31–33, 35, 38, 45, 48, 54, 57–58, 71, 71*n*51, 96, 127, 231–232, 234
  Culture and society sections 1.3 and 2.4
Peshawar 29, 32–33, 42, 48, 58, 78, 80, 96, 231
  Culture 50–51
  vikāra 18, 18*n*44

Rāja Gīrā 30*n*105
Rājagirī (Fort) 23, 29, 30, 30*n*105, 58, 78, 231
  Culture 51–52
Rājāwūri (Rajauri) 23, 23*n*66, 52

Ravi River 29*n*94
Ray 14–15, 57–58, 66, 72, 231

Salt Range 31, 49, 53, 56, 79
Sāwa River 50
Sialkot 29, 48*n*199
Sind 1*n*2, 21–22, 23, 32–33, 38, 78–79, 90*n*125
Somnāth 32, 38, 41, 42, 45, 58, 80, 81, 83–84
Srinagar 23*n*63
Sumatra 23*n*68
Suvarṇadvīpa, cf. Sumatra
Swāt (Valley) 30*n*105–106, 52*n*231, 56, 70

Udabhāṇḍa, cf. Wayhind
Uḍegrām 30*n*105

Varanasi 29, 43, 46, 81, 84
Volga 42

Wayhind 29, 32, 38, 49, 49*n*207, 51, 53–54, 78, 80, 96

Zalamkot 70

Personal Names, Dynastic and Ethnic Groups

Abhinavagupta 83, 115
Abū Aḥmad Ibn Ǧilaġtakīn 29*n*95
Abū l-Wafāʾ 14*n*11
Abū Maʿšar 62, 62*n*7
Abū Muḥammad al-Nāʾib al-Āmulī 64–65
Abū Naṣr ʿIrāq 15
Abū Sahl al-Masīḥī al-Ǧurǧānī 15
Achaemenids 17
Afrighids 13–15, 57
Alberonius 237
Alexander the Great 17
Alkhans 17
Aliboron, cf. Alberonius
Ali Huǧwīri 71
Alptigīn 15, 48, 53, 54
Ānandapāla 18, 48, 49*n*207, 77, 82
Aristotle (Aristotelian ideas) 86, 87, 132, 148–149
Āryabhaṭa 173
  Days of Āryabhaṭa (book) 63–64
Āsuri 108–109, 111

Balabhadra 29*n*95, 135, 168*n*43
(al-)Balāḏurī 21, 72
Barhatigīn 17, 19
Bayhaqī 8, 28, 39, 41, 42, 70
Bhīma(deva) 18, 54, 55, 56, 82
Bhīmapāla 18, 48
Bhoja 103, 162–163, 165, 175, 233
(al-)Bīrūnī
  Epistolary exchanges 44, 82, 139
  Literary review (in general) 4–6, (on his visits to al-Hind) 24–28
  Work's reception 237
  Sanskrit literature known to al-Bīrūnī 3–4, 47
Brahmagupta 1, 83, 135, 173
Buzurg Ibn Šahriyār 69

Durlabha 47, 83

Farruḫī 41, 72
Firdawsī 39

Galen 69, 132
Gardīzī 8, 41, 75–76
Gauḍa, cf. Gauḍapāda, the Sāṅkhya author
Gauḍapāda, the Advaita-Vedānta author 99*n*11, 111*n*76
Gauḍapāda, the Sāṅkhya author 99, 99*n*11, 111, 111*n*76, 197, 230
(al-)Ǧayhānnī 65, 66, 66*n*24, 75
Ghaznavids 57–58
  Court situation 39–43, 68–73, 95, 233
  Political agenda 26, 32–33, 37–39, 78–79, 163
  Relations with local rulers 77
  Studies on 5
Greek
  Literature known to al-Bīrūnī 132
  Greek philosophy 148–150, 164, 172

Hindu Šāhis, cf. Shahis
Hiraṇyagarbha 113, 116, 127, 191–193

Ibn Ḥawqal 14*n*13, 16*n*35, 21
Ibn Ḵurdāḏbah 21
Ibn Sīnā 14, 15, 139, 149, 172*n*61, 237
Ibn Zakariyyā' Rāzī 44, 89*n*119
Ibrāhīm al-Fazārī 1, 1*n*2
(al-)Īrānšahrī 89, 89*n*119
Īśvarakṛṣṇa 108*n*60, 109–111, 126, 197

Jaimini 91
Jayapāla 18, 48, 49*n*207, 50, 82
  Inscription 54
Jīvaśarman 44, 47, 52*n*231

Kallara 18
Kamalū 18
Kaniṣka (Kanik) 18, 19, 51
Kapila 91–92, 108–111, 113, 126, 134, 203, 247
Khalajs 19
Kidarites 17
Kuṣāṇas 17, 19
(al-)Ḵwarizmī 64, 237

Lagatūrmān 18

Maḥmūd (of Ghazna) 2, 16, 58, 70, 78–80, 84, 95, 98, 231
  Conquests to al-Hind 28–29, 30*n*106, 32, 52–53
  Court situation, cf. Ghaznavids
  Encounter with the Shahis 48
  Political agenda 32–34, 37–39
Ma'mūnids 14–15, 58
(al-)Muqaddasī 49*n*207
Mas'ūd (of Ghazna) 16, 26
Mas'ūd-I Sa'd-I Salmān 71

Nahuṣa 164–165, 188
Nandikeśvara (also Nandīśvara) 164–165, 188
Nezaks 17

Pañcaśikha 109, 111
Paramārtha, author of the **Suvarṇasaptati*¨ 99, 198–199
Parāśara, father of Vyāsa 117, 242
Pātanǧal (author and commentator of the *Kitāb Pātanǧal*) 93, 119–120, 121, 125, 136
  References to him in the *Taḥqīq mā li-l-Hind* 117–118
Patañjali, the yoga composer 113–116, 117, 123, 127, 165
Plato (Neo-Platonic ideas) 87, 132, 164
Ptolemy 63, 66*n*24, 132

Qābūs bin Wušmagīr bin Ziyār 14, 58–59
Qutayba b. Muslim al-Bāhilī 13

Rutbils 17

Šāhi of Kābul, or Kābulšāhs, cf. Early Shahis
Samanids 13, 15, 66*n*24
Sāmanta(deva) 18, 55
Śaṅkara 102, 102*n*35, 114, 175
Sebüktigīn 16, 32, 48
Shahis 17–20, 46, 48, 50–51, 53–58, 77–78, 80–82, 84, 94–96, 231–232, 234
  Encounter with Maḥmūd 32, 48–49, 52
  Literary review 5, 20*n*56
  Terminology 7*n*24
Śrīpāla 47
Syāvapala 45

Tilak 70
Trilocanapāla 18, 41, 48, 52, 77
Turki Šāhis, cf. Shahis
Turks 13, 17–18, 23, 42

(al-)ʿUtbī 8, 28, 41, 50, 51, 69
Utpala 47, 83

Vācaspatimiśra 99, 100, 101, 102, 111, 115, 116, 124, 175
Vārṣagaṇya 108*n*60, 109
Vateśvara 29*n*95, 47, 62*n*9
Vijayanandin 3, 47, 62*n*9, 83
Vijñānabhikṣu 115, 196–197
Vindhyavāsin 108*n*60
Vyāsa (Veda-) 113–117, 127, 242

Yaḥyā al-Barmakī 1
Yaʿqūb Ibn Ṭāriq 1, 1*n*2, 29*n*95, 64, 65

Sanskrit, Arabic and Persian Literature

Adab 139
Ādityapurāṇa 3, 45, 47, 84, 92, 93, 233
Agastimata 91
(Al-)arkand (Zīǧ) 62*n*9, 63–64
(Al-)āṯār al-bāqiya 59, 62, 65–68
  Date of composition 8

Bhagavadgītā 3, 5, 36–37, 47, 84, 97, 122, 160, 233, 236
Bhāraṭa (*Kitāb*) 117
Brāhmasphuṭasiddhānta, or *Brāhmasiddhānta* 1, 3, 47, 64*n*16, 66, 83, 135
Bṛhatsaṃhitā 3, 47
Buddhacarita 88, 97–98

Carakasaṃhitā or *Čaraka* (*Kitāb*) 1, 1*n*4, 3, 97–98

Dharma Pātañjala 139

Futūḥ al-buldān (*Kitāb*) 21

(Al-)ǧamāhir fī l-ǧawāhir 49*n*207, 72
Gauḍapādabhāṣya 100, 103, 110
  Commentarial tradition 197–199, 227–230
  Date of composition 99
Gītā (*Kitāb*) 3, 5, 36, 47, 68, 85, 125, 160, 211
(Al-)ġurra (*Kitāb*) 3*n*12, 64–65

Ḥadiṯs 139
Ḥudūd al-ʿālam 9, 13, 50

Jayamaṅgalā 103, 110
  Commentarial tradition 197–199, 227–230
  Date of composition 101

Karaṇasāra 29*n*95

Laghujātaka 3, 47

Mahābhārata 2, 47, 84, 88, 97, 109, 113, 116, 213, 215, 217
Maʾkaḏ al-mawāqīt (*Kitāb*) 65
(The)Marvels of India (عجايب الهند) 69
(Al-)masālik wa l-mamālik (*Kitāb*) 21
Māṭharavṛtti 100, 103, 110
  Commentarial tradition 197–199, 227–230
  Date of composition 99
Matsyapurāṇa 45, 47, 84, 92, 93, 233
Mokṣadharma 97, 109

Nyāyabhāṣya 91–92

Pātañjalayogaśāstra 103, 126, 127, 129–131, 161, 194–195, 233, 235–236
  Composition and authorship 113–116, 134
  Date of composition 102
Pātañjalayogaśāstravivaraṇa 98, 114–117, 159, 164, 175, 195
  Date of composition 102–103
Pātanǧal (*Kitāb*) 4, 4*n*18, 40, 47, 76, 84–86, 88, 91, 96–97, 129, 132–133, 195, 235–236, 247
  Date of composition 98
  Literary review 5–6
  References to it in the *Taḥqīq mā li-l-Hind* 1, 118, 125–127
  Work's status 9, 237
Pauliśasiddhānta 3, 47
Purāṇas 5, 47*n*196, 68, 76, 84, 85, 93, 125, 170, 171, 173, 233, 236

(Al-)qānūn al-Masʿūdī 5, 29, 29*n*96, 30
Quran 69, 139, 164*n*24

Rājamārtaṇḍa
  103, 115, 159, 175, 195, 233
  Date of composition 103
Rājataraṅgiṇī 9, 20, 30, 48*n*205, 56, 82

Sānk (*Kitāb*) 4, 4*n*18, 36, 47, 83–85, 88, 96–97, 129, 135, 230, 235–236
  Commentarial tradition 197–199, 227–230
  Date of composition 98
  Literary review 5–6
  References to it in the *Taḥqīq mā li-l-Hind* 1, 91, 111–112, 125–127, 202–203, Appendix
  Work's status 9, 237
Sāṅkhyakārikā 35, 85, 88, 98, 112, 234–235
  Commentarial tradition 197–199, 227–230
  Date of composition 99
Sāṅkhyapravacana (work by Vijñānabhikṣu) 115, 196
Sāṅkhyasaptativṛtti 100, 103, 110, 112
  Commentarial tradition 197–199, 227–230
  Date of composition 99–100
Sāṅkhyavṛtti 100, 103, 110, 112
  Commentarial tradition 197–199, 227–230
  Date of composition 99–100
Šašnāma 21
Ṣaṣṭitantra 109
(Al-)ṣaydana fī l-ṭibb (*Kitāb*) 43*n*169, 71
Sindhind 63–64
Ṣūrat al-arḍ 21
Suvarṇasaptati 100, 103, 110
  Commentarial tradition 197–199, 227–230
  Date of composition 99

Tafhīm 22*n*61, 33*n*119
  Date of composition 8
Taḥdīd al-amākin 13*n*2, 16, 26, 29, 30, 46*n*190, 49, 66*n*24
  Date of composition 8, 28
Taḥqīq mā li-l-Hind 3, 25, 59, 66–68, 80, 90, 232–234
  Date of composition 3, 3*n*10, 8, 28
  Work's status 237
Tattvakaumudī 103, 110, 196, 234
  Commentarial tradition 197–199, 227–230
  Date of composition 101
Tattvavaiśāradī 103, 115, 159, 165, 175, 195
  Date of composition 101

Upaniṣads 97, 97*n*4, 98, 126

Vāyupurāṇa 45, 47, 84, 86, 92, 93, 233
Vedas 47, 67, 84, 91, 117, 203*n*28, 211, 213, 214–215, 218, 228, 229, 233, 239
Viṣṇudharma or *Viṣṇudharmottarapurāṇa* 47, 92, 93, 160, 184, 233
Viṣṇupurāṇa 3, 45, 47, 84, 92, 93, 233
Vivaraṇa, cf. *Pātañjalayogaśāstravivaraṇa*

Yogabhāṣya 113, 116
Yogasūtra 113, 115, 116, 159
Yuktidīpikā 103, 110, 138*n*31, 196
  Commentarial tradition 197–199, 227–230
  Date of composition 100–101

Concepts related to Culture, Religion and Science

Actuality (Aristotelian concept) 85–87, 148, 240, 243
Advaita-Vedānta 90–91, 91*n*126, 99*n*11, 111–112*n*76
Allah 34–37, 40, 64*n*14, 84, 90–91*n*126, 92, 93, 150–155, 166, 192, 210–213, 239, 245, 247, 248
Bayt al-Ḥikma (or Dār al-Ḥikma) 15, 69
Brahminical, or Brahminism 58, 74, 77, 80, 129, 155, 170, 193, 232
  Terminology 11
Brahmin(s) 11, 18, 28*n*92, 44, 46, 50, 55, 57, 66*n*27, 70, 72, 74–77, 81–82, 84, 89–90,

Brahmin(s) (*cont.*) 91, 95, 101, 109, 139, 201, 224, 233–234, 235
Encounter with al-Bīrūnī 28*n*92, 43, 46–47, 58, 73–80, 94–95, 233
Buddhist, or Buddhism 19, 51, 21–22*n*58, 232
In the *Taḥqīq mā li-l-Hind* 89–90

Dār al-Ḥikma, cf. Bayt al-Ḥikma
Durgā 49, 55, 56

Falsafa, or Islamic philosophy 88, 132, 143–144, 148–149. 164
Festivals (Hindu) 28*n*92, 44, 56, 57, 59, 63–64, 68

God, cf. Allah or Īśvara

Hindu, or Hinduism 19, 20*n*56, 36, 40*n*151, 50, 51, 54–57, 57*n*256, 58, 74, 77, 80, 90, 125, 125*n*129, 231, 233
Terminology 11

Indian astronomers/astrologers, cf. Brahmins
Indian philosophers, cf. Brahmins
Indra 155*n*88, 206, 250, 250*n*33
Īśvara 35, 90–91*n*126, 123, 124*n*123, 150–155, 158, 166, 192, 210–213, 215, 215*n*66, 236

Lakṣmī 56
Lokāyata 91, 91*n*130

Ma'mūn Academy 15
Mazdeism 89
Meru (Mount) 67, 93, 146, 188–190
Mihrğān 64
Mīmāṃsā 90–91
Muslim, or Islam 38, 90, 232
Intellectual history 1–2, 15, 21–22, 62, 69–70, 95, 172
Terminology 11

Nārāyaṇa 67, 92, 151*n*71, 234*n*2
Nowrūz 63
Nyāya 90–91

Pārvatī 49
Polyglotism 33–37, 69–71
Potentiality 85, 86, 87, 148, 148–149*n*60, 240, 243

Sāṅkhya (philosophy) 111–112, 233, 235, 236, 237
Its knowledge in Panjab 35–37, 80–88, 234
Śiva 54–57, 139, 165, 234
Sufism 88

Tantra, or Tantric ideas 97, 166

Vaiśeṣika 90–92
Viṣṇu 51, 56–57, 234

Yoga (philosophy) 235–236
Its knowledge in Gandhāra and Panjab 80–88, 233–234

Zoroastrianism 90

Printed in the United States
by Baker & Taylor Publisher Services